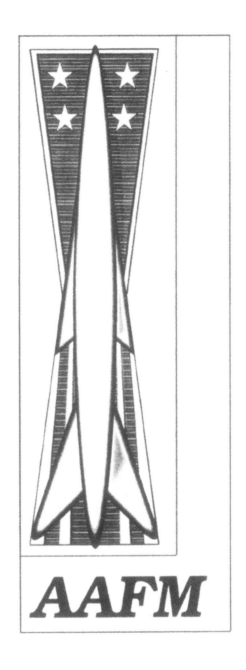

ASSOCIATION OF THE AIR FORCE MISSILEERS
"Victors in the Cold War"

TURNER PUBLISHING COMPANY

Titan I Missile topside at site.

TURNER PUBLISHING COMPANY

Turner Publishing Company Staff:
Publishing Consultant: Keith R. Steele
Production Coordinator: John Mark Jackson
Designer: Heather R. Warren

Featured Writer (Pages 12-33): David K. Stumpf

Library of Congress Catalog
Card Number: 98-86786
ISBN: 978-1-56311-455-7 (Hardcover)
ISBN: 978-1-68336-704-8 (Pbk)

Printed in the United States of America

Installing stage II handling ring - 568th SMS - "C" Complex - Launcher #1.

Table of Contents

Launch of BOMARC XY-4 intercepter. (Courtesy of Daniel Stachowiak.)

Foreword 4

Publisher's Message 5

The Missile Badge 6

Missileers 6

Association of the Air Force Missileers 7

Board of Directors 11

Air Force Missileers History 12

    Special Stories 34

    Biographies 75

Roster 124

Index 134

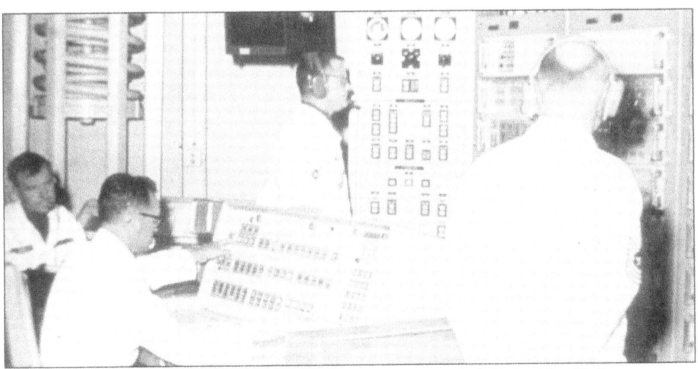

Major Dick Adams, at the console of a Titan II missile at Davis-Monthan Air Force Base. Inside of the Launch Control Facility.

Foreword

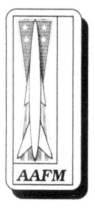

Association of Air Force Missileers

"Victors in The Cold War"

This book reflects the history of the missile systems, units and people who have served in the United States Air Force. The book was put together by the *Association of Air Force Missileers*, an organization of more than 1,800 members, and Turner Publishing, in conjunction with the 1998 National Meeting of *AAFM* at Cocoa Beach, Florida.

AAFM is dedicated to preserving the history of USAF missiles and the people who designed, developed, tested, fielded, operated, maintained and supported them. The histories and articles in this publication, along with the personal biogragphies of many missileers, are only part of the way that *AAFM* is meeting this goal.

The list of missileers included in the book, both the members of *AAFM* and other missileers, is only a part of the group of Air Force men and women who have served or now serve proudly as Air Force Missileers. We have served for over fifty years in a wide variety of ground and air launched missile systems, in locations around the world. The family of missileers, while relatively small in comparison to the whole of the U. S. Air Force, is a close and dedicated group of officers, noncommissioned officers, airmen and civilians who were truly "Victors in the Cold War" and continue to be an integral part of our nation's defense.

It is an honor to be part of this proud group of people.

Jay W. Kelley, Lieutenant General, USAF (Retired)
President

Post Office Box 5693 • Breckenridge, CO 80424 • (970) 453-0500
The Association of Air Force Missileers is a nonprofit, tax-exempt organization under Section 501(c) of the IRS Code.

The Missile Badge

Talk of a badge to recognize missileers goes back to as early as 1951, and rumors indicate that articles about a possible badge appeared in Air Force Times in the early and mid-1950s. The first record of an official proposal was memo dated 25 September 1956 from 1st Lt (Maj Gen (Retired)) Richard Boverie of the 11th Tactical Missile Squadron. Groundwork for the missile insignia was laid in 1957 and early 1958.

On 28 April 1958, AF Chief of Staff General Thomas White directed that, no later than 1 June 1958, "a distinctive badge for wear by missilemen will be designed and available for issue..." The basic guidelines for the Ad Ho committee that met that very day were: "The AF wants the design not to include wings of any type. The badge is to be of silver (no enamel). It is to be simple, yet tell the story of the missile. It should be no larger than the pilot wings and smaller designs are requested."

The missile is a generic missile so that no particular missile in the inventory would be represented. The vertical bands beside and beneath the missile portray vapor trails created by the missile during flight, a reminder of the lightning-quick speed with which it can strike the enemy. The star cluster (comprised of four stars, two on either side of the missile) is indicative of both the aerospace role and the traditional stars and stripes associated with our democracy. The hourglass shape signifies the weapon's round-the-clock readiness.

The first recipients of the Air Force's newest badge were Colonel William C. Erlenbush, commander of the 864th Strategic Missile Squadron (SAC) and Master Sergeant Jake Kindsfather, technical NCO of operations in the 4504th Tactical Missile Training Squadron (TAC). The two were selected by their commands as "representatives of airmen and officer missile specialists." Air Force Chief of Staff General Thomas White pinned on their badges in a special ceremony at the Pentagon in July, 1958.

Extracted from "The Missile Badge (A Not-so-brief History)", by Major Greg Ogletree, 1 June 1997 edition, available from the Association of Air Force Missileers. Greg's history traces the entire development, revisions and updates to the badge throughout its history.

Missileer

by Capt Robert A. Wyckoff

In vacant corners of our land, off rutted gravel trails,
There is a watchful breed of men, who see that peace prevails.
For them there are no waving flags, no blare of martial tune.
There is no romance in their job, no glory at high noon.

In an oft' repeated ritual, they casually hang their locks,
Where the wages of man's love and hate are restrained in a small, red box.
In a world of flickering, colored lights and endless robot din,
The missile crews will talk awhile, but soon will turn within.

To a flash of light or other-worldly tone, conditioned acts respond.
Behind each move, unspoken thoughts of the bombs beyond.
They live with patient waiting, with tactics, minds infused,
And the quiet murmur of the heart that hopes it's never used.

They feel the living throb of the mindless tool they run.
They hear the constant whir of a world that knows no sun.
Here light is ever present, no moon's nocturnal sway.
The clock's unnatural beat belies not night nor day.

Behind a concrete door slammed shut, no starlit skies of night,
No sun-bleached clouds in azure sky in which to dance in flight.
But certain as the rising sun these tacit warriors seldom see,
They're ever grimly ready, for someone has to be.
Beneath it all they're common men, who eat and sleep and dream,
But between them is a common bond of knowledge they're a team.
A group of men who love their land, who serve it long and well,
Who stand their thankless vigil on the brink of man-made hell.

In boredom fluxed with stress, encapsuled they reside.
They do their job without complaint of pleasures oft' denied.
For duty, honor, country and a matter of self-pride.

Association of Air Force Missileers

by Colonel (Retired) Charlie Simpson, Life Member L0001

Description:

The Association of Air Force Missileers (AAFM) is a tax-exempt, nonprofit organization, founded in 1993 and approved as an educational entity under Internal Revenue Service section 501 C 3. More than 1,800 missileers and those with an interest in missiles and missiles have joined since the organization began. The association is dedicated to the following goals:

Goal:

Preserving the heritage of United States Air Force missiles and the people who develop, test, deploy, operate, maintain and support them by:

Recognizing outstanding missileers

Encouraging and facilitating meetings and unit reunions

Keeping missileers informed

Serving as a central point of contact and locator for missileers

History of the Association

In the summer of 1991, a number of career missileers attended the 30th Anniversary Reunion of the 3901st Strategic Missile Evaluation Squadron in Las Vegas. During that reunion, these individuals discussed the need for an organization dedicated to Air Force missileers.

During the previous twenty or so years, there had been many attempts to form such an organization, usually with the Daedalian Society, the organization for military pilots, as a model. In each of the previous attempts, the organizations were formed and managed by active duty missileers, and each succeeded until the key players moved to another location. None lasted more than a few years, and most consisted of members at the locality where the organization began.

In the year and a half following the meeting in Las Vegas, Colonel (Retired) Charles G. Simpson, the current Executive Director of AAFM, continued to discuss a potential organization with other missileers. In late 1992, he submitted a proposal to a the group of missileers most active in these discussions, both active duty and retired. This initial cadre developed the goals, membership requirements, suggested donations for dues, the motto and logo for the new organization.

The organization was established as a nonprofit corporation in the state of Colorado in

January, 1993, and tax-exempt status was requested from the Internal Revenue Service. The organization selected section 501 (C) 3 status because of its goals, and that status was approved later in 1993. Using "Christmas card lists" from a number of the founders, brochures and applications were sent out to about 600 missileers. Within four months, more than 400 had become members of the association.

Simpson volunteered to serve as Executive Director, a committee of founding members nominated the initial board members and members were asked to elect the first board. The first officers were Major General (Retired) Ralph Spraker, President; Colonel (Retired) Dick Schoonmaker, Vice President; and Colonel (Retired) Pat Henry, Secretary/Treasurer. The other members of the board were Lieutenant General Dirk Jameson; Brigadier General Seb Coglitore; Colonel (Retired) Jack Lander; Lieutenant General Jay Kelley; Major General (Retired) Barry Horton; Brigadier General Lance Lord; Brigadier General Ron Gray; Chief Master Sergeant (Retired) Bob Kelchner and Chief Master Sergeant (Retired) Conrad Paquette.

Articles and notices in Air Force Magazine, The Retired Officers Association Magazine, the USAF Afterburner Newsletter for Retirees, and in several other national and local publications, including base papers at missile bases, resulted in a consistent early growth in membership. By the end of the first year, membership stood at 615.

Simpson represented the association with a traveling display at the 1993 Air Combat Command Missile Competition at Vandenberg Air Force Base, California, the first time that AAFM had a display at an official event. Over the next several months, Simpson took the display to most of the operational missile bases to spread the word about AAFM. Over the following four years, AAFM continued to grow and implement programs designed to accomplish its objectives. The association ended its fifth year with more than 1,650 members, and continues to grow both in membership and in the ways it meets its goals.

Membership

During the initial discussions among the founders of AAFM, a number of options were considered for membership. While a significant number of those who served or now serve in missile related career fields were "big missile" specialists - the Strategic Air Command intercontinental ballistic missiles (ICBMs) like Atlas, Titan and Minuteman, that

AAFM Executive Director Col. (Ret.) Charlie Simpson.

population represented only a portion of those who were Air Force missileers. It was decided early in the decision process that AAFM should include and represent all USAF missileers, whether they served in strategic, tactical, research or any other system. Many officers and enlisted members had earned the coveted missile badge while working on air-launched systems, space boosters and a host of other missile systems.

The discussion also considered the "who" of Air Force speciality. Many organizations limit their membership to a specific speciality - operators, maintainers, or some other narrow definition. Those of us who had served for most of our careers in missiles fully understood that we really were a closely knit, clearly identifiable family - and that family included officers, noncommissioned officers, airmen and civilians. We had worked on systems ranging from captured German missiles after World War II to the newest ICBM, space vehicle or air to air weapon. We had served as operators, maintainers, in test or development, in munitions, communications and many support fields, and in many major air commands. But we had one thing in common - we all called ourselves missileers.

In the end, membership was defined in two categories. Those who had earned the Air Force Missile Badge would be classified as "missileer" members. Others who had worked with missiles, or even had a strong interest in missiles, would be called "sponsor" members. The distinction is minor - only a single letter in the member number indicates the difference. It became quickly obvious to the AAFM board that the pride of those who served in USAF missile programs was strong, whether or not the individual earned the "pocket rocket." A look at the summary of membership later in this publication clearly shows that our "missileer family" has many branches on its family tree - but we are all missileers.

One requirement that was modified a couple of years after AAFM started was the requirement for the Missile Badge - when Air Force Chief of Staff General Tony McPeak did away with the badge, the Missile and Space Badge was issued to new missileers. The definition of a "missileer" member was modified to include this change, although the Missile Badge itself was reinstated when General Ronald Fogelman became Chief of Staff.

Since all of us who had served as missileers realized the great importance of the whole missile team, and especially the contributions of the young enlisted members, AAFM has emphasized the importance of encouraging membership for all ranks and levels of experience. AAFM is also not an organization solely for the career missileer. A significant number of our membership served only a single four year tour in the Air Force, either as young airmen in missile maintenance or other fields, or as young lieutenants on the crew force of one of our weapons systems.

The result has been a membership that runs the gamut - people who served from the lowest enlisted rank to four star general. Some served in the 1940s, before there was a missile badge, developing and testing systems like Navaho, Matador and Falcon. Others served in the 1950s in Mace, Genie and Jupiter and Thor. Members have served throughout the fifty years of USAF missile systems, in every conceivable role and

area of expertise. Our backgrounds are varied - but we are all in the same family - we are missileers.

AAFM Programs
Missile Heritage Grants

The AAFM Board of Directors developed a program in the first year aimed at assisting museums around the country in displaying missiles and missile related items. Each year, museums are invited to submit applications to the association on specific projects. A committee of board members reviews the applications and awards grants at the end of each year. In the first five years of operation, AAFM provided

$37,000 to museums for projects. Beginning in 1995, the association provided the grants in memory of members who had passed away during the previous year.

Air Force Space Command Awards

Each year, AAFM sponsors two awards through Headquarters, Air Force Space Command. The General Samuel Phillips Award for Operations is awarded to the best missile operations squadron each year. General Phillips was instrumental in the development of USAF intercontinental ballistic missiles. The Colonel Edward Payne Award for Maintenance is awarded to the missile maintenance squadron

National Atomic Museum's Thor.

Hill Museum's Snark.

that does the most to improve the quality of life for its members. Colonel Payne was the Deputy Commander for Maintenance in the 44th Strategic Missile Wing at Ellsworth AFB, and also served in other key maintenance positions at both Vandenberg AFB and Ellsworth AFB.

Space and Missile Competition (Guardian Challenge) Awards

Since 1994, AAFM has provided trophies to the winning team members from the missile units in operations, maintenance, communications, security and helicopter operations. These trophies are awarded to team members who do not receive individual trophies at the competition, but are named best in their respective specialties. In 1997, The Society of Strategic Air Command and AAFM jointly provided commemorative mugs to each Guardian Challenge participant. For 1998, AAFM provided commemorative coins to the participants. The Execute Director and President attend the competition each year, and the AAFM display is part of the Space and Missile Exposition.

Lifetime Honorary Memberships

In 1996, the association began a program to recognize the outstanding local-area supporters of our missile units. Each year, we ask commanders at each wing to name a member of the community who deserves recognition for his or her support of the people of the unit, the base and the Air Force.

AAFM Newsletter

This quarterly publication features historical articles and stories, current news about missiles and space, humorous articles by members, association news and information about events, meetings and reunions. The newsletter is provided to every member, to Air Force missile and space units, museums and other interested parties.

AAFM Web Page and Online Updates

The association web page features historical information and current news, as well as information about the association. Available at http://www.thebook.com/missileers, the page is visited over 600 times per month. The executive director also provides monthly e-mail updates to more than half the association membership.

National Meeting

Members of the association gather every two years for a meeting that includes tours of current and historical missile and space facilities,

local area tours and entertainment, a general membership meeting, an opportunity for members to meet and socialize and a banquet featuring a prominent missileer or space pioneer as speaker. In 1994, AAFM met in Colorado Springs, Colorado, and toured Space Command and North American Air Defence Command facilities. In 1996, the Santa Maria, California, gathering including tours of current space launch sites and historical Thor facilities. The 1998 meeting, at Cocoa Beach, Florida, includes tours of NASA and Air Force facilities at Cape Canaveral and Patrick Air Force Base. The association will return to Colorado Springs in the year 2000.

Local Area Meetings

The executive director of the association travels to missile bases around the country each year and conducts informal gatherings for members and other interested missileers. These meetings feature briefings on current missile programs by active duty people at the bases, and are often conducted in conjunction with some other event, such as the recent closing of some missile units, the competition and other events.

Assistance with Reunions and Meetings

The association has aided a number of missile organizations in conducting reunions. This assistance has included publicity for the event in the newsletter and web page, lists of members who served in the unit, guest speakers and displays. AAFM encourages units to combine their reunions with the AAFM National Meeting.

Member Directory and Locator

The member database tracks over 150 attributes for each member, including systems and units, job specialities, headquarters and other special assignment and other information. Each year, AAFM publishes a directory listing information about each member. This information includes current address, member lists by missile experience by system and unit and summaries of the demographics of the membership. Each new member is provided with a list of all other members in his or her state of residence, and members can request lists of members for a specific unit, system or specialty at any time. AAFM also maintains a list of more than 4,000 other missileers who are not members, and assists requestors in finding fellow missileers.

Historical Publications and Mementos

AAFM collects and makes available to members a variety of publications and

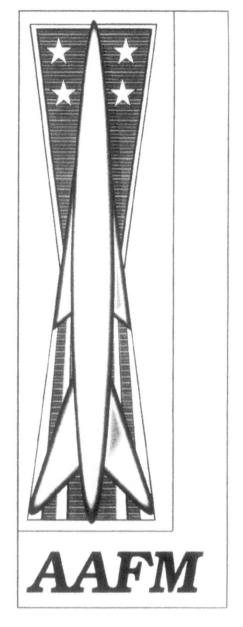

mementos. These include poems, cartoon books, a history of the missile badge, commemorative patches and books from missile units, other patches and pins and a number of items with the AAFM logo. Items are provided to members who donate to the Missile Heritage Fund.

AAFM Display and Slide Show

The association display appears at events around the country, and includes a computer controlled slide show featuring photos from throughout the history of Air Force missile programs. The executive director has taken the display to four to six events every year to help spread the word about the association.

1998 Association of Air Force Missileers Board of Directors

The current Board of Directors for the Association of Air Force Missileers consists of twelve members, with four directors elected every two years. Members serve a six year term as director. Officers are elected by the board every two years, following the general board election.

Officers

President - Lieutenant General (Retired) Jay Kelley - founding board member, Colorado Springs, Colorado, Titan II (381SMW, 390SMW), Minuteman (351SMW), Hq SAC, Hq AF Space Command, Joint Staff. Term 1996-2000.

Vice President - Colonel (Retired) Jim Burba - Bend, Oregon, Atlas F (556SMS), Minuteman (351SMW, 321SMW, 341SMW), 3901SMES, Space (Onizuka), Hq SAC, Joint Staff. Term 1996-2000.

Secretary - Master Sergeant (Retired) Dayna Castro - Lompoc, California, Minuteman (351SMW), Vandenberg (394MS, 1STRAD, 30SW, WSMC). Term 1996-2002.

Treasurer - Chief Master Sergeant (Retired) Bob Kelchner - Torrance, California, Minuteman (44SMW, 341SMW, 351SMW), 394MS, 3901SMES. Term 1996-2000.

Directors

Chief Master Sergeant (Retired) Joe Andrew - Woodbridge, Virginia, Matador, Minuteman (341SMW), 3901SMES. Term 1998-2004.

Brigadier General (Retired) Jim Crouch - Austin, Texas, Minuteman (44SMW, 321SMW, 341SMW), Hq SAC. Term 1998-2004. Past President.

Colonel (Retired) Dick Keen - Burnsville, Minnesota, Minuteman (44SMW, 90SMW, 341SMW), Hq SAC, Joint staff. Term 1996-2002.

Lieutenant Colonel Mike Lehnertz - Air Staff, the Pentagon, Minuteman (90SMW, 341SMW), ALCS (4ACCS), Hq SAC, Joint Staff. Term 1996-2002.

Lieutenant General Lance Lord - founding board member, Vice Commander, Air Force Space Command, Minuteman (321SMW, 341SMW, 90SMW), 3901SMES, 30SW, GLCM (Hq USAFE), Hq AF Space Command, Hq USAF. Term 1998-2004.

Major General Tom Neary - Hq USAF, Minuteman (90SMW, 341SMW, 351SMW), GLCM (485TMW), Hq SAC, Hq USAF. Term 1996-2002.

Major General (Retired) Bob Parker - San Antonio, Texas, Minuteman (44SMW, 91SMW, 321SMW, 341SMW), 4ACCS, 4315CCTS, Hq SAC, Hq USAF, Joint Staff, 20AF. Term 1998-2004.

Captain Julie Wittkoff , Ogden ALC, Hill AFB, Utah, Minuteman (90SMW, 91SMW). Term 1996-2000.

Executive Director - Colonel (Retired) Charlie Simpson, L001 - Breckenridge, Colorado, Titan I (569SMS), Minuteman (321SMW, 44SMW, 57AD), GLCM (487SMW), 3901SMES, HqSAC. The executive director is a volunteer manager who conducts AAFM business and represents the association at events, meetings and other activities.

Former Board Members

Colonel (Retired) Ron Bishop
Brigadier General (Retired) Seb Coglitore
Brigadier General (Retired) Ron Gray
Colonel (Retired) Pat Henry
Major General (Retired) Barry Horton (Deceased)
Lieutenant Colonel (Retired) Dirk Jameson
Colonel (Retired) Jim Knapp
Colonel (Retired) Jack Lander
Colonel (Retired) Ed Northrup
Colonel (Retired) Ed Osborne
Chief Master Sergeant (Retired) Conrad Paquette
Colonel (Retired) Bob Pepe
Colonel (Retired) Dick Schoonmaker
Major General (Retired) Ralph Spraker
Mister Jim Widlar

Air Force Missileers History

Air Force Missiles

by David K. Stumpf, Ph.D., Associate Member A1008

Many detailed and lengthy histories of Air Force missiles have been written. The purpose of this brief history is to place in perspective the many types of missiles that Air Force Missileers have been involved with, either in the form of research and development, operational test and evaluation or operational deployment. We have separated this history into broad categories: air-launched, ground-based cruise and ballistic missiles.

Air-launched Missiles

At the end of World War II (WWII), the United States Army Air Force was working on a number of projects for air-to-air and air-to-surface missiles, many in conjunction with the new ground-launched systems and others based on WWII projects from both the Allies and the Germans. In April, 1946, the air-launched portion of the USAAF guided missile program consisted of a number of projects.

Air-to-Surface projects included the MX-601, a Douglas Aircraft vertical bomb controlled in range and azimuth, called the Roc; the MX-674, a Bell Aircraft vertical bomb called the Tarzon; the MX-776, a Bell subsonic, 100 mile range subsonic missile called the Rascal; the MX-777, a McDonnell Aircraft 100 mile range supersonic missile; the MX-778, a Goodyear Aircraft 100 mile range subsonic missile; the MX-779, a Goodyear 100 mile range supersonic missile; and the Mastiff, a 300 mile supersonic missile with an atomic warhead.

At the same time a number of air-to-air projects were programmed. These included the MX-570, a Hughes Aircraft 9 mile range, subsonic, 50,000 foot altitude JB-3, called Tiamat;the MX-798, a Hughes 5 mile range version of MX-570; the MX-799, a Ryan Aeronautical fighter-launched subsonic missile, called Firebird; the MX-800, a M.W. Kellogg fighter-launched supersonic missile; the MX-801, a Bendix Aviation fighter-launched supersonic missile; and the MX-802, a General Electric bomber-launched supersonic missile, called Dragonfly.

Over the months following the April 1946 decision to pursue these projects, budget cuts, the emergence of a separate Air Force and interservice rivalry would all impact this list. Ground-launched systems were also being developed, and policy makers had to make some hard decisions as to the best use of limited resources.

By 1950, the USAF program for development of missiles had changed considerably from the original 1946 plan. The MX-674, or Bell Tarzon and the MX-776, or Bell Rascal I, to be followed by Rascal II with a 150 mile range were the only surviving air-to-surface projects. Only one air-to-air project, the MX-904, or Hughes Falcon for fighters, to be followed by a bomber-launched version, survived.

By June of 1953, the missile program had evolved into a National Guided Missile Program. The only air-launched missiles under USAF development at that time were the Rascal and the Falcon. Interestingly, the Sparrow and Sidewinder were being developed, but by the United States Navy.

The Missiles

Unlike ground based missile systems that are developed for specific basing, almost all air-launched missiles have been used in a variety of ways on a variety of aircraft. The missiles fall into several basic categories. Air-to-air missiles primarily are those used as weapons to down enemy aircraft. Air-to-ground missiles are used to destroy targets on the ground, and include both short-range and long-range systems. Some are designed for specific missions, the Short Range Attack Missile and the Hound Dog, for example, were nuclear armed strategic weapons. Others are designed to take out enemy radar systems, to penetrate hardened targets or other special purposes. Many air-launched missiles have been manufactured over long periods of time, by numerous manufacturers and in a variety of versions. Many have been used by the USAF, the Navy, Marine Corps and by the services of allied nations. Note that this history does not include some of the current development and test projects that are ongoing. It is intended to reflect the history of missiles that members of the association have worked on.

Air-to-Air Missiles

Genie (MB-1, AIR-2)

The Genie was a supersonic, unguided, free flight, air-to-air rocket designed to be carried and launched by interceptor aircraft from the Air Defense Command. Genie was also the first air-to-air missile armed with a nuclear weapon, the 1.5 kilo ton yield W-25. With this size warhead, Genie was designed for use against formations of enemy bombers by the interceptors controlled by the North American Air Defense Command.

The Genie program began in 1954 as the Douglas Aircraft MB-1 project and by 1956, when the flight test program was started, had been named Bird Dog, High Card and Ding-Dong as design changes were made. The first launches were by F-89Ds and the flight test prototype YF-102 Delta Dagger. Deployment began in 1957 as Genie (AIR-2).

The operational missile weighed 820 pounds, was 9 feet 7 inches in length with a wingspan of 18.5 inches and a diameter of 17.4 inches. Initially, the Genie's 40,000 pound thrust solid propellant rocket motor was manufactured by Aerojet-General, giving the Genie a range of 6 nautical miles and a top speed of Mach 3. In 1965, the Thiokol SR49-T-1 solid-fuel rocket motor, rated at 39,500 pounds of thrust replaced the Aerojet-General rocket motor. The Thiokol motor had better stability and operational temperature limits, something that had plagued the earlier deployed version.

Genie was an unguided missile. Since it was armed with a nuclear warhead, accuracy was not really necessary since the blast effects from the air burst of a 1.5 KT weapon meant that any aircraft within 3,000 feet would be severely damaged.

The Genie program had a unique distinction amongst the various nuclear armed air-to-air missiles deployed by the United States military forces; it was launched with a live nuclear warhead on 19 July 1957 from an F-89J, at 19,000 feet over the Nevada National Test Range. While the W-25 is listed as having a yield of 1.5 KT, the resulting explosion was measured at 2 kilotons.

By 1982 only 200 Genie missiles were still actively deployed on F-106 and CF-101 aircraft with the North American Air Command. When these aircraft were retired in favor of F-15, F-16 and F/A-18 fighters, there was no provision to fit them with Genie launch capability.

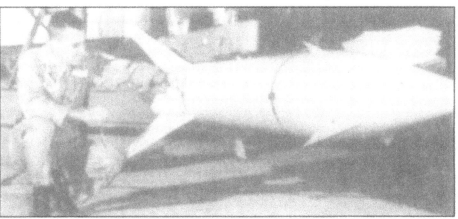

Genie.

Sidewinder (GAR-8, AIM-9A through AIM-9M)

The AIM-9 is a close range, infrared guided aircraft missile designed to be carried and fired from fighter planes against enemy aircraft, carried for both attack and defense. Developed in the 1950s, it continues in service today in many versions. The missile body is a heavy aluminum tube, with a length of 9 feet, 5 inches; over-all span, 25 inches; body diameter, 5 inches; and weight up to 160.1 pounds. The AIM-9 infrared system uses its target's heat signature for locating and tracking. A solid propellant motor accelerates the missile to speeds above Mach 2. Missiles are launched at distances measured from thousands of feet to more than 10 miles from the enemy aircraft. Launch may be made from off the line of flight of the target, and from above or below its flight altitude. Sidewinder was the first guided aircraft missile to down an enemy aircraft.

The AIM-9 was originally manufactured for the US Navy, which designed and developed the missile. Since the missile was first introduced, a number of contractors have manufactured various versions. The Sidewinder has been carried on a wide variety of Air Force, Navy and NATO aircraft.

Falcon (GAR-1, 2, 3, 4 and 11, AIM-4A through G, AIM-26B and AIM-47A)

Built primarily by the Hughes Aircraft Company but involving several other contractors, Falcon was a family of air-to-air guided missiles. Development began in 1947 under the name project Dragonfly, the Falcon was first designated XF-98, as it was a "pilotless interceptor." The Falcon was first tested in 1954 and became operational in 1955. Most versions were radar guided, and three versions (C, D and G) were infrared guided. The AIM-4F and G were introduced simultaneously in 1960 to provide reduced susceptibility to enemy

Falcon.

countermeasures and higher performance. The later version was the primary armament for F-106 but a number of aircraft carried the Falcon, including the F-89J, F-101, F-102, and F-4. The G versions were retired (with the F-106) in 1988. The AIM-26 was a nuclear-armed version. Unlike the Genie, the nuclear Falcon was guided to a specific target, rather than a general area. The AIM-47A was never operational - it was planned for use on the F-12A (which became the SR-71) and for other advanced interceptors that were canceled.

The Falcon was powered by a Thiokol single stage or two stage solid propellant motor. Generally, it was 7 feet 2 inches long and a half inches in diameter and weighed 150 pounds. About 48,000 were manufactured.

Sparrow (AIM 7- D through R)

The Sparrows are guided aircraft missiles designed to be carried and fired from fighter planes against enemy aircraft The Sparrow is a radar guided, all-weather, all-aspect capable missile. Manufactured by Raytheon and General Dynamics, more than 39,400 were produced. The various versions were developed to be carried by the F-4, F-111, F-104, F-14, F-15, F-16 and F-18.

Solid propellant motors are used for propulsion and to give the missile a speed above Mach 2. Missiles are launched at distances measured in thousands of feet to more than 25 miles from the enemy aircraft. The launch may be made from almost any angle off the line of flight of the enemy aircraft, and from above or below its flight altitude. The motors were manufactured by Aerojet-General and

Rocketdyne. Missiles are 11 feet, 10 inches long, 8 inches in diameter, with a wing span of 3 feet, 4 inches and weigh 504 pounds. Range is 25 miles and speed is Mach 3.5.

AMRAAM (AIM-120A and B)

The Advanced Medium Air-to-air Missile (AMRAMM) was developed jointly by the USAF and the Navy for use with the F-15, F-16, F-18, F-14, F-22 and NATO and allied fighters. The AIM-120 replaces the AIM-7 Sparrow, and is a medium range, look-down, shoot-down missile with fire-and-forget and multiple launch capabilities.

AMRAAM has inertial mid-course guidance and active radar terminal homing. The propulsion system was developed in 1979 by Alliant Techsystems, Hughes Aircraft and Raytheon. Testing was conducted in 1985 and 1986, and production began in 1987, with the first missile delivered in 1988. Total planned production is over 12,000 missiles. Propulsion is a boost-sustain propellant grain design solid motor with reduced smoke feature The missile weighs 345 pounds, is 75 12 feet long, 7 inches in diameter and a span of 2 feet, 1 inch. The range is 30 miles and speed is mach 4.

Air to Ground Missiles

Shrike (AGM-45A)

The Shrike is a short-range, air-to-ground missile used by tactical aircraft to locate and destroy radar transmitter stations. 'The missile

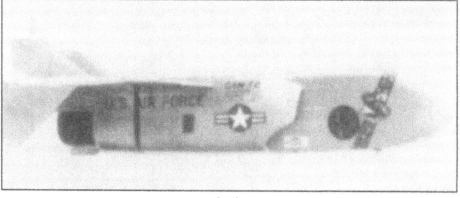

Quail.

Sidewinder.

contains a passive radar homing guidance unit capable of detecting radar transmissions, and of guiding the missile to the transmission source. The missile is launched at the discretion of the aircraft commander after the aircraft enters the applicable delivery envelope. Missile range depends on aircraft launch altitude and delivery mode. The missile was developed at Naval Ordnance Test Station China Lake, California. The guidance and control contractor is Texas Instruments.

The solid propellant motor is manufactured by Rocketdyne Division of North American Aviation. It contains almost 92 pounds of propellant, adequate for only a short powered flight. The momentum achieved during this period and the pull of gravity provide the high velocity the missile reaches during the terminal part of its free flight.

The missile is basically a cylinder which tapers to a point at the forward end. The assembled missile is 130 inches long, and has a body diameter of 8 inches. The overall span is approximately 36 inches. The total loaded weight is slightly more than 400 pounds.

Quail (GAM-72, ADM-20)

The Quail was a decoy missile carried by and launched from a B-52 bomber to confuse or dilute a hostile radar-controlled air defense system. Each aircraft could carry four missiles. The Quail had a range of more than 200 miles at nearly sonic speed, after launch from its carrier. It was manufactured by McDonnell Aircraft Corporation.

The Quail was powered by a General Electric J-58-GE-7 turbojet engine. The engine had an eight stage, axial flow compressor driven by a two stage turbine. Ethylene oxide was used for engine starting at extreme altitudes and fuel was JP-4 with Phillips 55MB additive. The additive prevented the formation of ice in the fuel at extremely low temperatures.

The missile airframe consisted of the forward body, aft body and wings. The forward body was made of glass fiber laminate-honeycomb composite and housed the offensive subsystems, the flight control system and the instrumentation system. The aft body, of conventional aluminum construction, housed the engine, fuel and oil systems; air turbine alternator and another offensive subsystems. The wings were conventional aircraft-type construction, a short-span modified delta planform design. Missiles were carried in the aircraft with the wings folded. In this position, each was 29 inches wide and 26 inches high. In the wing extended position, the missile was 5 feet 5 inches wide, 12 feet 11 inches long and 3 feet 4 inches high. It weighed slightly under 2000 pounds.

Before launch, the four missiles were housed aboard the aircraft in two carriage racks in the bomb bay. The carriages were lowered, the missile wings extended, and the engine started automatically. An interlock circuit prevented the extension of the launch gear while the bomb bay doors were closed. After launch, the lower carriages were jettisoned. Normally the upper carriages were retracted. If the missiles were not launched, they were retracted into the aircraft after the missile wings were refolded. If

Hound Dog.

B-52 Launching Hound Dog.

a launch gear failure prevented release or retraction of the missiles, thereby endangering the aircraft, the entire launch gear could be jettisoned. A fully extended missile would not clear the runway, if the bomber had to land with the launch gear down. Missiles were loaded into the aircraft by either the quick-load missile-package method or the single-missile sequence loading method. In the quick-load method, a package consisting of four missiles and the supporting launch gear components was built up in the maintenance area, and then loaded into the carrier aircraft as a unit. The sequential method consisted of loading a single missile onto a launch gear carriage which was already installed in the aircraft.

Hound Dog
(GAM-77, AGM-28A and B)

The North American Aviation Hound Dog, initially designated the GAM-77 and later the AGM-28A and B, was designed originally for a short three-year life span as a standoff weapon for the B-52. The missile was to have been replaced by the Skybolt air-launched ballistic missile, but the Skybolt program was canceled

in 1962. The missile would stay in service over 15 years before it was replaced by newer weapons like Short Range Attack Missile (SRAM) and Air-launched Cruise Missile (ALCM). North American was awarded the contract to build Hound Dog in August 1957, and they relied heavily on work done on the Navaho intercontinental cruise missile.

The first powered flight of Hound Dog occurred in April 1959, the first guided flight in October and the Air Force accepted the first production missile in December of that year. Over the next three and a half years, North American produced 722 missiles for SAC.

The missile was 42 feet 56 inches long, with a wing span of 12 feet. It weighed 12,000 pounds fully fueled with its single W-28 warhead, and was powered by a Pratt Whitney J-52 turbojet engine. North American's Autonetics Division developed the inertial guidance system in conjunction with a star tracker. One missile hung under each wing of a B-52 between the fuselage and the inboard engines. The missile had a range of about 700 miles, flew at Mach 2 plus and could evade enemy defenses by flying turns or doglegs to its target.

SAC activated Airborne Missile Maintenance Squadrons in 1962 at each of the

Hound Dog in flight.

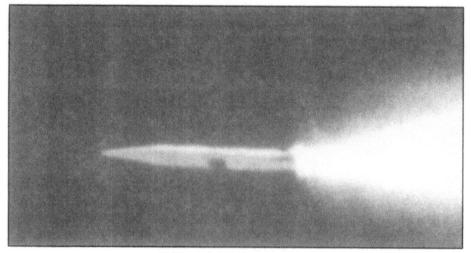

Skybolt air launched ballistic missile.

B-52 bases to provide the maintenance for the Hound Dog and its sister decoy missile, the Quail. These units had between 77 and 90 officers and airmen who had previously been assigned to Armament and Electronics Maintenance Squadrons.

During its service life, Hound Dogs were deployed at twenty-seven bases where SAC had B-52s. A total of 295 B-52 bombers were configured to carry the missile during its lifetime. In 1972, SAC began deploying the Short Range Attack Missile (SRAM), and began phasing out the Hound Dog. The last missile left service in June 1975.

Bullpup
(GAM-83A, AGM-12A, B, C, D and E)

The Bullpup was developed after Korea to fill the need for a high-performance, guided missile to match the capabilities of the new supersonic delivery aircraft. Built by Martin Marietta's Orlando Division, Bullpup was originally developed for the Navy, came in three basic versions, the AGM-12B (11 feet long and 571 pounds), the AGM-12C (13.6 feet long and 1785 pounds) and the AGM-12D and E (similar in size to the C). The B, C and E models carried conventional warheads and the D model was designed for use with a nuclear warhead. Range varied from 3 to 6 miles and propulsion was by prepackaged liquid rocket, although the original design used a solid motor. The missiles were radio guided by the launching aircraft. The

missiles were carried by the F-100 and F-105.

The missiles were extremely accurate and reliable, and could be launched from several fighter aircraft. They were designed for use against airfields, trains, truck convoys, bridges and other ground targets. Tracking flares in the tail allowed the pilot who launched the Bullpup to follow the missile visually while sending commands to guide the missile to the target. Bullpups were designed be treated like a "round of ammunition", so no pre-firing checkout was required and the missile could be loaded on an aircraft in about five minutes using standard bomb-handling equipment.

Skybolt (GAM-87A)

Designed to be a complement for, and then replace the Hound Dog cruise missile, the Skybolt was a highly mobile nuclear deterrent weapon planned to be carried by the B-52 and the British Vulcan bomber. The two stage, solid motor, inertially guided air-launched ballistic missile (ALBM) was built by Douglas Aircraft, and was first launched on April 19, 1962. The first launch was a partial success, because the second stage failed to ignite. The Skybolt had a range of 1,150 miles, was 33 feet long, 3 feet in diameter and weighed 11,000 pounds. A B-52 could carry up to four of the missiles, in addition to an internal bomb load. In a controversial decision that impacted relations between the United States and Great Britain, Secretary of

Defense McNamara canceled the Skybolt in the early 1960s.

Maverick (AGM-65A through G)

A tactical, rocket powered missile similar to the unpowered Walleye AGM-26A glide bomb, the Maverick uses launch and leave television guidance on early models or infrared guidance systems later, initially controlled by the aircraft crew, and was designed originally for the A-7, F-4 and F-111, and later used on the A-10, F-16 and F-15. Hughes Aircraft built the Maverick, designed to be used against pinpoint targets, like tanks and truck convoys. Over 100 Mavericks were used per day during the Gulf War.

More than 25,000 Mavericks were manufactured. The missile has a solid rocket motor, is 8 feet, 2 inches long, 1 foot in diameter, has a wingspan of 2 feet, 4 _ inches and weighs up to 662 pounds. Range is up to 14 miles.

HARM (AGM-88A through C)

The High Speed Anti-Radiation Missile (HARM) was developed as a follow-on to the Shrike to destroy enemy radar sites and systems. Proven in the Gulf War and used on the F-16 and F-15E, the HARM is built by Texas Instruments and powered by a Thiokol solid propellant motor. The passive homing guidance system homes in on enemy radar emissions. More than 10,000 HARMs have been delivered to the USAF and the Navy.

The missile is 13 feet, 8 _ inches long, 10 inches in diameter, has a wingspan of 3 feet, 8 _ inches and weighs 807 pounds. Its range is more than 10 miles.

Harpoon (AGM-84A)

The Harpoon, a Navy all-weather anti-ship cruise missile, was adapted in 1983 to be carried by the B-52G to support maritime anti-surface warfare operations. Each B-52G could carry eight missiles. When the use of the Harpoon was proposed, the USAF planned to equip B-52H models when the G models retired. The Harpoon uses sea-skimming cruise guidance monitored by a radar altimeter and active radar homing. The McDonnell Douglas missile is powered by a Teledyne turbojet engine, weighs 1145 pounds, is 12 feet, 7 inches long, a little over 1 foot in diameter and has a wing span of 3 feet.

ALCM (AGM-86B and C)

The Air-Launched Cruise Missile (ALCM) has a somewhat confusing pedigree. The initial operational requirement was as the replacement to the Quail decoy. With a longer range and more sophisticated electronics, this initial design was called the Subsonic Cruise Aircraft Decoy (SCAD). On 15 July 1970, two years after the initial requirement was identified, full scale development was authorized by the Department of Defense but refused by Congress. Slightly less then two years later, in February 1972, both the Defense Department and Congress agreed on the need for SCAD and Boeing Aircraft Company was awarded the full scale development contract. Development did not last long as Congress became alarmed at the rapidly escalating costs of the electronics systems for SCAD. With the simultaneous work on a Submarine-Launched Cruise Missile (SLCM) program for the Navy, the Department of Defense converted the SCAD program into a technology development study in 1973, effectively canceling the program.

The Air Force still needed a long range cruise missile to replace the Short Range Attack Missile (SRAM) and in 1974 Boeing was awarded a contract to utilize applicable SCAD technology toward this end. The AGM-86A was the first iteration of this new design, with seven times the range and three times the accuracy of SRAM. Best of all, it would fit onto the SRAM launch equipment, facilitating its rapid deployment into the B-52 fleet and upcoming B-1A fleet. Testing began in 1975 with drop tests for launcher compatibility and flight tests followed soon after. Completion of the flight tests in November 1976 led to the decision for full scale production in January 1977.

Elected with military cost cutting in mind, President Jimmy Carter formed the Joint Cruise Missiles Project which emphasized commonality of major components between the ALCM and Navy's SLCM. Additionally, the Defense Department directed the Air Force to build the extended range, AGM-86B, before the AGM-86A. Now the ALCM was to have a 1,550 nautical mile range. The AGM-86B was to compete with the Navy's SLCM, built by General Dynamics and called Tomahawk, to see whether Boeing or General Dynamics would produce the Air Force missile.

Five years after the awarding of the SCAD contract to Boeing, the first AGM-86B flew on 3 August 1979. After 10 flights, Boeing was awarded the Air Force contract. The first operational missiles were delivered on 23 April 1981 with the first B-52 launch of an operational missile, minus the nuclear warhead, taking place on 25 July 1981. Developmental testing was completed in 1982.

The ALCM is powered by a Williams or Teledyne turbofan engine, is 20 feet, 9 inches long, 2 feet in diameter and has a wingspan of 12 feet. It weighs 3,200 pounds and can be armed with either conventional or W-80 nuclear warhead. Conventionally armed ALCMs were used in large numbers during the Gulf War.

B-52G aircraft can carry 12 AGM-86Bs on their wing pylons. B-52Hs were likewise equipped originally but later modifications permitted eight AGM-86Bs to be carried internally.

ACM (AGM-129A)

The Advanced Cruise Missile (ACM) is deployed on the B-52H and is a low observable cruise missile with improved capabilities when compared to the ALCM. Developed by General Dynamics, with McDonnell Douglas selected as a second source, the first ACM was delivered in August 1993. A total of 461 missiles were ordered.

Powered by a Williams turbofan and using inertial guidance with a TERCOM update capability, the ACM was designed for a nuclear warhead. Range is 1,865 miles, and the missile is 20 feet, 10 inches long, 2 feet 3 3/4 inches wide and has wingspan of 10 feet, 2 inches.

Ground-Launched Cruise Missiles

This brief history of Air Force cruise missiles covers the major cruise missile developments in both surface-to-surface; Matador, Mace, Snark, and Gryphon; and surface-to-air, BOMARC. The Navaho cruise missile is included due to its extensive test program. Each missile is listed by its common name followed by known official designators.

Matador (TM-61A, TM-61C), Mace (TM-61B, TM-76A, TM-76B)

Immediately after the end of World War II, the Army Air Force began planning for the

ALCM.

Sram Rotary Launcher.

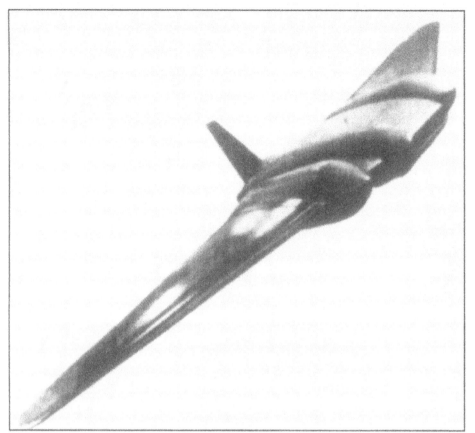

JB-1 FLYING BOMB.

The JB-2 (Buzz Bomb) with cradle still attached leaving the trailer ramp.

development of a new weapon in its arsenal, the guided missile. On 21 August 1945 the original characteristics for what was to become the Matador weapon system were promulgated by Headquarters Army Air Force in a document titled "Military Characteristics for Ground-to-Ground Guided Missiles (175-500 mile range).

The first nine test flights in the Matador program involved full-scale wooden models to test the launch technique. The first flight of a flight test vehicle was on 19 January 1949. The missile crashed 900 feet from the launcher due to booster misalignment. The second launch was successful and the missile reached its cruising altitude of 22,000 feet. The missile quickly out-ran its chase aircraft, reaching Mach 0.97. The

control program had been set for lower speeds, as a result the missile became divergently unstable in roll causing the horizontal stabilizer and elevators to pull off as well as both wings. Testing continued with 46 prototype missiles through March 1954 and then with 84 operational missiles from December 1952 to April 1954. Matador was launcher from a zero-length mobile launcher using a 57,000 pound thrust booster.

Matador's guidance system was a based on pulsed radar signals for right/left turn, climb or dive and terminal dive. The early system was limited to line-of-sight, which, at a cruise altitude of 40,000 feet was 200 nautical miles, well short of its 620 nautical mile range. In late 1954, a

new guidance system, Shanicle, was incorporated and the missile designator was changed to TM-61C. The Shanicle guidance system was based on a LORAN-type hyperbolic grid signal that required multiple precisely located ground stations which would be vulnerable under wartime conditions. Shanicle was also susceptible to electronic jamming. Seventy-four TM-61Cs were launched between April 1957 and September 1960; the resulting reliability was 71 percent with a circular error probable (CEP, the area in which fifty percent of the missiles launched could be expected to impact) of 2,700 feet. Instructor crew results during this same time frame gave a CEP of 1,600 feet.

On 9 March 1954 the 1st Pilotless Bomber Squadron (PBS) deployed with 50 TM-61A missiles to Bitburgh AFB, Germany. The 2nd PBS followed six months later with 50 missiles at Hahn AFB, Germany, for a final deployment of 100 TM-61A missiles. While this, in theory, greatly enhanced the nuclear umbrella available to the Tactical Air Command in Europe, in reality, Matador was clumsy to deploy. Twenty-five vehicles were needed to transport the missile in four pieces to the launch site, erect the launcher, build up the missile and place it on the launcher, a process which at best took 10 men close to 90 minutes. This was hardly an instantaneous response weapon but it did give TAC a weapon to use against heavily defended targets hundreds of miles inside the Soviet Union.

The TM-61C was deployed to Europe in 1957 with the 11th Tactical Missile Squadron (the Pilotless Bomber Squadron designator had been changed to Tactical Missile Squadron or TMS). TM-61Cs replaced the TM-61As at Hahn and Bitburg by 1958. Also in 1958, TM-61Cs were deployed to South Korea at Osan Air Force Base, Kimpo Airport in Seoul and Chinchon Ni, all under the 58th Tactical Missile Group.

The third guidance system for Matador, the Automatic Terrain Recognition and Navigation (ATRAN) radar map-matching system, was developed by Goodyear Aircraft Corporation in the 1948-52 period. Incorporation into Matador lead to redesignation as TM-61B and a name change to Mace. The changes between the two earlier configurations, TM-61A and TM-61C, included greater thrust and a more powerful 97,000 pound thrust booster. Mace, with the ATRAN guidance system first flew in 1956 and was given a final designation as TM-76A. The TM-76A top speed was Mach 0.7 to Mach 0.85 at 750 feet altitude and had a range of 540 nautical miles. Mace deployment began in 1959, eventually replacing Matador by 1962. At one point, six missile squadrons in the 38th Tactical Missile Wing, with a total of nearly 200 TM-61s and TM-76s were operational in Europe.

A lack of available radar maps of target areas limited the utility of the ATRAN system and it was soon replaced by the fourth and final modification, an inertial guidance system, TM-76AB. The inertial guidance systems of the TM-76B missiles required a fixed reference point so they were deployed in hardened above ground shelters giving the TM-76Bs an almost instantaneous response time compared to the early mode of deployment. By 1965, 50 TM-76B launchers were operational at Bitburg. Reliability and accuracy during operational test flights of Mace in Florida and Libya was found wanting and on 30 April 1969 the last Mace squadron in Europe, the 71st Tactical Missile Squadron, was deactivated and replaced by the Army's Pershing I system.

Deployment to Taiwan began in April 1957 with the 17th TMS at Tainan AFB. In June 1958, the 17th TMS was redesignated as the 868th TMS. Missiles were removed from Taiwan in 1962.

Deployment to Okinawa as the 498th Tactical Missile Group (TMG) began in 1961, with launch sites built at Bolo Point, Motobu Quarry and the U.S. Army Easely Range. TM-76B missiles deployed to Okinawa remained operational until 1971 when the 498th TMG was deactivated prior to the return of Okinawa to Japan in 1972.

Navaho (XSM-64)

The initial Navaho design was submitted to the Air Force in December 1945 by the Technical Research Laboratory of North American Aviation. The proposal was composed of three separate missile systems. The first was a V-2 with wings, to meet the 500 nautical mile range requirement; the second substituted the V-2 rocket engine with a turbojet-ramjet powerplant for 3,000 nautical mile range; and the third called for a rocket booster to be added to give the 5,000 nautical mile intercontinental range that the Air Force required. The Air Force adopted the 500 nautical mile range program as the MX-770 in 1946 and in 1947 added a 1,500 nautical mile variant of the turbojet-ramjet design. In 1950 the Navaho project was finalized to be a two part program. First was the design, construction and test of a turbojet test vehicle. A 3,600 nautical mile version of the test bed vehicle would be an interim weapon while the full range, 5,500 nautical mile operational weapon was being tested.

The turbojet test vehicle was designated as the X-10 and flew for the first time in October 1953. The missile was equipped with radio command control and landing gear to permit recovery and reuse of the test vehicles. Eleven missiles flew a total of 27 missions. On the 19th test, the test vehicle reached Mach 2.05, establishing a speed record for turbojet-powered aircraft.

The XSM-64 was longer then the X-10 and launched using a 76.5 foot long liquid propellant piggy back booster. Integrating the airframe and booster combination proved to be more difficult then expected. The first flight was to have been in September 1954 but did not occur until November 1956. Even when the XSM-64 was given priority nearly equal to the nascent ICBM and IRBM programs, it earned the sobriquet "Never go, Navaho," for its repeated failures at launch and in flight. Finally the Air Force canceled the program in July 1957.

Navaho did provide valuable data for the growing missile industry. The North American rocket engines used on the booster were conceptual forerunners to those on Redst0ne, Thor and Atlas. The first inertial guidance system used on submarines and on the Hound Dog missile were adapted from the Navaho guidance system.

585th Mace.

Navaho Cruise Missile.

Mace at Hahn.

Matador.

Snark (N-25, N-69A-E, SM-62)

The Snark program ended up being as unusual as its name. In March 1946, Northrup Aircraft Corporation was awarded a study contract for the development of two missiles, MX-777A and MX-777B. Dubbed by Jack Northrup, the company president, Snark and Boojum, respectively, the initial specifications were for 1,300 and 4,300 nautical mile ranges and a 5,000 pound payload.

The first Snark test flight vehicles were designated as the N-25 and first successfully flew on 16 April 1951. Powered by a J33 engine, the missile was designed to reach a top cruise speed of Mach 0.85. To reduce test flight program costs, the N-25 was equipped with a drag chute and a tricycle skid arrangement for recovery on the runway at Holloman AFB, New Mexico. Launch was from a sled using rocket boosters. The first test program ended after 21 flights with five of the original sixteen missiles remaining.

From the beginning of the program, Snark was to have an inertial guidance system. Tested on B-29 and B-45 aircraft, the one-ton guidance system worked, but not long enough for the proposed two-hour flight time of a typical operational mission.

In June 1950, the Air Force made radical changes to the Snark program, requiring an extensive redesign of the missile. Range was increased to 5,500 nautical miles with a supersonic dash at the end of the mission; the payload was increased to 7,000 pounds and the CEP was now 1,500 feet. Now designated as the N-69, the fuselage was stretched, the nose contour was sharpened, the external air scoop was flush mounted and the wing shape was changed to include leading edge extensions on the outer half of wing. Eventually the N-69 was produced in four variants. The N-69A,B and C were powered by a J71; the N-69D and E models were powered by the J57. The N-69 missile was launched from a zero-length launcher using twin solid rocket boosters. The final deployment version used two four second duration 130,000 pound thrust boosters.

The first four attempts to launch N-69A missiles were failures. Finally 13 months after the first attempt, the missile flew for three and one-half hours but exploded on recovery. The first successful recovery did not take place until October 1956. Since Snark's tentative activation date had originally been April 1953, with full operational status by October 1953, the program was in trouble.

A particularly disturbing problem in the test flight program was the discovery that the original terminal dive to impact weapon system delivery was not possible due to inadequate pitch control. In July 1955, Northrup proposed that the nose compartment carrying the warhead be detached from the missile in level flight and follow a ballistic trajectory towards the target. The first flight test of this concept was on a modified N-69C on 26 September 1955. With this relatively unsophisticated delivery system, the Air Force and enlarged the required CEP to 8,000 feet!

Beginning with the N-69D model, Snark used a stellar guided inertial guidance system that was beset with problems. Interestingly enough, the first full-range flight was successful

Bomarc.

and revealed that current maps had misplaced Ascension Island by several miles. Later flights were not as productive with the CEP for a 2,100 nautical mile flight being 20 nautical miles. Even the W-39 thermonuclear warhead with a low megaton yield could not make up for this kind of miss distance. In all fairness, the most accurate of the full-range flights impacted 4.2 nautical miles left and 0.3 nautical miles short of its target. By February 1960, test results showed a less then 50 percent chance of the guidance system performing to specifications.

In perhaps one of the most unreasoned deployment decisions during the Cold War, the Air Force decided to deploy Snark at Presque Isle, Maine, on 21 March 1957. Construction began in May 1958. The 556th Strategic Missile Squadron (SMS) was activated in December 1957 at Patrick AFB, Florida for test launch and missile handling training proposes. Just before deactivation of the 556th SMS and its functions assigned to the 702nd Strategic Missile Wing (SMW), the first Air Force launch of a Snark missile took on in June 1958.

In January 1959, the 702nd SMW was activated at Presque Isle, receiving its first Snark SM-62. The launch site at Presque Isle had six assembly and checkout buildings, each with two launch pads. The missiles were stored indoors and wheeled to the launch pad and fired. The 702nd SMS was eventually equipped with 30 missiles. In November 1959, SAC recommended deactivation of the program. This was rejected and the first Snark went on strategic alert on 18 March 1960. The 702nd SMW became fully operational in February 1961 only to be deactivated four months later, on 25 June 1961, as the newly elected Kennedy Administration attempted to weed out weapon systems that were obsolete.

BOMARC (IM-99A and B/CIM-10A and B

BOMARC has several unique distinctions. First, it was the first and only surface-to-air missile developed by the Air Force. Second, the

Snark Launch.

acronym for its name is probably the most unusual for any weapon system developed by the United States. BOMARC is the acronym for Boeing and Michigan Aeronautical Research Center.

In 1949 the Air Force authorized Boeing Aircraft to improve upon a number of limited test programs that demonstrated the utility of ramjets as powerplants for interceptor missiles against both missiles and bombers. Boeing teamed with the Michigan Aeronautical Research Center for the development of BOMARC. Full development of the missile began in 1951 with Marquardt Company manufacturing the ramjets and Aerojet-General supplying the liquid rocket vertical takeoff booster. To facilitate prolonged storage but quick reaction time, the booster propellants were hypergolic, igniting on contact. Unfortunately, the highly corrosive oxidizer, nitric acid, caused sufficient storage problems that the booster propellants were loaded at the last minute in the operational missiles. Test flight models were designated as XIM-99, the test bed for rocket booster development, with neither a guidance system or the ramjet powerplant installed. XIM-99 flight tests began in 1955 and ended after 38 flights in 1957 with the delivery of the first guided and ramjet powered test vehicles, designated as YIM-99A. On 16 May 1957, Boeing received the first production contract and on 7 November 1957, the production of 23 IM-99A missiles began.

BOMARC A (IM-99A) became operational on 19 September 1959 with the 46th Air Defense Missile Squadron (ADMS) at McGuire AFB, New Jersey. One year later, all 28 missiles at McGuire AFB were operational, along with four more squadrons; the 6th ADMS at Suffolk County, New York; the 22nd ADMS at Langley AFB, Virginia; the 26th ADMS at Otis, AFB, Massachusetts and the 30th ADMS at Dow AFB, Maine; for a total of 140 missiles. With a range of 250 nautical miles, CIM-10A served to cover the northern half of the Atlantic seaboard.

The BOMARC guidance system was an interesting hybrid. After reaching a cruise altitude of 60,000 feet, guidance was switched to the Semi-Automatic Ground Environment (SAGE) air defense network that was used to coordinate the Air Defense Command fighters. Terminal guidance was turned on when the missiles approached to within 10 miles of their targets.

Problems with the hypergolic propellants culminated in a accident on 7 June 1960 at McGuire AFB when a high pressure helium tank failed, rupturing the fuel tank and causing a fire that engulfed the missile, including its W-40 thermonuclear warhead. The warhead's high explosives did not detonate and contamination was limited to the dispersion of the fire fighting water.

The BOMARC flight test and crew training program utilized the Navy's Regulus II cruise missile as a supersonic target. Since the BOMARC radar was designed to detect clouds of bombers, the Regulus II had to be fit with radar reflectors to increase its radar signature. The first launch against Regulus II took place on 9 September 1959 and was recorded as a near miss. With a limited number of Regulus II missiles available as targets, direct hits were not always the desired outcome. On 17 September 1959, the first direct hit of a supersonic target by a BOMARC missile was achieved. After 27 months and 46 Regulus II target launches, QF-104s and QB-47s became available and the Regulus II target program ended on 31 September 1961.

The Air Force was well aware of the problems with the liquid rocket boosters and with the breakthroughs in solid rocket propellant technology that took place in the late 1950s, began the design of the BOMARC B (IM-99B) which incorporated a single Thiokol 50,000 pound thrust solid propellant booster. Incorporation of the booster allowed launch in 30 seconds, a significant improvement over the two minutes necessary to load the liquid rocket propellants used in the IM-99A. Range was increased to 440 nautical miles and cruise altitude could now reach 100,000 feet.

On 1 June 1961, the first IM-99B site became operational with the 37th DMS at Kincheloe AFB, Michigan. Over the next two and one-half years, operational sites included: 74th ADMS, Duluth Municipal Airport, Minnesota; the 35th ADMS at Niagara Falls; the first Canadian sites at 446th Surface to Air Missile (SAM), North Bay, Ontario; and the 447th SAM La Macaz, Quebec. IM-99Bs were eventually deployed at all of the IM-99A sites except the 6th ADMS and the 30th ADMS. IM-99A training ended on 10 March 1962 and on 1 December 1964 both the 6th ADMS and 30th ADMS were closed. In June 1963, the BOMARC A (IM-99A) and BOMARC B (IM-99B) designators were changed to CIM-10A and CIM-10B, respectively, standing for Coffin-launched Interceptor Missile, as part of the uniform designation system adopted by the Pentagon.

BOMARC reached a deployment maximum of 242 missiles but this was not long lived. By June 1969 only 22 missiles were still operational. With the realization that the Soviet Union was now relying much more heavily on ICBMs and submarine-launched ballistic missiles (SLBMs), the need for BOMARC was lessened. On 1 October 1972 the last BOMARC was deactivated at McQuire AFB.

Gryphon (GLCM, BGM-109G)

The story of the ground launched cruise missile (GLCM) is somewhat convoluted. GLCM was born from the realization that after the cancellation of the SCAD program, future cruise missile development would require cooperation between the Air Force and the Navy. The Air Force was directed to support the Navy's terrain contour mapping (TERCOM) and begin an effort to integrate the developing Navy missile for aircraft use. While the ALCM system previously described evolved separately from this point, the air launch option for the submarine launched cruise missile (SLCM) was still being pursued by the Navy. By 1977, the initial testing of the SLCM had involved aircraft launches and the Navy continued to propose replacement of the Boeing ALCM with a version of SLCM. While only six SLCMs could be installed on the B-52 SRAM launcher, instead of the eight Boeing ALCMs, cost savings were apparent to the Pentagon. Nonetheless, the desire to maintain two approaches to cruise missile design and performance led to the ALCM program proceeding with the Boeing design while the Navy was directed to develop ship launched, submarine launched and ground launched versions of the newly named Tomahawk cruise missile.

GLCM had a range of 1,500 miles and superb accuracy. Published photographs show it squarely hitting a target screen no larger then a football goal post set. The first launch of the GLCM from the transporter-erector-launcher (TEL) took place on 16 May 1980 and was successful.

For the USAF mission, the system was a ground mobile system deployed in flights of sixteen missiles apiece. Each flight consisted of four TELs, two launch control centers and 16 support vehicles and was manned by 69 people, including a flight commander, four launch officers, 19 maintenance personnel, one medical technician and 44 security personnel. Personnel were from both the USAF and the host NATO nation. The flight, normally secured in hardened Quick Reaction Alert shelters in the base GLCM Alert and Maintenance Area, could be relocated to remote, off-road locations selected by the flight commander to dig in, conceal equipment and prepare launch in wartime.

The launch sequence began when the TEL was elevated for launch, and the missile boosted from the TEL by a solid-fuel rocket motor. At four seconds into flight the fins extended and at 10 seconds the wings and jet engine inlet deployed. At 13 seconds, the booster was jettisoned and the small air-breathing cruise engine started. The missile then used TERCOM guidance for low-level flight to ensure highly accurate, all-weather penetration to the target.

In December, 1979, the NATO Foreign and Defense Ministers decided that the time had come to counter the Warsaw Pact's buildup of theater nuclear forces. The Soviets were deploying medium range SS-20s targeted against Western Europe, and attempts at Intermediate Nuclear Forces (INF) negotiations limiting these weapons were not fruitful. The NATO ministers decided that GLCM would be deployed in Europe, a total of 464 missiles at six locations. Two sites in England, and one each in Italy, Germany, Belgium and Holland.

Planning and construction began in the early 1980's at each of the six selected sites, all of which required considerable construction for mission and support facilities. For example, at Comiso, Sicily, the site selected was Magliocco Air Base, an old Italian and German fighter-bomber base built in 1935 and virtually destroyed by the Americans in July, 1943, when Patton

GLCM.

invaded that region of Sicily. Little was done to repair the damage from World War II, so a major task in 1982 was the destruction and removal of old bombed out buildings so construction could begin. Each site that was selected was unique in its previous use and the amount of construction required.

The GLCM program was one of the key elements that brought an end to the Cold War. After deployment of this highly accurate and visible weapon system, the Soviets became serious about arms reduction talks, and finally agreed in the late 1980's to an INF treaty that resulted in elimination of the Pershing and GLCM missiles in exchange for like reductions in Soviet systems.

Ballistic Missiles

In April 1946, Consolidated Vultee Aircraft was awarded the first development study contract for a 5,000 nautical mile range intercontinental ballistic missile (ICBM), the MX-774. Slightly more then one year later, the program was canceled as part of military austerity programs. Consolidated Vultee Aircraft continued research on its own funds and flight tested three MX-774 missiles. Several concepts that were carried into later ICBMs were developed with the MX-774, including using gas/liquid pressure to support the airframe, swiveling engines and a separating warhead.

After the 1947 cancellation of MX-774, funding for development of ICBMs was scarce until January 1951, when the AF funded Consolidated Vultee Aircraft for a development program for ballistic and cruise options for a 5,000 mile range missile carrying one of the current heavy nuclear warheads, with a goal of a CEP of 1,500 feet. The cruise, or glide, option was dropped later that year, and funding and development were limited until 1955. In 1954, the Western Development Division of the Air Force Air Research and Development Command was created with Brigadier General B. A. Schriever in command. This new division was formed to focus energy and resources on developing the concept of the ICBM into a weapon system as rapidly as feasible. In 1955, President Eisenhower, following advisory committee recommendations to accelerate ICBM work, as well as develop an intermediate range ballistic missile (IRBM) as in interim solution, assigned the highest national priority to the dual track Atlas and Titan ICBM programs.

The initial operating capacity for both systems was originally defined to be able to launch 25 percent the assigned 40 missiles at each of the three planned bases within 15 minutes, with another 10 launched within two hours. The mix of Atlas, Titan ICBMs and Thor and Jupiter IRBMs varied over the years between 1955 and 1957, with plans for varying totals, mixes and locations for missiles. The October 1957 launch of Sputnik changed all of the plans, with a great acceleration of all programs and an increase to the number of squadrons and missiles.

The concept of concurrent development was critical for the successful development and rapid deployment of the ICBM systems. Basing modes were decided upon and construction well underway before the operational testing program

for Atlas, Titan I and II and Minuteman were complete. The concurrency concept foreshortened the time needed to bring each system to initial operating capacity.

Intermediate Range Ballistic Missiles

In 1955, during the development process for the early ICBM systems, a key decision was to develop an IRBM weapon system. The Army, under the leadership of Werner von Braun at the Redstone Arsenal, began developing its own medium range missile, Jupiter. The Air Force began work on the Thor. Since the range of the IRBM was 20 to 30 percent of the range of an ICBM, the currently available inertial guidance systems provided sufficient accuracy of the relatively short distances.

In November 1957, the Secretary of Defense made the decision to place both the Thor and Jupiter in production, with both to have initial operational capability by June 1958. The real surprise was a decision to have the Air Force deploy both missiles. The Jupiter would not be an Army-manned system, since missiles with ranges greater than 200 nautical miles were assigned to the Air Force. The Air Force Chief of

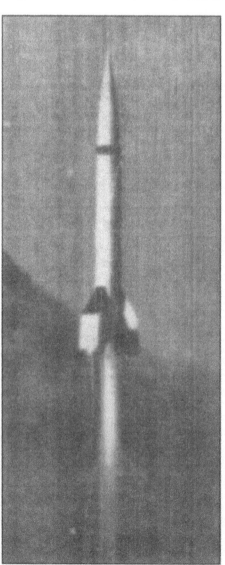

T-MX 774 launch.

Staff strongly pushed for elimination of the Jupiter and increase in Thor production, but the decision by Secretary of Defense McElroy stood and both Thor and Jupiter would be deployed in Europe.

Jupiter (SM-78, WS 315A-2)

The Jupiter missile, manufactured by Chrysler Corporation, was a single stage missile, 58 feet long and 105 inches in diameter. Weighing 110,00 pounds at launch, Jupiter was powered by a single Rocketdyne S-3D liquid propellant engine which used liquid oxygen and kerosene (RP-1), producing 150,000 pounds of

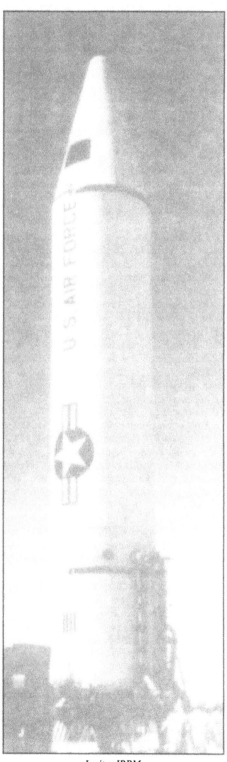

Jupiter IRBM.

thrust. A Ford Instrument Company inertial guidance was used and the warhead was a single W-49 thermonuclear weapon, housed in the first deployed ablative heatshield reentry vehicle. Range was 1,500 miles.

Flight testing of components began on 14 March 1956 with the launch of a modified Redstone rocket called Jupiter C. The Army was still assigned the task and funding responsibility for Jupiter and conducted the first successful launch of a Jupiter research and development missile on 31 May 1957. Five months later the first Jupiter IRBM prototype was launched on 22 October 1957. The first launch from a tactical launcher took place three years later, due to problems with the equipment.

Jupiter stood on alert, loaded with RP-1, on a launch ring surrounded at the base by an environmental shelter. Liquid oxygen was loaded with high capacity pumps just before launch. No effort was made to protect the Jupiter missile or its support equipment from the effects of a nearby nuclear or conventional blast. While considered mobile, the twenty vehicles, including, the 6000 gallon fuel tanker and 4000 gallon liquid oxygen trailer, and two cranes, made Jupiter's mobility more of an academic point than a reality. Jupiter missiles were deployed in a 3x5 configuration, three missiles per launch site, five sites.

The 864th, 865th and 866th Strategic Missile Squadrons (SMSs) were activated by SAC to support the deployment of Jupiter to Turkey and Italy in January 1958. The first launch by an Italian crew took place in April 1961; by a Turkish crew in April 1962. The first fielded missiles were manned by SAC crews due to delays in NATO crew training. Two squadrons of 15 missiles each were deployed to Gioia del Colle in southern Italy with the first site becoming operational on 15 July 1960 and the last site being turned over to the Italians on 20 June 1961. One squadron was deployed to Cigli, Turkey, reaching full operational status with Turkish crews on 28 February 1962. The Jupiter warheads were maintained under SAC custody even though the three training squadrons were deactivated after February 1962.

The Jupiter missile played a key role in the Cuban Missile Crisis of October 1962. President Kennedy privately agreed to complete the deactivation of the Jupiter missiles as part of the negotiations to resolve the crisis. On 17 January 1963, the phaseout of weapon system in Italy was announced. On 23 January 1963, the Turkish government announced that its single squadron was also to be deactivated and all missiles were off alert by 1964.

Thor (SM-75, WS 315A-1)

The single stage Thor missile was 65 feet long, 8 feet in diameter and weighed 110,000 pounds at launch. Powered by a 160,000 pound thrust Rocketdyne MB-3 engine using liquid oxygen and RP-1, Thor used two vernier rocket engines for roll control as well as added thrust at lift off. The vernier engines also gave Thor a variable range capability, from 300 to 1500 nautical miles. Guidance was an all inertial system originally developed by A.C. Spark Plug

Thor in England.

for the Atlas program. The warhead was the W-49 housed in a heatsink configuration Mark 2 reentry vehicle, or in a Mark 3 ablative heatshield reentry vehicle.

The Thor missile has the distinction of being the first Air Force missile designed and built without a prototype. Production began at the Douglas Aircraft Company facilities in Santa Monica, California in March 1956 and the first missile was received by the Air Force on 26 October 1956. Unfortunately, the lack of a research and development program resulted in less than satisfactory results at the first and second launch attempts; both were failures. On the seventh launch, 20 September 1957, a range of 1,100 nautical miles was achieved. Two months later, Thor was being under full production. The first operationally configured missile was launched on 26 November 1958.

Four Thor squadrons were deployed in England, the 77 RAF SMS at Feltwell, the 97th RAF SMS at Hemswell, the 98th RAF SMS at Driffield and the 144 RAF SMS at North Luffenham. The first squadron was turned over to the RAF on 22 June 1959 and the last on 22 April 1960. SAC established the 705th Strategic Missile Wing (SMW) at Lakenheath RAF Station on 20 February 1958 to support Thor deployment. The 705th SMW was later moved to South Ruislip and merged with the 7th Air Division.

Each Thor squadron had a 3x5 configuration, three missiles at each of five sites. Once the RAF personnel were trained, the only remaining SAC personnel were the launch officers assigned to each site. SAC retained custody of the nuclear warhead. Unlike Jupiter, Thor was completely covered by an environmental shelter and stored empty in a horizontal position. Upon receipt of a launch order, the shelter was rolled back, the missile erected and loaded with propellants. Typical time to launch was 15 to 20 minutes. On 3 June 1960 the first warhead was mated to a Thor missile at Feltwell and Thor joined the strategic alert forces of Great Britain and the United States. On 1 May 1962, the United States informed Britain that support of the Thor deployment would cease on 31 October 1964. Britain chose not to support Thor on its own and deactivation began on 29 November 1962. On 15 August 1963 the last Thor was removed from alert.

The Thor missile was used in a variety of

other Air Force programs, including serving as satellite launch vehicle. Thor was also one of only three operational ballistic missiles, including the Navy's Polaris A-2 and the Army's Redstone, used to launch live nuclear weapons during the nuclear weapon test programs in the Pacific in the early 1960's. Seven Thor launches resulted in three successful weapon detonations. After two unsuccessful attempts, on 8 July 1962 a Thor missile was launched from Johnston Island. Nineteen miles downrange and 248 miles in altitude, the explosion had a yield of 1.4 megatons. The primary objective was to monitor high-altitude air burst effects on radio communications and radar operation. An unexpected result was the tremendous electromagnetic pulse experienced in the Hawaiian Islands. Not only was the overcast night sky turned into day for six minutes, the pulse of electrons set off burglar alarms and tripped circuit breakers over a wide area.

The next Thor launch in the test series was a spectacular disaster as the missile had to be destroyed immediately after engine ignition. The resulting damage and radioactive contamination required a massive effort to make Johnston Island usable again. The next two tests were failures as the missile had to be destroyed. The final two launches were highly successful.

A second operational life for the Thor missile was with the 10th Aerospace Defense Squadron (ADS) of the Air Defense Command. Located at Johnston Island, the 10th ADS was declared operational on 29 May 1964 and remained so, 24 hours day, for eight years. Two launchers at Johnston Island and two at Vandenberg AFB were used in the program. On 19 August 1972 the program was placed on reserve status and three years later deactivated as the W-49 warheads were withdrawn from service.

Intercontinental Ballistic Missiles

The First Generation

Atlas (SM-65, PGM-16D, CGM-16D, CGM-16E, HGM-16F)

Atlas, built by the Convair Division of General Dynamics, was derived from the MX-774 program that had been canceled in 1947. The pressure stabilized structure, separable warhead and gimbaling engine concepts developed with MX-774 were incorporated in Atlas from the beginning. The original Atlas design required five main and two vernier engines and was 90 feet long and 12 feet in diameter. With the development of lightweight thermonuclear weapons, the design was modified to use three main engines.

Atlas was a one and one-half stage missile, a unique feature not found on any other ballistic missile in the United States arsenal. All three main engines and the two verniers ignited at launch but the staging event dropped two of the three main engines, leaving the third (called the sustainer engine) engine and two vernier operating through the remainder of powered flight.

The operationally configured Altas varied in length from 75 feet to 82 feet 6 inches, depending on the reentry vehicle being carried. Atlas was 10 feet in diameter and weighed 244,000 pounds at launch. Three main engines and two vernier engines generated at total of 360,000 pounds of thrust. The pressured tank design allowed the stainless steel skin to be the

Atlas.

Atlas on site at Warren.

Atlas E (Coffin Launcher) Vandenberg AFB 1961.

Atlas F.

thickness of a dime. While this meant a significant savings in airframe weight, it also meant that if any leaks occurred when the missile was not fully loaded with propellants, the airframe would collapse under the weight of the payload. The first Atlas D squadron carried the General Electric Mark 2 heatsink reentry vehicle, armed with a W-49 warhead. All remaining Atlas D missiles carried the General Electric ablative heatshield Mark 3 reentry vehicle, armed with a W-49. Atlas E and F missiles carried the Avco ablative heatshield Mark 4 reentry vehicle armed with a W-38 warhead. The Atlas range was 5,500 nautical miles. Atlas D using radio command guidance while the E and F models used an inertial guidance system.

Atlas A, B and C models were used in the research and development program. Atlas A carried a dummy nosecone and sustainer engine. Designed to test for major flaws in the basic design, Atlas A was first launched on 11 June 1957. After approximately one minute of flight, one of the booster engines failed and the missile was destroyed by the range safety officer. Three months later, the second launch was successful but again powered flight was cut short due to engine malfunction. On 17 December 1957 the third Atlas A test was a complete success. The Atlas B and C models utilized all three main engines with a total of 25 Atlas A, B and C flights before the first operationally configured Atlas D flight. Atlas D testing began on 29 July 1959. On 9 September 1959, Vandenberg AFB launched its first Atlas D. Six weeks later, on 31 October 1959, the first Atlas D was placed on alert at Vandenberg AFB Launch Complex 576A-1. Atlas E, powered by improved engines and carrying the Mark 4 reentry vehicle, was first tested on 8 March 1960 and Atlas F with minor

modifications to permit silo-lift launcher operation, was first tested on 1 August 1962.

Atlas was deployed in four shelter configurations. The first operational and training sites at Vandenberg AFB, armed with Atlas D, had the missiles maintained vertically in their gantry launchers (PGM-16D). The Air Force realized that this was an extremely vulnerable configuration but wanted three missiles on alert as soon as possible. Atlas Ds were also deployed in above ground "coffin" shelters which offered better environmental protection but virtually no blast protection (CGM-16D). Atlas E was deployed in below ground coffins with improve blast protection (CGM-16E) but only with the deployment of Atlas F in the silo-lift configuration (HGM-16F) was true hardening against nuclear blast achieved.

With the successful depolyment of the first Minuteman I squadron in 1963, the Air Force realized that the Atlas D and E series missiles were obsolete. Deactivation began on 15 May 1964 and was completed on 17 February 1965. Atlas F was soon to follow, with deactivation beginning on 30 December 1964 and completed on 12 March 1965.

Atlas missiles were heavily involved in the manned space program as the launch vehicles for both the Mercury Program and Agena docking targets for the Gemini Program. A total of 112 Altas E/F missiles with reentry vehicle payloads were flown from 11 October 1960 to 13 October 1974. Of these, 54 were used for development and operational training and the other 58 used for research and development testing of advanced reentry vehicle systems and for launching target reentry vehicles for the Nike-Zeus anti-ballistic missile system

Titan I (SM68, HGM-25A)

In 1954 the Air Force began to study a recommendation of the Atlas Scientific Advisory Committee that a second ICBM design be developed. The original rationale for the one and one-half stage design of Atlas had been the concern about ignition of rocket engines in the near vacuum at high altitudes. This was no longer considered an unsurmountable problem and so a true two stage missile should now be developed. One obvious and often overlooked advantage was that such a missile could be transported in two pieces, something that was not feasible for Atlas and made for some logistical difficulties. An additional compelling reason was the simple need for competition in the growing field of ICBM development.

On 27 October 1955, the Glenn L. Martin Company design for the alternative ICBM was selected for design and development. Not only was Martin Company responsible for the missile, it was also to play a key role in the complete development of the weapon system basing equipment. From the onset the new missile was to be housed in the silo-lift configuration with 3 missiles per launch complex, three launch complexes per squadron.

Titan I was 98 feet long, weighed 220,000 pounds at launch, with a 10 foot diameter for Stage I and an 8 foot diameter for Stage II. Martin Company designers internalized the support structure, saving the weight of a dedicated exterior skeleton yet still making the structure self-supporting. Innovative chemical milling and aluminum welding procedures were developed during the Titan I design and development program. Stage I was powered by two Aerojet LR87-AJ-3 engines burning liquid oxygen and

Titan I installation crane.

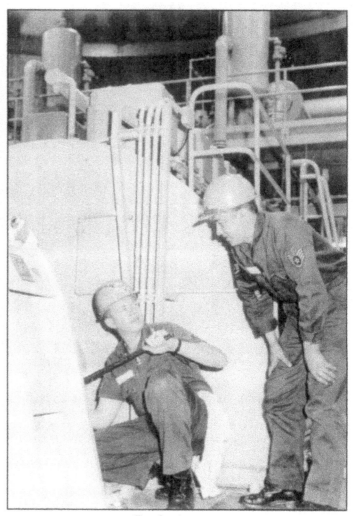

Titan I maintenance.

3 Titan I topside.

Titan I and B-52.

RP-1, producing 300,000 pounds of thrust. Stage II was powered by a single LR91-AJ-3 engine that burned the same propellants and produced 80,000 pounds of thrust. The Avco Mark 4 reentry vehicle carried a single W-38 or W-49 warhead. Range was 5,500 to 6,300 nautical miles. Although initially programmed for inertial guidance, the Titan I inertial guidance system was switched to the Atlas E and F missiles, leaving Titan I with a radio command guidance system similar to that of Atlas D.

Flight testing began in December 1958 with the first of Lot A missiles configured with a live Stage I and simple radio guidance but with a dummy Stage II loaded with water. The first launch took place on 6 February 1959. On 4 May 1958, successful staging was achieved, as the two Stage II translation rockets pulled the dummy Stage II away from Stage I. The Lot B missiles had the first fully powered Stage II though engine duration was limited. The Lot B flight test program was marred by static test fire stand explosions at Denver and on-pad explosions at Cape Canaveral. The first successful flight of a Lot B missile took place on 27 January 1960, impacting 2,020 nautical miles down range. Lot C testing began with a spectacular explosion as the missile's range destruct package inadvertently fired due to the shock of the hold down bolts releasing. Flights of the Lot G and Lot J missiles were highly successful and the Titan I program was ready for deployment.

The first Titan I squadron was declared operational on 18 April 1962 and the missiles were placed on alert on 20 April 1962 and all six squadrons were operational by 5 May 1963. Two squadrons of nine missiles were located at Lowry AFB, Colorado and four squadrons of nine missiles each were located at Mountain Home AFB, Idaho; Beale AFB, California; Larson AFB, Washington; and Ellsworth AFB, South Dakota. Deactivation of Titan I took place simultaneously with the Atlas program. The first Titan I was taken off alert on 4 January 1965, the last on 1 April 1965.

The Second Generation

Atlas and Titan I represent the first general of ICBMs. Propulsion technology at the time of the inception of their design required the use of RP-1 and liquid oxygen as fuel and oxidizer, respectively. Storable liquid propellants were not yet developed and solid rocket propellant, at the time, was not powerful enough to lift the initial atomic and thermonuclear weapon designs.

The Air Force, realizing that Atlas and Titan I were but the initial part of the ultimate blend of weapons for the strategic missile force, continued research aimed at developing a more easily maintained and more rapidly launched ICBM force. On a normal day, over 80 operations, maintenance, support, security and food service occupied a Titan I site, all to operate 3 ICBMs. The complexes were only hardened to 100 psi, and response time was at least 15 minutes, with the missiles sitting in the open for the last few minutes before launch. Even though both Atlas and Titan I were undergoing major

modifications to improve reliability, all were deactivated by June 1965. Two developments permitted this massive retirement of the first generation of ICBMs.

The first was the breakthrough in solid rocket propellant technology that resulted in sufficient power to launch the newest and smallest thermonuclear warheads. This breakthrough resulted in the Minuteman and Peacekeeper series of ICBMs. The second was the development of a hypergolic, ignition on contact, liquid propellant combination which lead to the Titan II ICBM.

Titan II (SM-68B, LGM-25C)

Titan II evolved from the Titan I program as a result of a study conducted to simplify Titan I operations. In April 1960 the first Titan II development plan was released. With the launch of a modified Titan I, VS-1 (live Stage I, water ballasted Stage II) from the Silo Launch Test

Facility, Vandenberg AFB, California, in May, 1961, the Air Force confirmed that a large ICBM could survive the acoustical environment of a confined launch within an underground silo structure. This was the culmination of several years of research into silo launch concepts involving Aerojet Corporation and Martin Company. Development of the Titan II missile began in June 1960, one month after Martin Company signed a contract with the Air Force for development of the new missile.

Titan II was 103 feet long and 10 feet in diameter, weighing 330,000 pounds at launch. Fabrication of the Titan II airframe was nearly identical to that of Titan I, a feature that sped the development of the Titan II program considerably. Stage I was powered by an Aerojet LR87-AJ-5 engine composed of two thrust chambers, capable of generating 430,000 pounds of thrust. Stage II was powered by one Aerojet LR91-AJ-5 engine, capable of generating 80,000 pounds of thrust. These engines represented a

Titan II Crew 5-116, Little Rock.

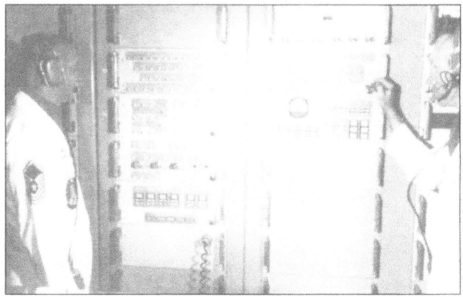

Titan II BMAT and MFT.

Titan II complex.

Titan II in Silo.

concerted effort by Aerojet to simplify engine operation and control, significant reducing the number of control features used on the Titan II engines when compared to the Titan I engines. The payload was a General Electric Mark 6 ablative heatshield reentry vehicle that carried a W-53 thermonuclear warhead, the largest yield nuclear weapon in the strategic missile force. Titan II had a range of 5,500 nautical miles and carried an A.C. Sparkplug all inertial guidance system.

The first launch of Titan II took place on 16 March 1962 with the successful launch of N-2 from Cape Canaveral, Florida. N-2 was a fully configured missile with both stages capable of fully powered flight. Slightly less then one year and 10 launches later, the first launch of a Titan II from its silo took place. While missile N-7 cleared the silo at Launch Complex 395-C, Vandenberg AFB, demonstrating that the in-silo launch was completely successful, failure of several umbilicals to pull free caused the guidance system to be inoperative. The missile blew up at 18,000 feet due to inadvertent stage separation, showering the beach and ocean near Launch Complex 395-C with debris. Of the remaining 21 research and development launches, of which 8 were from the silo facilities at Vandenberg AFB, 17 were completely successful, including the last 12. The first operationally configured missile, B-15, was launched successfully on 17 February 1964 from Launch Complex 395-B, Vandenberg AFB.

Titan II was deployed in the 1x9 configuration, nine missiles per squadron, two squadrons per missile wing. A total of 54 missiles were deployed, 18 at the 308th SMW, Little Rock AFB, Arkansas; 18 at the 381st SMW, McConnell AFB, Kansas and 18 at the 390th SMW, Davis-Monthan AFB, Tucson, Arizona. The Titan II system was declared fully operational on 31 December 1963.

The decision to deactivate Titan II came as a result of two accidents and the desire of President Ronald Reagan to modernize our strategic nuclear forces. The oxidizer spill at Launch Complex 533-7, near Rock, Kansas in August 1978 and the catastrophic explosion

at Launch Complex 374-7 near Damascus, Arkansas in September 1980 were caused by human error, not poor facility or missile design. Nonetheless, opponents argued that Titan II was clearly more expensive to maintain than the Minuteman system. With only 52 missiles and hence 52 targets no longer covered, deactivation represented an opportunity to retire an older system with the potential to begin development of a newer modern system.

It must be kept in mind that the Titan II program had been designed to last for 10 years and at the point of deactivation was nearly 20 years old. An upgrade to the guidance system in 1976-78 had eliminated the single most serious logistical problem, that of a guidance system with few remaining spare parts. Improvements in oxidizer tank seals had eliminated most of the corrosion problems and an upgrade in system facilities was underway at the time of deactivation.

The deactivation process began at Davis-Monthan AFB with the first launch complex taken off alert on 2 July 1982 and ended with the last launch complex taken off alert at Little Rock AFB on 3 May 1987.

Minuteman (SM80, LGM-30A,B,G,F)

While the Minuteman program was initially funded on 9 October 1958 with the selection of Boeing Company Aerospace Division as the prime contractor, critical research to optimize large diameter solid propellant rocket engines had begun in September 1955. Aerojet-General Corporation, Grand Central/Lockheed and Phillips Petroleum/Rocketdyne were part of the Air Force Large Scale Solid Rocket Feasibility Program. After one year, Aerojet-General was selected to remain in the program and Thiokol and Hercules were added to the project. While many aspects of solid propellant rocket engines needed to be worked out, one in particular was the need to generate improved range. This was accomplished by both improving thrust performance and by optimizing the trajectory of

the missile. While liquid propellant missiles had relatively delicate airframes and could not be pitched over into the optimum trajectory quickly, solid propellant missiles were far more robust. Early and rapid pitch-over optimized the flight profile and created extra range.

Unlike Atlas and Titan, much of the early testing of Minuteman took place at Edwards AFB, California, where test models with limited propellant loads were used to finalize the in-silo launcher configuration in a series of eight test "flights." The first launch of a operationally configured Minuteman I took place on 1 February 1961 at Cape Canaveral and was successful, flying the full 4,600 nautical miles.

Ironically, the Minuteman I test flight program mirrored the Titan II program in that the first launch was successful but the first silo launch, on 30 June 1961, ended with a spectacular explosion just as the missile cleared the silo. The Stage II motor ignited prematurely and while the silo was not damaged, the largest explosion seen at the Cape to date was the result. The first successful silo launch on 17 November 1961 was successful. On 29 June 1962 the first SAC crew launched a Minuteman IA missile. On 3 July 1963 the first Minuteman I squadron became operational.

Minuteman is a three stage missile and was deployed in four models; Minuteman I (LGM-30A and LGM-30B), Minuteman II (LGM-30F) and Minuteman III (LGM-30G). All used an inertial guidance system built by the Autonetics Division of Rockwell International. The greatest diameter for all Minuteman models was 5 feet 6 inches. Minuteman IA, 53 feet 8 inches long, carried a single Avco Mark 5 reentry vehicle containing a W-59 warhead with a range of 6,000 nautical miles. Stage I thrust was 210,000 pounds; Stage II, 60,000 pounds; and Stage III 35,000 pounds. Minuteman IA weighed 65,500 pounds. Minuteman IB, 55 feet 11 inches long, still carried only one reentry vehicle, the Avco Mark 11 containing a W-56 warhead but range was increased to 6,300 nautical miles. Stage I, II and III thrusts were the same as Minuteman IA as was total missile weight.

Minuteman II, 59 feet 9 inches long, also carried the Avco Mark 11 reentry vehicle and W-56 warhead but had a range of over 7,000 nautical miles and utilized an improved inertial guidance system that increased its targeting capability. Stage I thrust was 210,000 pounds; Stage II 60,300 pounds; and Stage III 35,000 pounds. Minuteman II weighed 73,000 pounds.

Minuteman III, 59 feet 9 inches long, was redesigned from Stage II up to permit the addition of a multiple independently targetable reentry vehicle system capable of carrying three General Electric Mark 12 or Mark 12A reentry vehicles each containing a single W-64 or W-78 warhead respectively. Stage I thrust was 210,000 pounds; Stage II 60,300 pounds and Stage III 34,300 pounds. Minuteman III weighed 78,000 pounds. A post-boost vehicle was used to position the reentry vehicles and associated decoys on different trajectories.

Another important feature of the Minuteman program was the upgrading of targeting capabilities throughout the life of the program. Minuteman IA was limited to one target for its guidance system. Minuteman IB was expanded to two targets, Minuteman II to eight targets and Minuteman III could accept three different sets of targets for the cluster of reentry vehicles.

Early in Minuteman development, two basing modes were considered, in-silo launch and mobile basing using the nation's rail system. The in-silo launch system was chosen for deployment but serious consideration and equipment evaluation of the mobile launch mode took place over a two year period, only to be canceled on 1 December 1961. At one point, the Air Force requested funding for 10,000 Minuteman missiles but this was soon cut back to 800 and then increased to the final 1,000 that

Minuteman II (W5133B) Deputy Console Charlie Simpson and Darrell Downing.

Minuteman dual launch.

Minuteman LCC at Ellsworth.

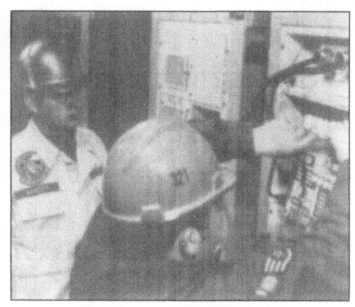

321st Maintenance with 3901 SMES evaluation.

Minuteman III Maintenance.

Minuteman modernized LCC.

were deployed. In 1974, an airborne deployment mode, where the missile would be extracted by parachute out of a C-5 aircraft at 20,000 feet and stabilized by parachutes until it reached 8,000 feet and ignited, was tested but not adopted.

The actual deployment mode of the Minuteman missile was also unlike any of the previous systems. In Atlas and Titan I and II, the launch control center was in close proximity to the missile silo. With Minuteman, since maintenance was greatly reduced due its solid propellant, each launch control center was responsible for 10 widely dispersed missiles, each missile was at least five miles from a launch control center and three miles from another missile. Each squadron had five "flights" of 10 missiles each, with each wing composed of three squadrons. This made for a widely dispersed missile field, difficult to target completely.

A description of Minuteman deployment is necessarily complicated due not only the large number of missiles involved but also the replacement of Minuteman IA and IB missiles by Minuteman II and replacement of Minuteman II by Minuteman III. Each base is listed in order of activation. A summary table is not included.

Malmstrom AFB, Montana

Minuteman IA was deployed only at the 341st SMW in three squadrons, the 10th SMS, the 12th SMS and the 490th SMS. The first flight of Minuteman I missiles (Alpha Flight, 10th SMS) were placed on modified alert status during the Cuban Missile Crisis of October 1962. The 341st SMW became fully operational in October 1963. In July 1964, the Air Force announced that the twentieth and final Minuteman squadron would be stationed at Malmstrom, composed of 50 Minuteman II missiles with the 564th SMS. New sites were built and the 564th SMS was declared fully operational in May 1967.

The Minuteman Force Modernization Program (MFMP) called for the replacement of all Minuteman IA and IB missiles with Minuteman IIs once the new Minuteman II sites were completed. The 10th, 12th and 490th SMS facilities were modernized to accept the Minuteman IIs advance guidance and targeting capabilities and by 27 May 1969, this activity was complete as the 341st SMW was now completely converted to Minuteman II. Malmstrom AFB was now host to 200

Minuteman II missiles. In 1975, the 50 Minuteman II missiles of the 564th SMW were replaced with the upgraded Minuteman III missile and the 341st SMW returned to fully operational status on 8 July 1975. Malmstrom AFB now had 150 Minuteman II and 50 Minuteman III missiles on strategic alert.

Ellsworth AFB, South Dakota

Minuteman IB was deployed in the 44th SMW composed of the 66th, 67th and 68th SMS. Each squadron had 50 missiles and the wing was declared fully operational on 24 October 1963. The MFMP replaced all 150 Minuteman IB missiles with Minuteman II, with the replacement completed 13 March 1973. The 44th SMW was deactivated in July 1994.

Minot AFB, North Dakota

The 455th SMW, composed of the 740th, 741st and 742nd SMS, received 150 Minuteman IB missiles, becoming fully operational 19 March 1964. On 25 June 1968, the 455th SMW was deactivated and its assets assigned to the 91st SMW. The MFMP replaced all Minuteman

IB missiles with Minuteman III during 1970 and 1971. On 13 December 1971, the 90th SMW was fully operational with 150 Minuteman III missiles.

Whiteman AFB, Montana

The 351st SMW, composed of the 508th, 509th and 510th SMS, received 150 Minuteman IB missiles, becoming fully operational on 30 June 1964. The MFMP replaced all Minuteman IBs with Minuteman IIIs in 1966-67 with the 351st SMW becoming fully operational with Minuteman IIIs on 3 October 1967.

F.E. Warren AFB, Wyoming

The 90th SMW, composed of the 319th, 320th, 321st, and 400th SMS received 200 Minuteman IB missiles, becoming fully operational on 30 June 1965. The MFMP replaced all Minuteman IBs with Minuteman IIIs in 1973-74, with the 90th SMW becoming fully operational with Minuteman IIIs on 26 January 1975.

Grand Forks AFB, North Dakota

The 321st SMW, composed of the 446th, 447th and 448th SMS, received 150 Minuteman II missiles, becoming fully operational on 22 November 1966. The MFMP replaced the Minuteman II missiles with Minuteman III and the 321st SMW became fully operational with Minuteman IIIs on 8 March 1973.

Third Generation

Peacekeeper (LGM-118A)

In 1974, development began on a new ICBM, called Missile-X, or MX. Over the next seven years, a wide variety of basing modes were evaluated, including rail, ground vehicle and air mobile, using a number of different shelter concepts. In 1981, the decision was made to deploy the MX, now called Peacekeeper, in existing Minuteman III silos.

Several technological advances make Peacekeeper unique. First was the use of Kevlar epoxy composite instead of stainless steel, aluminum or titanium as in Minuteman. This greatly reduced the weight of the Peacekeeper airframe, permitting greater range and a larger payload. The second innovation was the use of an extendable nozzle for the second and third stages. This permitted installation in existing Minuteman III silos. The third innovation is the use of the cold launch technique. Using the lessons learned in development the Polaris, Poseidon and Trident missile launch systems from ballistic missile submarines, Peacekeeper is propelled out of its silo using a steam generator at the base of its shipping cannister, to an altitude of 150-300 feet where Stage I ignites and powered flight begins.

Peacekeeper was built by Martin Marietta Company and is 70 feet long and 7 feet 8 inches in diameter and weighs 192,300 pounds. Inertial guidance is used to accurately target up to 10 independently targetable Avco Mark 21 reentry vehicles carrying one W-87 warhead. Stage I produces 500,000 pounds of thrust; Stage II 300,000 pounds of thrust and Stage III 125,000 pounds of thrust.

Initially, 114 Peacekeepers in silos were planned, but deployment was capped at 50 in 1990, deployed in fifty modified Minuteman III silo of the 400th SMS, 90th SMW at F.E. Warren AFB. Installation began in June 1986. The first missile was placed on alert on 10 October 1986. The fifty Peacekeepers will be removed as part of START II by 2004.

Missile Competition
by Colonel (Retired) Charles G. Simpson, Member Number L0001, and a competitor in 1969 and 1970

The Air Force has conducted competitions almost throughout its history. Our leadership learned a long time ago the competitive events hone the combat capabilities of our forces. William Tell, Gunsmoke, Bomb Comp, Peacekeeper Challenge; these are just some of the highly competitive events that have been conducted over the years to test our interceptors, fighters, bombers, security forces and almost every other specialty in the USAF.

In late 1966, the Commander in Chief, Strategic Air Command announced plans for the first SAC Missile Combat Competition, to be called Curtain Raiser. This competition would be conducted primarily at Vandenberg AFB, California, with some parts of the new Minuteman II competition taking place at Malmstrom AFB, Montana. The missile competition started small. Curtain Raiser competitors included two combat crews and one target and alignment team from each of the six Minuteman and three Titan II wings. The big prize was the Blanchard Trophy, awarded to the wing with best overall score. The trophy was named for General William H. Blanchard, a former Vice Chief of Staff of the Air Force. General Blanchard was Director of Operations for SAC when the first ICBMs entered the Air Force inventory.

Due to the pace of the war in Viet Nam, SAC canceled the 1968 competition, but resumed in 1969 with Olympic Arena, the name that would identify the Missile Combat Competition until Air Force Space Command renamed the event in 1993.

In 1969, each Minuteman wing sent two-man missile combat crews and two maintenance teams, a three-man combat targeting team and a five man missile maintenance team. The Titan teams included two four-man combat crews, a three-man ordnance team, a three-man target alignment team and a seven-man reentry vehicle team. Three numbered air forces were represented, 2nd, 8th and 15th Air Force. The competition lasted a week in the early days - with six full days of competition. Each missile crew went through three missile procedure trainer rides, and each maintenance team took part in multiple events. The 321st Strategic Missile Wing from Grand Forks took the Blanchard home in 1969, with the second narrowest margin in competition history, winning by only 1.5 points out of a 4,800 point total.

The scope of the competition remained basically unchanged until 1972, when the number of participants was significantly increased. The

A Peacekeeper ICBM, like the one above, was launched at 1:01 a.m. from Vandenberg AFB, California and travelled 4200 miles to its target in the Marshall Islands on September 11.

number of combat crews was doubled, Titan maintenance added electronic laboratory and missile handling teams and Minuteman added electro-mechanical and missile handling teams. In 1973, Minuteman I competition functions were conducted at Warren AFB, Wyoming, since there was no longer a Minuteman I silo available at Vandenberg AFB. In 1975, the first female competitors participated, both as missile crewmembers and security police team members. The size of the competition teams grew again the next year, when the competition was expanded to include communications, civil engineering and vehicle support teams. The security police presence was changed from three to seven team members. Size of the teams shrunk in 1978 (budgets affect competitions, too), with fewer competitors and elimination of the vehicle support teams. For 1978, 1979 and 1980, the Titan maintenance events were conducted at McConnell AFB, Kansas, because a Titan II site wasn't available at Vandenberg AFB.

The Olympic Arena title was dropped for 1982-1985, with the event known only (at least officially) as the SAC Missile Combat Competition. In 1982, the 44th Strategic Missile Wing from Ellsworth edged the 381st Strategic Missile Wing from McConnell by a single point out of a 3,000 point total.

In 1983, Titan wings finished 1, 2, 3, and the 381st Strategic Missile Wing from McConnell joined the 351st Strategic Missile Wing from Whiteman as a four-time winner of the Blanchard Trophy. The Olympic Arena title returned in 1986, and it was the final year for Titan II in the competition. Peacekeeper entered the competition in 1988, and the first place wing was victorious by a single point. In 1991, the 341st Strategic Missile Wing from Malmstrom

1969 Competition Headquarters.

321 SMW Team, 1969.

AFB garnered the first ever back-to-back win. The security police were not represented because of the Gulf War.

1992 marked the 25th anniversary of the competition, and the last for the Strategic Air Command. In 1992 and 1993, units that were competing for the last time, the 44th Strategic Missile Wing from Ellsworth and the 351st Strategic Missile Wing from Whiteman 1993 also marked the one year that Air Combat Command was the host command since the ICBM force was in transition after the death of SAC. 1994 meant a new command, a new name and a significant change in the competition. For the first time, the event integrated ICBMs and space forces, and was called Guardian Challenge. The initial Space and Missile Competition included four Minuteman units and space units from around the world. The format changed to a "squadron" competition (similar to Gunsmoke), with the Blanchard trophy going to the winning Missile unit. Missile teams included combat crews, four teams from missile maintenance, security police, communications and helicopter operations teams. In 1998, the competition returned to a wing format.

The names, the format and the scope of the annual competition has changed through its first thirty years, but the basic goals have remained the same. For those who have forgotten the reasons for the event, they are to:

- recognize superior people

- enhance esprit de corps and strengthen teamwork

- improve readiness and combat capabilities through preparation, innovation, competition and teamwork

- create competition-tough crews

- send the message that the command is prepared

Blanchard Trophy Winners

1967 - 351SMW, Whiteman (MM)
1969 - 321SMW, Grand Forks (MM)
1970 - 44SMW, Ellsworth (MM)
1971 - 351SMW, Whiteman (MM)
1972 - 381SMW, McConnell (Titan)
1973 - 90SMW, Warren (MM)
1974 - 321SMW, Grand Forks (MM)
1975 - 381SMW, McConnell (Titan)
1976 - 341SMW, Malmstrom (MM)
1977- 351SMW, Whiteman (MM)
1978 - 91SMW, Minot (MM)
1979 - 390SMW, Davis Monthan (Titan)
1980 - 381SMW, McConnell (Titan)
1981 - 351SMW, Whiteman (MM)
1982 - 44 SMW, Ellsworth (MM)
1983 - 381SMW, McConnell (Titan)
1984 - 90SMW, Warren (MM)
1985 - 308SMW, Little Rock (Titan)
1986 - 341SMW, Malmstrom (MM)
1987 - 321SMW, Grand Forks (MM)
1988 - 91SMW, Minot (MM)
1990 - 341SMW, Malmstrom (MM)
1991 - 341SMW, Malmstrom (MM)
1992 - 44 MW, Ellsworth (MM)
1993 - 351MW, Whiteman (MM)
1994 - 742 Missile Squadron, Minot
1995 - 10th Missile Squadron, Malmstrom
1996 - 319th Missile Squadron, Warren
1997 - 320th Missile Squadron, Warren
1998 - 341st Space Wing, Malmstrom

Bibliography

"Administrative History of Ballistic Systems Division, 1 July to 31 December 1963," K243.012, AFHRA, Maxwell AFB, AL. This document is classified as SECRET. The information used is unclassified.

"Aerojet, the Creative Company," edited by the Aerojet History Group, 1995, S.F. Cooper Company, Los Angeles, California.

"A History of the U.S. Air Force Ballistic Missiles," E.G. Schwiebert, 1965, F.A Praeger, New York, New York.

"Air Force Missile Patches, Official and Unofficial, 1954-Present," G. Olgetree, 1997, Patriot Press, Lompoc, California.

"Ballistic Missiles in the USAF, 1945-1960," 1989, Jacob Neufeld, Office of Air Force History, United States Air Force, Washington, D.C.

"From Snark to Peacekeeper: A Pictorial History of Strategic Air Command Missiles," 1990, Office of the Historian, HQ Strategic Air Command, Offutt AFB, NE.

"Ground-Launched Cruise Missile Fact Sheet, 83-14," United States Air Force.

"Missileers' Heritage," F.X. Ruggiero, Report Number 2065-81, Air Command And Staff College, Air University, Maxwell AFB, AL

"SAC Missile Chronology, 1939-1988," Office of the Historian, HQ Strategic Air Command, Offutt AFB, NE.

"Strategic Air Command: People, Aircraft and Missiles, 2nd Edition," edited by N. Polmar and T. Laur, The Nautical and Aviation Publishing Company of America, Baltimore, MD.

"The 1966 Aerospace Year Book," 1962, Spartan Books, Washington, D.C.

"The Air Officers Guide," 1962, Stackpole Company, Harrisburg, PA.

"The Development of the SM-68 Titan," by Warren E. Greene, 1962, Historical Office, Air Force Systems Command.

"The Evolution of the Cruise Missile," Kenneth P. Werrell, 1985, Air University Press, Maxwell AFB, Alabama.

"The History of the U.S. Nuclear Arsenal," J.N. Gibson, 1989, Brompton Books Corporation, Greenwich, Connecticut.

"Tomahawk Cruise Missile," N. Macknight, 1995, Motorbooks, Osceola, WI.

"U.S. Nuclear Weapons, The Secret History," C. Hansen, 1988, Aerofax, Inc., Arlington, Texas.

Individuals

Chief Master Sergeant (Retired) Walter Kundis, AAFM Member Number A0640
Major Gregory Ogletree, AAFM Member Number LOO49
Colonel (Retired) R. Douglas Livingston, AAFM Member Number A1138
Colonel (Retired) John Bacs

What Do We Call Them?

What do we call them? Initially, missiles were designated like aircraft - as bombers or fighters. Atlas was the B-65, Matador the B-61 and BOMARC the F-99. Then, in the 1950s, somebody realized that missile were were unmanned vehicles and they became Strategic Missiles (SM-65 for Atlas), Tactical Missiles (TM-61 for Matador) and Interceptor Missiles (IM-99 for BOMARC). In the early 1960s, the Department of Defense (DOD) developed the standard DOD designation system for all aircraft and missiles. Every missile now had three prefix letters and a suffix. The first letter indicated the "launcher" (C for coffin, H for silo/elevated to launch and L for silo launched). The second letter indicated the mission (G for surface target, A for aerial target) and the third letter was M for missile. The suffix was the series - the first generation was the A, the second B, and so on. Therefore, the Atlas E became the CGM-16E, the Titan I the HGM-25A and Minuteman II the LGM-30F.

AAFM Vice President Col (Ret) Jim Burbo as 1973 "Voice of the Big Board."

General Bruce Holloway presents 1969 Blanchard Trophey.

Air Force Missileers
Special Stories

A Minuteman intercontinental ballistic missile is silhouetted against the setting sun as it lifts from its underground silo at Vandenberg. The solid fueled Minuteman has an effective range of more than 5,000 miles.

Special Stories

The Role Of The Surveyor
by Victor J. Haas

As a Geodetic Surveyor in the USAF, I feel that it is important to give the members of the AAFM a little history on the key role of the Geodetic Surveyor and the various missiles in the Air Force. I have been retired from the Air Force since 1967, but I will submit an article on the involvement and importance we played in the overall missile programs.

My first assignment to an Air Force Missile Unit was in 1955, when I was assigned to the 1st Pilotless Bomber Squadron, which later was redesignated as the 1st Tactical Missile Group at Bitburg AFB, Germany. I was assigned as NCOIC of the Target Data Section which was under the Missile Operations Center. Our primary function in this unit was to deploy our survey crews to various areas in Germany and to perform surveys of the Missile Guidance Sites as well as possible launch sites. We always maintained a ready status by constantly maintaining the missile target data current, which involved constantly receiving and updating missile data based on current weather information as deemed necessary. I continued my career in the MATADOR while being assigned to the 587th Tactical Missile Group in Tainan, Taiwan. The same duties were performed in this unit as with the 585th Tactical Missile Group.

In 1958, I was assigned to a detachment at Orlando AFB, FL. This detachment was under the 1370th Photo Mapping Wing. Turner AFB, GA. A short time later we became the 1381st Geodetic Survey Squadron (MAC), Orlando AFB, FL. This organization was immediately involved in performing very precise Geodetic Surveys for the Thor Missile Sites in England. The personnel in this organization were constantly deployed to perform, not only the original surveys, but also check surveys for the Thor Sites. The 1381st Geodetic Survey Squadron later also became very much a part of many of the AF Missile Sites requiring very precise surveys. These surveys included the Minuteman, Atlas, Titan, Jupiter, GAM-77 and other related required surveys. This was the primary function of the 1381st GSS. Our military personnel as well as civilian advisors, not only performed the Geodetic Control Surveys, but also were required to perform precise Gravity. Astronomic as well as the Geodetic Surveys of these various missile facilities. As NCOIC of the Missile Site Survey Branch of the 1381st GSS, I was very proud of having the privilege of serving the Air Force in this organization.

We were a rare breed of Air Force personnel as we were the only Geodetic Survey Squadron in the USAF. We were a very proud, close knit family, always looking out for each other and their families as our assignments were always on TDY status.

US Naval Academy Class of 1954 "Cold War Warriors"

The following story was compiled from material submitted by Major General Richard Boverie, USAF (RET) and Colonel Michael A. Nassr, USAF (RET). Colonel Nassr was the editor for the US Naval Academy Class of 1954, 40th Anniversary History, titled "Cold War Warriors." Lieutenant General Aloysius G. Casey, USAF (RET) also was instrumental in the writing and development of the history.

It was a watershed year. Not only did 1950 open the second half of the 20th century, it

Minuteman III launch.

35

marked the beginning of profound changes to the American scene. As the first members of the US Naval Academy Class of 1954 enrolled on 14 June 1950 (Flag Day), the country was shifting into high gear following World War II. The West was challenged by the Soviets in 1948 and had to launch a massive airlift to break the Berlin blockade. A new kind of war (the Cold War) unfolded. In 1949 Chiang Kai-shek left Mainland China to the Communists, and the Soviets caught us by surprise with its first atomic bomb. That same year the North Atlantic Treaty Organization (NATO) was established to counter growing threats of Communist aggression.

As entering midshipmen in 1950, we had no way of knowing that our nation's future faced four decades of prolonged vigilance, brinkmanship and outright hostilities. But by the time our last members were retiring at the end of the 80s, the Cold War would be won and the threats of communism and a nuclear holocaust would have suddenly faded away.

Transporter-Erector enroute to site-91st SMW Minot.

1969 Strategic Air Command Missile Combat Competition Vandenberg Air Force Base, California. Lt. Gen. Jack J. Catton, 15th Air Force commander, third from left, and Col. B.H. Davidson, 91st Strategic Missile Wing commander, salute the top combat crew and the crew with the highest single score in the 1969 SAC Missile Combat Competition. Crew S-102, Capt. Dick D. Ingram, left, and 1st Lt. Joseph A. Pate III, second from left, made the highest score so far with 378.5 points out of the possible 400. Crew s-006 1st Lt. Elbert J. Leuty Jr., second from right, and 2nd Lt. Thomas M. Banning III, are the top crew in the competition with a remarkable 92.3 percent of the possible 1200 points. Both crews are members of the 91st SMW, Minot AFB, ND.

Missile and Space Activities

Space and missile systems had a very significant role in the Cold War. They became the strongest capability of the Soviet Union and a major factor in the final demise of its empire. In the period that the "Class of 54" served our nation, these systems were designed, developed, deployed and, in the minds of many became the decisive edge which finally persuaded the Soviet Union it could not win the Cold War.

The first attempt by the Air Force in the early 50s to field a surface-to-surface missile was called "a pilotless aircraft" because the budgeteers had no category for missiles. It was crude in today's terms, basically and aircraft without a cockpit which was launched from a rail and guided by radar to a point where it dove to the approximate vicinity of the target. It came to be known as the TM61 "Matador," a supersonic missile able to carry 3,000 pounds of explosives with a range of about 500 miles.

Twelve of our classmates entered guided missile training after graduation and took on all the associated tasks of flying this bird. Involved in this pioneering effort were Jack Appel, Dick Boverie, Ken Burns, Al Casey, Ed Defede, Dale Gramley, John Miller, Stan Prahalis, Dick Proffitt, Chuck Prohaska, Dick Sassone and Carl Soreco. After schooling in Denver, they reported to Orlando, FL, where operational units were assembled for deployment overseas. Carl Soreco recalls conducting live firings at Cape Canaveral from a wooden blockhouse that was sandblown by the rocket booster as the Matador left the launcher.

To their credit and that of others involved, the TM-61 was deployed and served as an early nuclear missile without major mishap. In 1956 many of our classmates set up one of the first tactical missile squadrons at Sembach Air Base in Germany, where they began conducting nuclear warhead check-outs, Red Alerts and tactical deployments. They arrived in time to respond to the Hungarian Revolution crisis of that year by deploying to secret wartime launch sites with their nuclear warheads pointing east. Humorously, the "secret wartime sites" were under the close scrutiny of high-ranking Russian liaison officers in West Germany who circled the sites while standing in their jeeps and observing operations through binoculars. Later, pilots Tann Lightsey, Jim Foster and Bill Miller served in a similar tactical missile squadron in Okinawa.

Of these original dozen officers, three became general officers. Al Casey went on to manage the Peacekeeper (MX) program and serve as Commander of Air Force Systems Command's Space Division as a lieutenant general. Dick Boverie served on the National Security Council Staffs of Presidents Nixon, Ford and Reagan and completed his USAF service as a major general and deputy assistant secretary of defense. The third is Ken Burns who became a pilot and won his stars commanding primarily flying organizations.

Classmates had active development and operational roles in the Air Force's many follow-on missile and space booster programs. Fred Clark worked on the Skybolt program, Creight Bricker on Skybolt and BOMARC and Al Freer became a combat crew commander for both the

A Transporter-Erector positions a Minuteman Intercontinental Ballistic Missile for loading into its underground launch silo at Vandenberg AFB, California. These specially built vehicles transport the missiles to the launch facility, erect them to a vertical position and lower them into the silo.

10 May 1966 Complex 374-3. Team play crew s-131. MCCC, Capt. Pillman; DMCCC, Lt. Kissel; BMAT, SMSGT Kundis; MFT, SMSGT Blaydes. Failed missile verification-replace actuaton.

Titan I Operational Launch Complex.

SNARK and Atlas missiles. Harvey Taffet became Director of Safety. AF Eastern Test Range. Pentagon leadership realized that Space Systems were key to all of the services, for military intelligence, missile warning communications, navigation and weather. This led to the formation of US Space Command in 1985 and classmate Bob Herres, as a four star, became its first commander.

Origin Of The Stump
TSgt Cliff Worthy, SSgt Larry Sutton, Sgt John Fowler, Sgt Jerry Spooner, Sgt Ron Wisniewski, A1/c Michael Hutstetler

In the early years of Olympic Arena, the annual Strategic Air Command Missile Competition, the nightly score posting ceremony became a competition of its own with each missile wing's cheering section attempting to drown out all others. Each group was constantly searching for that perfect mascot, icon or logo to best depict their superiority during score posting.

Mascots soon started to appear. There were cowboys, space invaders, vikings and even Teddy Roosevelt, but none to compare with the Arkansas Razorback waving a tree root and chanting "Root-Root-Root." None that is until the 1976 competition when six members of the 351st Missile Maintenance Teams Section (MMT) conspired to locate and procure the ultimate, the "Whiteman Stump."

There it was in a Eucalyptus grove next to Airfield Road on Vandenberg AFB. The MMT crew quickly set to work liberating the stump, cutting it to the proper size, adding a coat of polish and a pair of brass handles. At score posting that night, all other competitors rightly feared that they had been permanently trumped.

The Whiteman Stump will soon retire, a little shy of a full 20 year hitch, but this often decorated veteran deserves early retirement considering his leadership in helping Whiteman win more trophies than any other Missile Wing in Olympic Arena History.

Launch Of The "Blue Gander" Door
by LtCol.Robert E. Marsh

It was a Saturday night and our squadron, the 395th, had planned a party at the Officer's Club for that evening. Harold Korbel, who was my back up as GCO (guidance control officer), and I had been flying all day getting our time in for the month. When we got back to base operations they had a message for me to go out to OSTF (Operational System Test Facility) right away as they were going to have an unscheduled night test. We had been doing lots of tests of various functions of the system for a long time and this was to be the final test before launch, if everything went all right.

OSTF was the special test launch site for Titan I missiles so it did not have all of the underground tunnels that the hard sites would have. In fact, the Launch Control Center had a lot of extra monitoring equipment and even had a glassed in gallery for VIPs to watch what was going on.

The test went off very well with all of the various things that had to be done and I ran the guidance programs and talked to the bird and did a lot of surveillance of what was going on up topside with the TV cameras that I had.

Everything went well and our MMT team had gone down into the Silo to check and see that it was clear to lower the missile. This was of course a fully loaded missile with LOX and RP-1. Our team checked in that everything looked good in the silo and they would be clear of the second blast door in one minute (as they were still in the lock, the telephone on the safe side of the blast door wouldn't work). The test conductor held the count until they had time to get through and secure the last door on the blast lock. The lower launcher button was pressed and the bird started down. We could see everything that was going on as there were TV cameras at various places in the silo. The first thing that I noticed was that the missile was going down much faster than it normally did. It didn't take long to realize that we had a runaway. We all looked at each other and wondered how big of a blast it was going to be?

It was a big one. In fact we found out later that the explosion was seen in San Franciso. It was figured at two kilo tons!

Naturally we all stayed put and started checkups to see that everyone was all right, as they had search lights topside with people manning them for photographing as well as the fire trucks that were on site. I of course had run out of film in the cameras that I had on the antenna so didn't get any pictures.

It was quite a while before we were able to let our families know we were all right. The families living right on the base heard the explosion and experienced great anxiety until a phone message got through saying that nobody was hurt.

It wasn't until the next day that we really got to see what had happened. It cratered about the top 60 feet of the silo and took four floors of the five story underground equipment rooms up through the overbearing. The biggest thing we found was a 100-ton concrete silo door about a half mile away near the golf course. The other door was never found so we claimed the world record for putting something big into space.

We later found out the cause was a casting in the elevator hydraulic system broke and it did not have any pressure left to retard the elevator while lowering the missile. When the first stage LOX and RP-1 tanks hit bottom, they broke and exploded. Of course it was only seconds before the second stage did the same.

We were all fortunate that no one was seriously hurt. In fact one of the fellows topside who was manning one of the search lights had obviously been trying to light a cigarette when it started to drop. There was a string of cigarettes from where he dropped the matches to where he dove under the search light generator for protection, and just got a few scratches.

The Story Behind Thor - 151
From the files of Bob Carpenter

Circa 1958: Shortly after he assumed command, General David Wade expressed the desire that the first Thor IRBM launch from Vandenberg AFB be an Air Force, rather than a contractor launch.

A first firing was a requirement placed on the contractor by the Air Force to demonstrate that the weapon system was in order and functioning properly. Now, on the basis of General Wade's request, arrangements were made for intensive launch crew training of six personnel from the 392nd Missile Training Squadron.

Through the summer and fall of 1958, while these men were studying missile launch techniques and procedures, construction was

A picture of the last Atlas E missile to leave the 548 Strategic Missile Squadron at Forbes AFB, Kansas, in December 1964. Then, AIC Stan Bieleski is in the back row, against the missile without a hat.

speeded on the Thor emplacements. When the contractor at last reported he could have at least one pad in a near-operating configuration toward the end of the year, the Division initiated final planning for the launch to be conducted, first in November and then in December, when installation difficulties cropped up.

As the December launch date approached, a number of dress rehearsals were held for the special crew, who included Capt Bennie Castillow, launch control officer; Capt John C. Bon Tempo, systems monitor; SMSgt Charles E. Gifford, launch monitor control officer; MSgt Max L. Meyer, guidance alignment technician; MSgt William L. Hodges, missile technician; and MSgt Michael J. Aueri, missile technician.

All personnel involved in the exercise recognized the importance of a first successful launch. Thus it was with much relief and satisfaction when, on 16 December 1958, the launch crew assumed control of the weapon system for the final 15 minute countdown and successfully fired Thor Missile 151 some 1,500 miles over the Pacific Ocean. Lift-off was at 1544:45 PST.

Gen Wade listed a number of historic "firsts" from the launch: 1st ballistic missile fired from VAFB westward into the Pacific Missile Range; 1st known operation where new equipment, a new missile, new personnel and a new organization were integrated into a successful operation; 1st completely automatic launch of a fully operational ballistic missile using normal IOC launching procedures; and 1st ballistic missile launched by a SAC operational crew.

Was It Charmin?

by MSgt (Ret) Harold Renninger

The incident described here took place at Site One, 564th SMS, in the 389th SMW, the Atlas D and E Wing at Warren AFB, WY. The launch site was located 25 miles north and seven miles west on I-25 from Warren, on land now part of the Whitaker Ranch. The 564th SMS was the first combat-ready ICBM squadron in the USAF (the 576th was operational but not combat-ready). Our activities were closely monitored by SAC, the Pentagon and congress. My launch crew, Crew 3, was chosen to conduct a DPL (dual propellant loading) for a team of visitors from Washington, including members of the Joint Chiefs and several congressmen. Our missile was stored horizontally in a concrete bunker called a coffin complex. When the LCO (launch control officer), Lt Jason F. Mayhew (now LtCol (Ret) and an AAFM member) attempted to initiate the erection sequence with everyone watching, we had a red light on the erection bar. The maintenance crew at the pad, including me, began troubleshooting and discovered a faulty missile contact switch. In order to proceed with the erection, we needed a kluge (a term coined to indicate a by-pass for a problem). We decided that the cardboard tube in the center of a roll of toilet paper would be the perfect temporary fix since it could be compressed and would not damage the tank section if it fell out.

After installing this highly technical modification between the re-entry vehicle

November 1963, Crew S-01, 551 SMS Lincoln AFB, Nebraska. L to R: L/C John Campbell MCCC, Cpt Van Hunnm, DMCCC, SSGt Harold Wood BMAT, TSGT Rodney Fohr MFT, SSGT Lance Shirley EPPT.

adapter and the missile contact switch, we got a green light and the erection proceeded, normally. We told Lt Mayhew not to open the nose clamp and rotate the boom back nine degrees from the normal perpendicular launch position because the kluge would fall out and stop the normal DPL sequencing. The DPL was completed and the missile stood fully loaded with RP-1 and liquid nitrogen (we used LN2 in place of liquid oxygen (LOX) for safety during exercises). The nose clamp closed and the boom stayed firmly attached. Those of us on the maintenance crew at the launch pad are not really sure what went on in the LCC, nor how Lt Mayhew explained why the missile was not in launch configurations, but we heard that he told the visitors that it "was too windy to take the boom back." One of the congressman stated, "I hope we don't have to go to war on a windy day." LtCol Mayhew may want to offer us a view from "inside the blockhouse" - I would sure like hear his story.

One More Time

by Col (Ret) Larry Hasbrouck

As the senior instructor for the ALCS at Ellsworth (1963-67) I was tagged as the launch director for airborne launches from Vandenberg. The following occurred on a launch during that time frame:

We had a routine flight from Ellsworth to Vandenberg, ran some tests with the ground launch director and got into the launch phase without any trouble. The ALCS had a command called the "auto command." This combined the enable and launch command into the key turn. We were testing this capability.

The launch director gave me the command "initiate the auto command." I instructed the ALCS crew, Maj Jack Farley was the MCCC, to

do the same. Jack and his deputy turned keys. Nothing, no signal left the aircraft and a red fault light appeared. Jack, true to his checklist, said, "Deputy, Fault checklist." I knew that we had about one minute left and if we didn't succeed the ground crew would turn keys (Heaven forbid). Having found the airborne system not to be true to the tech data and capable of doing just about anything, said "Jack turn the damn keys again!" Jack hesitated, I put my hand on Jack's and gave the countdown (very quickly) to his obedient deputy and we (the three of us) key turned again. It worked, the missile fuse lit and it went its merry way down the weather test range.

Many questions came from the ground, all answered with, "please don't compare the air with the ground." However, true to the SAC tradition we put in a TO change saying, "If you don't succeed the first time try again." Obviously the TO said it in more scientific terms and with better TO jargon. This, by the way, was a characteristic of the entire ALCS system not discovered until that particular launch.

Transportation And Missileers

by John J. (Pat) Napolitano

Throughout the more than 50 years that make up the history of USAF missile systems, we have traveled by about every means known to man. From jeeps, 6x6 trucks and tracked vehicles to aircraft, we have used a wide variety of ground, air and water vehicles to transport missiles and missileers.

We have moved our big missiles in C-133s, C-141s, on trailers and on train cars. Small missiles have traveled in even more varied ways; remember the recent story of the commercial trucker who disappeared with a

Orlando AFB, 1956. 17th TMS.

Flight maintenance at Wueschheim GAMA. Bravo Flight transporter, erector launcher (TEL) unloaded. August 1988.

truckload of air-to-air missiles on a cross country trip? We have moved missileers in even more different ways.

In my days in Titan I, I traveled by pickup truck, big blue diesel bus, H-19 helicopter, my own Jaguar sports car (the long paved access roads in Idaho were great for a "high speed approach") and in convoys of several other vehicles transporting the two stages of the missile to the site. In Minuteman, we rode in a variety of vans, trucks, cars, station wagon and helicopters. The current missile force travels in Expeditions, new transporter-erectors and reliable choppers.

Those of us who served in the Ground Launched Cruise Missile System worked with one of the most varied off-road transportation system in USAF history, from the big MAN tractors that towed the transporter-erector-launcher and launch control centers to the variety of trucks, Hummers and other vehicles we used to get to the flight area. If it rolled, flew or floated, we probably used it. This issue includes a number

of stories and articles about our travels and it only scratches the surface.

Moving Missileers in Montana

The 341st Strategic Missile Wing at Malmstrom AFB with four squadrons (10th, 12th, 490th and 564th), 200 Launch Facilities (LF) and 20 Launch Control Facilities (LCF) now called Missile Alert Facility (MAF) cover a large part of the west central Montana prairie. The wing's missile complex stretches from about 50 miles west and north to 150 miles east and south of Great Falls. The 490SMS had the wing's farthest LCFs ranging from 90 miles for Mike-1 to 150 road miles for Oscar-1. With the farthest sites, the 490SMS probably experienced the widest variety of transportation methods an crew changeover schemes ever devised by clever and cunning wing staff members. As a member of the 490SMS from 1964-67, I experienced most of these methods and schemes. Here is a thumbnail of my missile crew transportation memories:

Planes: A C-47 Gooney Bird flew the 100 miles from Malmstrom to Lewiston Airport and carried all five 490SMS crews plus Charlie, Delta and Echo crews from the 10SMS. I remember one pilot, a short, old 105MS Operations Branch Officer (OBO), who used to scare the crews to death with white knuckle landings, and choppy, bumpy flights. At Lewiston, crews would pick up prepositioned station wagons, drive to LCFs, and changeover. Relieved crews would return to Lewiston, board a C-47 for trip back to Malmstrom, and another white knuckle landing. Eventually, crews refused to fly with the old 10SMS OBO. Another problem: crews going to sites closest to Lewiston had to wait on the crews from the farthest sites, the cause of much griping. Flights suddenly stop; no qualified pilots available.

Buses: By fare the slowest, most boring change-over method was to be bused out to Lewiston, do the car thing to the various LCFs, return to Lewiston and bus back to Malmstrom. Luckily this didn't last long - the crews revolted!

Choppers: I remember well the H-19, an old piston engine Sikorsky chopper that held only two crews. It was a bulbous looking contraption that looked anything but flyable. The pilots sat up high in the cockpit, and passengers sat below and behind. The interior layout was such that I can remember staring through the pilot's legs at the pedal controls. The H-19 was very noisy, slow, and reeked of exhaust fumes and burnt oil. The seats were oil stained and oil dripped from the gearbox onto our white (house painters) crew uniforms. I think I ruined a uniform on each flight. Eventually, the H-19s were traded in on the...

CH-3: (Cargo version of the Vietnam War HH-3E Jolly Green Giant). Much bigger and faster than the H-19, held 20+ pax, a regular cattle car. The 341st had two of these monsters. It was a relatively new aircraft, but like the H-19, it was noisy, smelly and oily. The pax sat under the huge rotor and gear box which made a terrible graunching sound. Too big to land at most LCFs; it dropped crews at a central point like the C-47. I hated flying in it and it took longer to get out and get home due to the wait for the farther out LCFs. I recall wishing we could get rid of them. Be careful what you wish for, one crashed and the other caught fire; both were completely destroyed. Luckily, with no loss of life or injuries. Next came the...

UH-1F (Huey): Much better, clean, quiet (everything is relative) and more efficient at changeover of crews. Only drawback: same for all the air transports, they only flew in good weather; when things went to pot you drove. Worst situation: weather closed in after you got relieved grounding the choppers at LCFs; stranding you, sometimes in blizzard conditions, for days. Best situation: riding the copilot seat on a clear day, the pilot letting you try to fly this very unstable flying machine. Or the pilot, newly returned from Vietnam, practiced strafing runs on trains, chased herds of antelope all over central Montana or surprised the crew members in the back by demonstrating that a helicopter could really perform 90 degree banked turns like a fighter.

Cars: In the end, when the weather turned bad, we drove Ford station wagons. All my

Titan II Re-entry Vehicle.

Titan II maintenance team.

Titan II re-entry vehicle.

deputies were from warm climates (Alabama, California and Arkansas) with zero snow driving experience. I let one drive in the snow once and didn't get two miles from base when we spun off the road. Being from Pennsylvania, a die-hard car nut and self-proclaimed excellent driver, especially in snow, I did all the driving from then on. In real bad weather, the squadron would have Transportation install tire chains on all the vehicles and tell us not to exceed 20mph. Naturally, we hated chains, too slow and uncomfortable, and we would remove them immediately once off the base. Although the wing limited its vehicles to 55 mph, Montana's speed limit was "reasonable and prudent." I can recall telling wide-eyed deputies, while going 65 mph on snowpack, "We will be OK Dep unless we have to stop or turn." I think I held the record for the fastest time to Kilo-1 in the snow. By the way, you would have to wait until after changeover to report to the base that you arrived; otherwise they may have started questioning how you drove 125 miles in an hour and 40 minutes at 55 mph! The best part of driving out to the 490SMS was the breakfast stop in the tiny town of Stanford at a restaurant that had the best cinnamon rolls in Montana. Of course, when the wing staff caught wind of this crew perk, they quickly came with numerous reasons as to why crews should not stop. I can remember hearing a commander's words "I better not get any reports of blue vehicles parked at that restaurant in Stanford." Cooler heads prevailed, however, and it was deemed better for safety that crews take a rest break. That is unless you were carrying classified materials out to the LCF. Something, you will recall we did with agonizing regularity.

After I left the crew force, I believe most of the wings acquired big, suburban type vehicles, some with 4-wheel drive, and used these exclusively. This system probably gave some stability to the crew changeover procedure and did away with the hit or miss weather dependent, chopper flights. Although I bet it wasn't near as exciting, interesting or as much fun.

Boating To BOMARC
by Richard A. Rice

Stationed with the 4751st ADSM at Hurlburt Field (Eglin AF Auxiliary Field 9, Florida in the late 1950s and early 1960s. I rode a ferryboat to work. The BOMARC launch site was on Santa Rosa Island and the only other way to get there was a considerable longer route through Fort Walton Beach. Air Force sailors manned the boat which took us from the base across Santa Rosa Sound. Then it was a bus or auto trip of several miles to the site. Occasionally a power outage on shore kept the ramp from working and we were stranded on the boat, but never for very long.

How We Got Back From Work
by Larry Hansbrouck

My deputy, Mike Babbidge, and I had finished a tour at Echo, the 66th SMS Command Post. A chopper (Huey) had brought the relief crews from Bravo and Delta as well, so we had stops at both on the way back to Ellsworth, about 110 road miles away.

The weather wasn't good, and was getting worse. Sleet, snow, freezing rain, but not enough to force the chopper to sit down at one of the control centers. We left Bravo for the last leg to the base and things got worse. I always enjoyed sitting up front with the pilot (being non-rated I guess it gave me a thrill). In any event the defroster-deicer for the front windshield wasn't doing well, especially on my side. The pilot said if we could sit down for a couple of minutes the engine would warm up enough to clear the windshield and we would be on our way. My question was why not? No way! You had to get 15th AF permission to land any place other than the base or a missile site. We knew we didn't have time or the courage to go through that ritual.

The pilot said "Well, I'm going to get as close to the ground as possible to see if that would do any good." It didn't. My memory escapes me but I think they carried an ice scraper on board.

I won't blame or take the blame for this idea but the next thing I remember was standing on the helicopter skid holding on with one arm and scraping the windshield with the other at about five feet above the ground. I did a good enough job so that the pilot's visibility was sufficient to get us back to the base.

What lengths we went to not to have to stay one minute longer than we had to at a control center especially an extra night or two.

Arizona Highways
by Dick Adams

As a MCCC instructor and later, a stanboard evaluator for the 390th SMW (Titan II) at Davis-Monthan AFB, AZ, 1965-68, we often flew in Huey helicopters to the missile sites to complete instruction or evaluations. Some of our helicopter pilots had recently returned from Vietnam and some were prone to continue their combat flying by "attacking" 18-wheeler convoys on the interstate highways on the way to the missile sites.

One flight that I will always remember was when a helicopter instructor placed a visor on the student pilot, placed the helicopter in a 45 degree bank and then turned the controls over to the student. The student was to right the helicopter, and he did, but it took a few corrections and us non-flight status missileers recognized very quickly what the term "white knuckle flight" meant.

The Colonel's Car
by Charles C. Simpson

The 1965 Ford Station Wagon belonged to our squadron commander, at least Air Force had assigned it to him. My deputy, our facility manager, and I were heading to a hotel from Grand Forks AFB on the gravel county road leading north from Highway 2 to the LCF. It was January, normally cold for North Dakota, and the roads were, as usual, snowpacked. I corrected for the large lump in the road and sailed into the roadside ditch,

Minuteman site construction scenes.

Minuteman site construction scenes.

which was filled with windblown snow. The Ford slowly settled to the frame, we were really stuck. Our FM walked to a nearby farmhouse and the farmer showed up with a pre-war Farmall, the Ford didn't budge. The farmer said "I'll be right back." In a couple of minutes the largest new green John Deere you have ever seen came chugging down the road. It only took a couple of minutes to have the Ford back on the road. The farmer refused an offer for payment, he said, "This is the most fun I have had all winter."

Lox, Atlas and Steam
by Elmer Brooks

During my Air Force career, I had the dubious distinction of being responsible for two of the most cantankerous "beasts" in SAC's missile fleet, the liquid-fueled Atlas F and Titan II. Around 1963, as an instructor missile combat crew commander in the 551st Strategic Missile Squadron, Lincoln AFB, NE, my crew was undergoing a Stanboard Check during a Dual Propellant Loading Exercise. This was before SAC became sadder and wiser about using RP-1 and LOX (instead of liquid nitrogen) for exercises, and before the tech data emergency procedures (especially those pertaining to AGE) did more than scratch the surface of the catastrophic things that could happen and how to recover from them.

After loading Lox, in the middle of the launcher platform ascent sequence, we experienced a major malfunction. The primary diesel power generator shut down automatically, and we switched to the standby generator. Shortly, we noticed that it was overheating rapidly, and we had to shut it down also. This caused the launcher platform to stop rising, before the liquid oxygen boil-off valve had cleared the silo cap. The super-cooled gaseous oxygen began to boil-off and vent through the valve, impinging on the steel launcher platform suspension cables. That could cause the cables to snap, dropping launcher platform and loaded missile into the bottom of the silo. A potential catastrophe was in the making.

Of course, our trusty "Dash One" didn't cover this predicament. And, when I turned to the Stanboard crew commander to rescue us, I could tell by the pained expression on his face that he didn't have the answer either. Fortunately. I had an outstanding crew and we had spent many hours of alert duty studying the AGE systems and how they interact with one another. We knew that this complex had been experiencing problems with its automatic water valves servicing the cooling tower. We panned the topside TV camera to the water cooling tower and, sure enough, there was no steam cloud rising. Necessity being the mother of invention, I decided to improvise. After a quick review of the water plumbing schematics, we devised a plan that the Stanboard crew commander bought off. I sent the missile facilities technician and the ballistic missile analyst technician into the silo to reposition some water valves manually in order to reroute cooling water to the diesel generators. Thank God, it worked. We restarted the generators and successfully backed out of the PLX.

Watch Out - Falling BOMARC
by Richard A. Rice

Stationed at Hurlburt Field, FL in the early 1960s, my Air Force unit was test launching BOMARC antiaircraft missiles over the Gulf of Mexico from a launch site on nearby Santa Rosa Island. The missile, about 50 feet long, 3 feet in diameter, with sharp wing and tail surfaces, looked more like a supersonic fighter without a cockpit than a guided missile. Loaded with highly toxic fuels and an explosive device designed to reduce it to fine pieces of scrap metal in the event that it strayed off course, the BOMARC presented a variety of potentially lethal hazards in the event of a mishap. Just such a catastrophe occurred one day when instead of arcing over the Gulf, an 1800 guidance error sent it toward the mainland of northwest Florida. The explosive device did its job right over the administrative area of the launch site and parts began raining down toward those of us who had been evacuated there to observe the launch.

Three young airmen had driven a military 3/4 ton truck down the road a few hundred feet closer to the launch pad to get a better view. When the blast occurred they all ran for the safety of the truck, however, all did not share the same concept of what type of safety the truck afforded. Two of them dove underneath while airman #3 jumped behind the wheel and drove off. Fortunately no one was hurt but it is a good example of how different we humans react under stress.

The Missileer Eye Doctor And The Russians

Doctor Doug Smith, a member of AAFM, served four years as a missile launch officer at Malmstrom AFB, MT from 1970-74. Like many of us, he pulled many alerts beneath the Montana plains. During the peak of the Cold War, he manned 10 Minuteman missiles targeted for the Soviet Union.

Doug left the Air Force, earned a degree in Optometry from Pacific University in 1979, and began practice in Oregon. Early this year, he visited Petropavlovsk, a former Soviet nuclear naval port in Siberia. He visited there as part of a group of Rotarians from Oregon and Alaska, working to connect medical clinics and the city's library to the Internet.

Doug is studying Russian and is raising funds to purchase low-vision eye care equipment and a laser for eye surgery for the many diabetes patients living in that region of Russia. He expects to meet some of his former adversaries, Russian missileers, on his next visit.

My First Day At Cooke AFB, CA
by Jerry Strong

On 4 October 1957, Dallas, TX: Radar Bomb Scoring Detachment 1 of the 10th Radar Bomb Scoring Squadron at Love Field. We hear the news that Russia has launched "Sputnik One" into orbit. That's all the talk, anyone you see and talk with, all the radio and TV stations, Sputnik, Sputnik, beep beep, beep. Everyone is trying to sound like the signal sent by Sputnik. The whole world is changed by this launch of the first artificial satellite. Education policy changes, loans and grants are made to science and math teachers, student loans are started (engineering, science and math are the majors). Americans' response to hear that Russia beat us into space - is that this is unthinkable. Everyone's interests turns to missiles, rockets and space.

Back at Love Field, we are supporting Convair with the testing of the ECM pod for the new B-58 bomber. I am full time maintenance on the radar and plotting board equipment, but operate the radar on special missions. Next comes Chance Vought's request for support on flying the guidance system for the Regulus II missile. This is a cruise type missile to be carried and launched from a submarine. It is the backup/ stopgap in case the tube launched Polaris missile is a failure. The Chance Vought engineer is a Japanese fellow who had worked on the Mitsubishi "Zero" fighter plane during WWII. We ask some questions about the Zero; however, most are about the Regulus guidance system. The test is a success with the system mounted in a Douglas B-26 aircraft. The Chance Vought engineer takes the radio microphone and gives them the results, "bulls eye, bulls eye."

A request came in for our radar AFSC to go to the 704th Instrumentation Squadron at Cooke AFB, CA. I put my name in but nobody else from Det 1 volunteered. Orders arrived to report on or about 20 April 1958 with delay en route approved. Next: what is the 704th Instrumentation Squadron? Where is Cooke AFB? How do I get there?

Carswell AFB is our headquarters and support base. The transportation office can't tell me where Cooke AFB is. I don't have an automobile, so the best suggestion is to take the train to Monterey, CA, and find out "where to go." It is a two day train trip from Dallas to Monterey. Part of my baggage doesn't arrive at Monterey, and I still don't know where Cooke AFB is. After a couple of days, I went back to the train station to check on my duffel bag. It's still not in. The freight agent says he will forward it when it comes in and asks where I want it sent. I replied, "Cooke AFB, but nobody knows where that is." Then I said, "That was Camp Cooke, US Army." That solved all my problems, he knew that the railroad station was at Surf and the base was between Santa Maria and Lompoc.

I went to the bus station, bought a ticket based on that information and left for Cooke AFB. The bus left after dark and we rode all night and into the early morning. We left Santa Maria and rode a long way in the country. The bus pulled over to the side of the road and stopped. We are at Cooke AFB. Two of us got off. The driver gives us our baggage from the bus compartment and drives away. What do we see? A small wood building just off the road for the main gate guard shack. About three quarters of a mile from the road is the first building. That building is the 704th Strategic Missile Wing Headquarters where we sign in. The sign-in book is numbered and we are between 450 and 475 to sign in to Cooke AFB. The other fellow is also going to the 704th Instrumentation Squadron. He is a supply person. We got our base processing papers, an information pamphlet and a map. We

Minuteman construction scenes.

Representatives from the various wings. Col. (Ret.) William A. Knapp is second from the left.

find our squadron in the old hospital area. There are three officers and four enlisted men (now 6).

The squadron commander is LtCol Perry, former range safety officer at both White Sands and Cape Canaveral missile ranges. He was selected for this job because of his ability and reputation. One of his several degrees is in civil engineering. He is spending most of his time checking the work of the contractors on base.

The two of us that reported in today are

single. We are assigned to the barracks and shown the mess hall. The one barracks has people from five or six squadrons, with each squadron having only a few people assigned. The mess hall is an old Army company type. Everyone shares the mess hall. The aide to Gen Wade stands in line with the rest of us. The old mess sergeant greets everyone. "What else would you like to have served? What do you like or dislike about the food or service?" This would stay the

1970 Missile combat competition.

The Strategic missile wing commanders draw for their wings' competition start times at the 1979 SAC Missile Combat Competition-Olympic Arena, Vandenburg AFB, CA.

same until the new, large consolidated mess hall opens.

Later in the afternoon we hear that a C-47 has crashed short of the runway. Several of us go to look. The plane landed short of the runway, went between two trees and knocked both wings off outside the engines. The runway was closed until further notice and air operations moved to the Santa Maria Airport. I would learn more about Santa Maria Airport in a few weeks.

Many things happened that first day at Cooke AFB, and much, much more in the 14 months that I was at Cooke/Vandenberg. I don't remember now, but I am sure that on that first day I must have said, "Why, why did I ever leave Dallas?"

Eight months later the first Thor was launched from Vandenberg. In the 704th Instrumentation Squadron, we were 100% Air Force, no civilians were in control of any

operations. In eight months, we went from no people, no buildings, no equipment, no training or experience to the first launch. Each year, as we get older, the feats get larger, but I think maybe we won't live long enough to make this accomplishment too large. The crack was in the curtain, the missile gap was already over.

The First Sergeant

We had been at the base for three or four weeks. The squadron is starting to fill up with people. There are still only three officers, but we now have a first sergeant. He is from the "old Army" with 26 to 28 years of service, the "First Soldier" type. He is six foot four, a big, rawboned man, who is in charge and everybody knows it.

Gen Power, CINCSAC, is coming to the base and the base must be cleaned up. Squadron commanders and first sergeants are exempt from the detail. Everybody else, from lieutenant colonel on down, will be on the clean-up detail. One of the officers is a first lieutenant, an Annapolis graduate waiting to get out to go to dental school. The first sergeant has the lieutenant as his jeep driver to keep him out of the work detail. So he and the first sergeant bring us coffee and ice water and make sure that we have regular breaks. Only the 704th Instrumentation Squadron has a first sergeant that looks after his men on the clean-up detail, the "Old First Soldier."

Cooke AFB had been Camp Cooke, a US Army training base. Part of our clean-up area is around the old tank garages. Some of these held three tanks and the others had space for five tanks. Each garage also had an office and a maintenance area which are like miniature time capsules. The maintenance NCOs had laid out their shops and each is different. Some have tool boards with painted outlines or shadows, others just the boards and hooks. Names and dates are written or carved in the boards on the wall. It makes us wonder what happened to them after leaving their tank units.

After three days or so, all the areas that Gen Power is going to visit are cleaned up - not bad for a hard detail, and working together we get to know each member of the squadron better. As to the lieutenant, everyone thinks that the first sergeant pulled a neat trick by making him his driver, and nobody resented the assignment.

The "Old First Soldier" only stays with us a few more months. When the first ever senior master sergeant promotions are announced, he isn't selected. So he decides that it is time to retire from the Air Force. Sadly, I don't remember his name, but I will never forget our "First Soldier" in the 704th.

GLCM Guidance System

Submitted by Col (Ret) George Grill, who was with General Dynamics at Greenham Common AB, England. It may not the the first time you have seen this - it seems to apply to all guidance systems.

The missile knows where it is at all times. It knows this because it knows where it isn't. By subtacting where it is from where it isn't, or where it isn't from where it is (whichever is greater), it attains a difference or deviation. The guidance system uses deviations to generate

SAC Missile Combat Competition Olympic Arena 1979. "Best Operations Crew" L to R: 1/LT Carlos Cherry, MCCC, 2/LT Mike Kelly, DMCCC, SSGT Bruce Cook, BMFT, MSGT Beryle Farmer.

corrective commands to drive the missile from a position where it is to a position where it isn't, arriving at a position where it wasn't but is now. Consequently, the position where it is, is the position where it wasn't. So it follows that the position where it was is the position where it isn't. In the event that the position where it is now is not the position where it wasn't, the system has acquired a variation. The variation being the difference betwen where the missile is and where the missile wasn't. If the variation is considered to be a significant factor, the missile guidance logic system will allow for the variation, provided that the missile knows where it was or is not now. Due to the variation modifying some of the information obtained by the missile, it is not sure where it is. However, the thought process of the missile is that it is sure where it isn't, and it knows where it was. In now subracts where it should be from where it wasn't and adds the variable obtrained by subtacting where it isn't from where it was. In guidnace system "language" this is called error, or the difference between deviation and variation found in the algebraic difference between where the missile shouldn't be and where it is. Simple.

MOCAM
by Jack Roberts

I was a missile maintenance technician (MMT) with the 548SMS (Atlas E) at Forbes AFB from mid-1962 until deactivation in 1965. I was first assigned to operations, but before I could be sent to Vandenberg for launch training, I was reassigned to the Mobile Maintenance and Check-Out (MOCAM) section. As soon as I had my 5 level, I became the MMT on MOCAM Team 1. The team (usually consisting of five or six specialists) traveled in old worn out 48 passenger buses. These buses worked okay as there was plenty of room for our tools, check-out equipment, etc. We took out some of the seats in the back to make them even more convenient. One could get a relatively good nap on them as

well. Every member of the MOCAM team got a bus license, so we took turns driving. Once, our bus started giving us trouble on the highway west of Topeka. It was clear that we would not make it to the site, so we pulled over in a rest area. We had no radio, so all we could do was wait for someone on the way to or from the site to stop and take word back that we were stranded, or for someone to notice that we had not checked in at the site. We scrounged some firewood and built a fire in a BBQ pit in the rest area to stay warm. It was several hours before we were "rescued." We built a makeshift oven, warmed up our foilpak lunches, and had dinner. As I recall, the wrecker and another bus got there just in time for us to go back to the base. Our CO, Col Clugston, was at one of the sites one day to observe something we were doing. One of the other specialists on the site was needed at another site, so the colonel told him to take his staff car and go. The colonel told him that if he did not get back in time, that he, the colonel, would ride back to the base with the MOCAM crew. Col Clugston was like that.

By the time the MOCAM team was ready to go, it was dark, the wind was howling, and it had begun to snow. Of course, the bus would not start. This was not uncommon, so we always parked it so that we could push it if necessary. Picture this! We all got off to push the bus, and the colonel was right there beside me helping push. Within a week of the bus incident, the buses disappeared. In their place were brand new crew cab four wheel drive Dodge pickups, two for each MOCAM crew. The G2000 engine service carts used on the Atlas engines were about the size of a golf cart but I estimate they weighed 1500 pounds or more. Even though the carts were on wheels, they were not made to be pulled on the highway so the base transportation squadron had to haul them to the site on a low-boy trailer, unload them, then haul them back when we were finished with them.

Someone got the idea to mount the carts on a truck so they could be more easily moved around to the sites. It was decided that the

MOCAM teams would get checked out in the trucks and do our own transportation of the G2000 carts. Good idea. However, the only truck chassis available were two old worn-out International dump trucks. The dump beds were removed and the G2000 carts were mounted on the truck frames. This was okay except that these trucks had electric clutches. That's right, electric clutches. They had a standard 5-speed floor shift, but no clutch pedal. Instead, there was a button on the gear shift lever that you depressed when you wanted to change gears. The transmission had a torque converter on it, so that you just held the brake, depressed the shift button, shifted into low, released the button, released the brake and went on your way. There was a sort of "clunk" when you released the button to get started, but otherwise it worked okay. I still have my AF driver's license which lists "5-ton truck with electric clutch" as well as pick-up, staff car, and 49 passenger bus.

The launch crews used station wagons, almost exclusively Chevys, as I recall. Helicopters were used as base-site transportation, but rarely. Near the end of the life span of the 548th, small STOL aircraft were used more frequently as base-site transportation. Short runways were prepared near the site, just outside the outer security fence. I recall that these planes (I think they were called Otters) had a big radial engine and swivel landing gear so that crosswind landings could be made more successfully. The pilots had a reputation similar to the "bush" pilots in Alaska. The security guards used these planes frequently.

The Missile Badge
Submitted by Duffy Liebenguth

About two years ago when the Air Force conducted a major review of uniform policies, there was a suggestion made to eliminate the Missile Badge. MajGen McGinty, who was the Chief of Personnel for USAF at the time and chairman of the uniform board, ruled in favor of keeping the Missile Badge for wear on the uniform.

MajGen McGinty stated, "A lot of people are very proud of the fact they were missileers. During the Cold War they saved the world."

This statement was published in a copy of *Air Force Times* magazine with the article on uniform changes mandated by Gen Fogelman right after he became Chief of Staff of the Air Force.

Nation Watches Missilery Come To Life at VAFB
by Gene Wright, Valleys News Service Correspondent and submitted by Max L. Meyer

Lompoc, 17 December: With an awe-inspiring roar, missilery came to life before the eyes of over 150 reporters yesterday afternoon at Vandenberg AFB.

This reporter felt lucky, indeed, to be one of the select group that had gathered from every corner of the land to watch the launching of a Thor, an intermediate range ballistic missile, from the SAC base.

The occasion marked several firsts in this field: it was the first launching of a ballistic missile from this base and, consequently, the West Coast; it was the first launching of a Thor by operational equipment, meaning it was being launched as a weapon and not just in a test; and, it was the first one launched by an all-Air Force crew.

Two Sites

Following a briefing, conducted by Maj Jim Marquis in the main cantonment area, the newsmen, radio and TV reporters, cameramen and photographers were transported in two groups to the observation sites: one for reporters and one for television and photography.

Maj Marquis had announced during his briefing that the planned time of firing was 12:15 p.m. and we arrived at 11 o'clock at the site where portable bleachers had been erected, about 8,200 feet from the launching pad.

Due west through a break in a grove of trees we could see the shiny metal shed which housed the missile. We had been informed that the Thor lays on its side until just before firing time when the shed moves out of the way on tracks and the missile is raise to an upright position to be fueled and then fired.

Awaiting Move

Everyone eagerly awaited the sight of the shed moving back. While waiting, there was ample time to look around the area and point out points of interest to those who had not visited the base before.

Directly behind the bleachers to the east we could see the three Atlas sites with their massive gantry towers outlined against the hills. Near the center pad lay an Atlas missile, shrouded on its transporting trailer.

To the south against the skyline we could see the Navy's Mt. Tranquillon with its squatty radar building.

At 12:30 came the announcement of the first delay. Later the shed slowly moved back from the missile and the excitement of the crowd began to grow. But their hopes were dashed when the shed later rolled back to again cover the Thor and it was announced that the countdown had begun anew.

After several more delays, caused by minor technical difficulties, the arrival of some unplanned freight trains on the railroad which lay between us and site, and the fact that two ships had ventured unknowingly into the flight

area, it was announced at 3:20 that the final countdown had begun.

Rolled Away

In a few more minutes the shed rolled away for the second time and interest heightened again among the spectators. All eyes were glued on the area and those fortunate and wise enough to have brought field glasses passed on comments to the others.

Then at 3:30 the Thor slowly rose into the air like a giant black finger to stand silent and grim, outlined against the offshore fog bank, as the delicate fueling operations began.

At launch time minus two minutes, part of the elevating and fueling apparatus slowly lowered out of sight. As the seconds ticked by, the talk of the crowd grew still as everyone watched for the flame.

Then with a shout, someone announced, "Fire in the tail," and the flames began to spew from the base of the missile.

The flame grew brighter and the Thor rose slowly and, so far, silently into the air, wavering slightly as it lifted from the concrete pad.

Then a great roar which grew in intensity

Vice President Humphrey, Capt. John W. Haley III Commander (S-117), Col. Ry D. Sampson, commander 390th SMW at Davis-Monthan AFB, AZ site 571-6 (currently a national historical site south of Tucson.

enveloped us as the missile began to climb into the hazy sky. The noise grew from a whisper to a roar and then began to diminish as the weapon climbed higher and higher, gaining more speed with each passing second.

Only Pin Point

It continued to rise until only the pin point of the flame was visible and it appeared to be directly over the grandstand. Then it turned to head out to sea to its destination in the vast Pacific Missile Range. In a few more seconds it had disappeared.

Back on the launching pad, huge clouds of smoke and water vapor continued to billow into the air in mute testimony to the tremendous heat generated in the blast.

The faces of the Air Force men, which had grown a little grim with the passing hours and delays, broke into smiles which grew broader and broader as the Thor rose higher and higher. Their first attempt was a success.

And there was no more grumbling in the crowd. Forgotten was the long wait on the hard wooden bleachers, for we all agreed that it had been a sight worth waiting to see.

Alert

by Paul F. Murphy

Oh I have slipped beneath the surface dirt,
Downward crept and not felt hurt.
Blue suited I've entered my capsule tomb,
And sat for hours in a concrete womb.

Chances are you have never seen my place,
Or done my job sitting in inner space.
But all your flights soaring through the air,
Would not be, were I not there.

Minutes creep by in a slow unending parade,
Lights flick on, glow brightly, then fade.
A missile sits in deadly rest;
a key remains secure;
Above me a world goes on because of
boredom - I endure.

My T.O.

Submitted by George R. Nagy

This is my T.O. There are many like it, but this one is mine. My T.O. is my best friend. It is my life. I must master it as I must master my life. My T.O., without me it is useless. Without my T.O., I am useless. I must process my checklists well. I must process better than my enemy who is trying to kill me. I must process my checklists before my enemy processes his. I will...

My T.O. and I know that what counts in war is not the PLCs we run, the missiles we enable, nor the sorties we launch. We know it is the hits that count. We will hit...

My T.O. is human, even as I, because it is my life. Thus I will learn it as a brother. I will learn its weaknesses, its strengths, its notes, its warnings, its amplifications, an its VBs. I will ever guard it against the ravages of weather and damage. I will keep my T.O. clean and posted, even as I am clean and posted. We will become part of each other. We will...

Before God I swear this creed. My T.O. and I are the defenders of my country. We are the

Gate guard in front of HQ training building Denver, Lowery AFB May 1961.

Crew briefing, 91st SMW.

masters of our enemy. We are the saviors of my life. So be it, until there is no enemy, but only Peace.

The Creed Of The Missileer

(with apologies to the USMC)
Submitted by Greg Ogletree

For nearly four decades, missileers have fought and won a battle greater than any other. They have not soared through endless skies, or danced on silvered wings. They have not scored their hits or felt the thrill of victory. They hear no bugles call or mournful note of taps, they wear few medals and receive little praise from those they serve each day.

Buried deep beneath the earth or crawling across its surface, they wait for an event that should it occur, will mean deterrence has failed. They're carefully trained and tested to ensure they can do the one thing their very existence is meant to prevent. They know as long as the missiles are ready to launch they most likely never will. Every hour of every day they wait, every hour of every day they win a war they hope they will never fight.

For the vigilant missileers each hour of deterrence is their personal victory, every moment of freedom their personal conquest, and every day that passes their personal prize. What they fight is a war of wills. What they win is time. Time for a child to grow, learn, love and flourish. Time for a nation to break new barriers, heal old wounds, reach new heights, and serve its people. Time for this world to solve its problems, feed its hungry, cure its sick and find peace. Time for mankind to seek out new challenges and achieve its ultimate potential.

Time then is the prize for today's missileers. They have fought and won the cold war, but having won they must fight on. Yesterday's deterrence is gone, today's is being won right now, and tomorrow's must forever be earned. For the blue-suited missileers the victory is never final, the battle never finished, and the war never over. They work and wait, their war fighting is constant, and the peace we enjoy today is a tribute to their skill.

Glory Trip
by Greg Ogletree

Transported from its strategic perch
Where poised for quick ascent,
This bird's now ready for a special trip;
Declawed and preened to heart's content.

She's groomed from nose to nozzle,
All readings in the green;
The count toward "Mark" has begun,
All eyes and ears are keen.

Well-trained hands twist in sync
To send the magic spark;
Tons of concrete glide toward the south,
Then fire lights up the dark.

A ring of smoke heralds Minuteman,
Floating halo-like o'er head;
While Peacekeeper's egress is absent fire,
Being freed by steam instead.

Trailing tongues of flame and smoke
The missile climbs ever higher,
Pitching slightly toward the sea,
As it paints a tall white spire.

It flies down-range with lightning speed
And comes in like a shooting star,
Reminding all who witness this,
No target is too far.

The glory's brief, the moment's spent,
But the message is loud and clear:
We're ever ready to defend
All that we hold dear.

They Also Serve Who Only Sit
by Capt. Greg Ogletree

In the current generation
While peace-niks talk and pray,
Some sentinel saved the nation,
Some crew dog won the day.

At base gates some chant and linger,
Acting the Spartan type.
A few will raise the finger;
While others joint paint their gripe.

But nowadays let voices praise,
With vigor and with power,
Those ever-ready missileers who keep
A watch upon our nation's sleep

While sitting beneath the prairies deep
For a dollar sixty-five an hour.

Come drink to them in bitters,
Or toast their fame in stout;
The ever-vigilant waiting sitters
Who sit while we dine out.

They leave before the day breaks,
Not knowing when they'll return,
And drive past valleys, hills and lakes,
To sites whose names they learn.

They cast their looks on sundry books
While entombed within the pit.
Deprived of beer or brandy,
At abstinence they're handy;
But still, with me it's dandy,
So long as they will sit.

For the Dove is found within the church,
Preaching to the choir;
And the Hawk is balanced on his perch,
Calling him a liar.

The Left and Right give me the jitters,
Telling tales of woe;
Thank God we have the missile sitters
To keep the status quo!

No orator's rage can they assuage;
No view can they espouse;
But in their high-back they can sit
And turn the TV dial a bit
To ease our conscience while we flit
To someone else's house.

They watch the pitchers pitch no-hitters
And they wear down the telephone;
But ah, those useful sitters
Willingly sit while we are prone.

Others may join the ranks of quitters,
Join all the faithless crew;
You can bet the combat ready missile sitters
Will see the crisis through.

Launching of Lunar Probe #1.

Orlando AFB, Florida. September 30, 1955.

509th SMS, Whiteman AFB. July 1971.

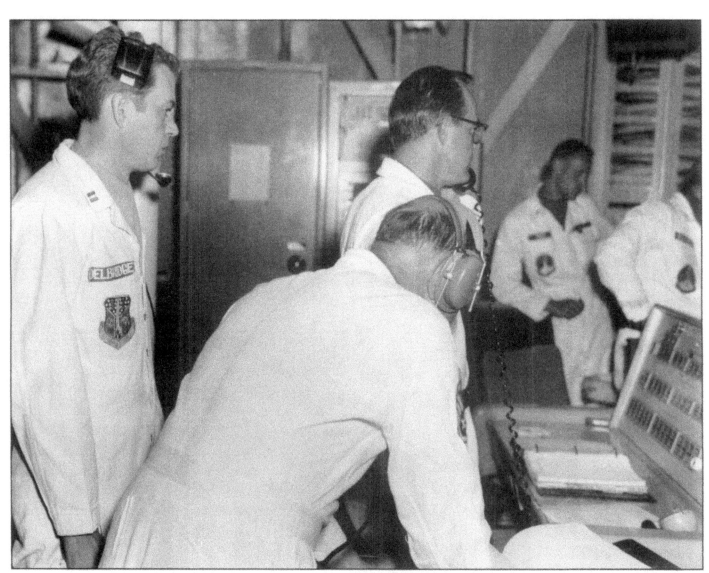

10 May 1966, Complex 374-3. Team play crew S-131. MCCC-Capt Pillman, DMCCC-Lt. Kissel, BMAT-SMSGT Kundis, MFT-SMSGT Blaydes.

No Lone Zone

by Dan Partner and submitted by Fred H O'Hern Jr.

Of all combatants in the cold war, none is more carefully selected, screened, trained, tested, evaluated, observed, and confined than the missilemen handling nuclear weapons. They are responsible for assembling, testing and mating of man's most destructive devices to the multimillion-dollar missiles designed to lob the bombs to targets continents and oceans away.

These demanding tasks require intimate knowledge of the nuclear weapon and of the intricate electronic and electrical systems in what is formally called the re-entry vehicle.

This marvelously engineered package, popularly known as the warhead or nose cone, is the business end of the missile which kicks loose from the carrier at 500-700 miles high and arcs down to the target at more than 15,000 miles an hour.

The Strategic Air Command which controls 90% of the Free World's nuclear weapons, maintains these versatile warhead crews at Atlas, Titan and Minuteman bases covering more than 100,000 square miles in 18 midwestern and western states.

One of these crews is based at Lowry AFB where the 451st Strategic Missile Wing attends 18 Titan I missiles buried under concrete and steel east of Denver. Each is capped with an A-bomb warhead ticketed for a specific target more than 6,000 miles away.

The A-bombs are flown to Lowry in C-124 cargo planes and moved to a heavily guarded, isolated area manned by the Re-entry Vehicle (RV) Branch of the 451st Missile Maintenance Squadron. LtCol Bruce D. Smith commands the RV force of two officers and 32 highly rated airmen.

Combat defense force personnel constantly patrol the area in maintaining strictest security. Only about a half dozen other men in the 1,200 man wing have top secret security clearance to enter the area.

The RV Branch has a four-part mission: Receive, store, maintain and account for all warheads and related ground equipment.

Attach 100% reliable warheads of the Titan missiles. This means that the product must work perfectly at all times, not just 99.99% of the time.

Test, store and issue Titan pyrotechnics. These include explosive bolts, engine igniters, and rockets used in separation of the missile stages while in flight.

Maintain the capability to remove the warhead from the missile in an emergency. This would be necessary if something went seriously wrong in the missile systems or in event of fire in the underground silos.

The warheads are received in three parts, the flare section, containing the mechanical separation mechanism; the cylinder containing the atomic bomb; and the nose cone containing vital electrical components.

Assembling these components, a secret procedure, and testing the systems is done by four 4-man crews. The No. 1 team serves as an evaluation board for the other teams and demands a high degree of performance.

Evaluation teams from SAC headquarters make periodic inspections.

Never is a man allowed individual access to the vehicle. Regulations prescribe "no lone zones" where the presence of only one man is prohibited. This precludes any impulsive attempt to damage or activate the systems.

The vehicles are taken to the missile sites on special truck-drawn transporters under heavy guard. A crane gently lowers the warhead to the blunt end of the Titan's second stage. There it is bolted and sealed by a maintenance team.

Periodic inspections are made to detect deterioration or damage while in the air-conditioned silos. New re-entry vehicles, or improved nuclear devices designed by the Atomic Energy Commission, are installed on the missiles as they become available.

Operation "Glory Trip" 16 April 1968. First launch of operational Minuteman missile.

Presenting safety award.

51

The Sky's No Limit
by MSgt Edison T. Blair

Where a black-topped road dead-ends in weeds and palmettos in an abandoned hangar area at Orlando AFB, FL, a TM-61 Matador guided missile slants its nose skyward, vibrating with the thrust of its jet engine blasting against the roadway. Suddenly, a T-33 jet trainer flashes overhead and noses up in the direction the Matador points. For guidance training purposes the T-33 becomes the missile and the Matador's engine whines to a stop.

Several airmen crouching behind tractors and vans stand up. They remove their earplugs and tuck them into little plastic cylinders worn on their dog tag chains. There's a smell of hot metal and jet fuel in the air.

"We did all right. Hope those guidance boys do as well" SSgt Bob Emmerling says, Bob is chief of the launch team and they just finished their part of a "count down." From now on it is up to guidance.

Several miles away guidance teams track the T-33 with radar. They radio signals to the pilot, guiding him as they would the missile toward a theoretical target. He follows directions and notes their errors. Later, he will tell them how close they came to the target.

Meanwhile, inside a fenced compound on the other side of the base, a simulated Missile Operations Center has been set up in a classroom. Groups of airmen are computing target data, weather and winds between the launch site and the target. They determine the altitude, speed and flying time needed to put the bird on the target and relay this information to guidance, control, armament and launch teams in the field.

Each team must perform certain operations at a specific time. The Missile Operations Center "keeps the count." They coordinate every move and relay all information through a communications net of telephone, teletype and radio. There's a back-up system in case part of the net goes out. If one of the teams has trouble, they stop the count until it is cleared up.

This is how TAC trained the Matador groups that have been on 24-hour alert for several years close to where the Iron Curtain stretches across Europe. Sgt Emmerling's outfit is being readied for such a vigil, wherever needed.

"In this outfit everybody and his job is important," he declares. "We can't do our job unless guidance and controls check out. They can't do theirs unless the assembly team is on the ball. None of us can do anything unless the truck drivers, supply men, transportation people, and just about everybody you can think of have done their job right, and at the right time. That's why we have teams instead of crews."

A rocket booster kicks the Matador into the air from a zero-length launcher mounted on a trailer. It is guided at better than 650 miles an hour until it dives on the target, then it cracks the sound barrier. Armed with either a conventional or nuclear warhead, its mission is interdiction, knocking out troop concentrations, supply lines and communications immediately behind the battle line without risking the lives of fighter-bomber pilots who would otherwise do the job. This is the Air Forces' first operational missile. Bigger, better and far more potent birds are under development to

1976 SAC Missile Competition winners. 341st SMW, Malmstrom AFB.

91st SMW display at 1969 Missile Combat Crew Competition.

strengthen its strategic, tactical and interception capabilities.

The Matador is highly mobile. Everything rolls on trucks, vans and trailers. The birds have a range of several hundred miles and can be launched equally well from camouflaged pads, highways or country lanes. The missiles are shipped broken down into components in special crates. Airmen uncrate, assemble, check them out and launch them in the field. Components are interchangeable between missiles.

To help prevent snafus, team chiefs in the 588th Tactical Missile Groups have cross-trained their men to drive trucks, tractors, and operate the crane. A crane operator doesn't sit around an hour or two just to lift a bird on or off the launcher. A member of the team does it.

"The tech schools do a good job of training these men," SSgt Salem Wilt, another launch team chief, says, "And so do the Mobile Training Detachments sent out by Training Command. All we have to do is teach them to work as a team. They all pull their weight and don't gripe, even when they have to tear it apart and put it back in storage. All our maintenance is preventive. We don't sign it off until it can fly by itself. Not like when you finish an airplane inspection and the crew takes it up to find the bugs you've overlooked. We can't leave bugs, it's gotta be right the first time. It only flies once."

Most of the men at Orlando get a starry look in their eyes when they talk about some of the new missiles under development. A1/c Nick Kopac is an ex-infantryman and a jet engine mechanic by AFSC. But he spends about half his time on the flight's ground power equipment and helping the airframe men.

"I had to study pretty darned hard to stay at the top of my class in tech school," Kopac says. "That's the only way I could get my choice of assignment and I wanted missiles. Now I hope to train for the Snark when I finish this tour."

His buddy, a guidance specialist, has his heart set on getting into a fighter squadron equipped with Falcon air-to-air missiles. He is fascinated by what he knows about their homing guidance system.

Orlando is the home of the Air Force's first Guided Missile Wing which has had its share of growing pains. There was no previous missile experience to draw on in preparing Tables of Organization, establishing supply levels and logistic requirements for the new hardware. But they were worked out, sometimes by trial and error. Later missile wings will benefit from their experience.

But then, growing pains aren't new to the Air Force. It has been only 50 years since that small group of men, whose vision was mistaken for foolhardiness, talked Congress and the War Department into buying the first military airplane. They began a revolution in equipment and weapons systems that has changed the whole concept of military tactics. Missiles are another phase of the revolution started by men like Arnold, Spaatz and Mitchell.

Some think that missiles will completely replace the airplane. They won't in the foreseeable future. They are a category of weapons that make up an effective but complex weapons systems. Even with their marvelously fast reactions and all-seeing electronic eyes they cannot think for themselves. Men must make the decisions and push the buttons. The Air Force expects the human pilot to continue to play a leading part in warfare for years to come.

However, what is good today might not be good tomorrow, Gen Twining warned cadets at the Air Force Academy recently.

"Airmen have developed a motto," he said, "which is in essence, 'to proceed unhampered by tradition.' We must fight the tendency to hang on too long to favorite weapons and tactics. Missiles have created the need for entirely new structures and we must integrate them into our combat force without losing effectiveness."

If the sky has a limit, guided missile development will certainly help us find it. Research has already started on atomic-electrical-ion propulsion systems that will push our civilization out into space. But even in the workaday world, the research being done is enough to fire the imagination and ambitions of people like Kopac, Emmerling and a lot of others.

At Holloman AFB in the New Mexico desert, research teams are burning up the wind with a two-stage rocket known in the trade as Hypersonic Test Vehicle. It scorches along at 5,000 feet on March 7, seven times the speed of sound, while 15 pounds of instruments in the nose cone gather data. The information it acquires can be projected to provide data on how missiles would operate at far higher speeds at greater altitudes.

The National Advisory Committee for Aeronautics has fired small model rockets at speeds of Mach 10.4 (6800 mph). In their laboratories they've duplicated conditions matching Mach 20 (13,200 mph) or more. Another test vehicle, the Lockheed X-17, stands as high as a four-story building and weighs in at better than six tons. Rockets hurl it into the ionosphere. Then it dives at incredible speed back through the blanket of air surrounding the earth.

All this is being done to find out what happens when the warhead of an Intercontinental Ballistic Missile (ICBM) re-enters the earth's atmosphere. The problem is a tough one; of the falling stars you're likely to see on a clear moonless night probably not more than one in 10 million ever hits the ground, because heat generated by their friction with air molecules vaporizes them. We have to overcome that problem if our missile's warhead is to reach the target.

The two ICBMs now under development, Atlas and Titan, have caused as much comment as the A-bomb. Their range will be measured in thousands of miles, their altitudes in hundreds. The thrust of the rocket motors may be a million pounds or better. That's more horsepower than 10,000 freight locomotives can develop together. They will carry their own oxygen supply to burn the fuel. Speeds will be greater than 10,000 miles per hour; their entire flight will take only 30-40 minutes. Not much time for intercept. It is easy to see why it presents re-entry problems; shock-wave temperatures may reach 20,000 degrees Fahrenheit.

But even with a thermonuclear warhead the ICBM cannot be called an "ultimate weapon." This is just the first of such missiles. "The compelling motive for the development of space technology is national defense." MajGen Bernard A. Schriever, boss of the ICBM project, states flatly.

These ballistic missiles, designed to wipe out industrial complexes and destroy the enemy's production of war materials, will be part of Strategic Air Command's weapon system. A single ICBM with a thermonuclear warhead will do more damage than thousands of WWII bombers.

Sixteen prime contractors supported by 200 subcontractors have hired 75,000 people to work on the ICBM project. The whole program, bigger than that organized to produce the first A-bomb, comes under the Air Research and Development Command's AF Ballistic Missile Division. How soon the ICBM will be operational and exactly what its capabilities are, the enemy would like

Home of the 91st SMW.

Col Gary E. Marsh, 308th SMW, Commander briefing Gen. Allen, Air Force Chief of Staff at Missile Complex 373-7, Little Rock AFB, AR.

Fighting Viking painting located above the LCC tunnel at the 446th SMS A-0.

to know. However, AFBMD's Commander, MajGen Schriever, reports: "All the major mileposts in their development have been passed on schedule."

A ballistic missile is powered and guided only during the early part of its flight. Rocket motors will fall away as they are spent until only the warhead is left, which will then follow a ballistic path depending on its shape, size, density and direction of air flow and the force of gravity. In a sense it starts as an aircraft that consumes all of itself except the warhead, which it tosses at the target.

Wind velocities, air density and temperature at every level, from the launch site, hundreds of miles out into space, and back again to the target, must be computed into the guidance. Even the speed and direction of the earth's rotation with relation to the missile must be fed into the computer if the bird is to be accurate. All in all, it will take some pretty sharp airmen with fantastic machines to man the missile site of the future.

To bridge the gap until the ICBMs are ready, ARDC and its contractors have readied the Snark, an air-breathing pilotless bomber that has all the range, accuracy and load-carrying capacity of the strategic ICBM. Launched from a zero-length launcher, the subsonic Snark is powered with turbojet engines and has a self-contained guidance system. It has already passed most of its tests and will soon be operational.

Only slightly behind the Snark in development is the Navaho, a rocket-boosted ramjet supersonic pilotless bomber that flies at very high altitudes. It will be equipped with a more efficient and more accurate guidance system than the Snark. Both these strategic birds are as deadly as the ICBM, though they are more vulnerable to interception.

Missiles are in the works to handle other assignments as well. The Rascal is a long-range air-to-surface pilotless bomber carried part way to the target by one of SAC's jet bombers. Once released, well outside detection and interception range, it is driven at high speeds by rocket motors

to the heart of an enemy's war plants or shipping centers.

In the Intermediate Range Ballistic (Missile) Field, Douglas Aircraft Company and its subcontractors are working on the Thor. It is a surface-to-surface strategic missile driven by the same rocket motor that will boost the ICBM toward the ionosphere. It is intended for use against both strategic and tactical targets up to 1500 miles from the launch site.

One of the most promising surface-to-air interceptor birds under development is the Bomarc. Powered with ramjet engines, it will develop supersonic speed and is designed to seek out and destroy enemy aircraft at very great distances from the launch site.

The GAR-1 Falcon is a rocket-powered, supersonic, air-to-air missile that is part of ADC's F-102A and F-89H weapons system. Once the pilot makes his intercept and locks onto the target, he closes in. When the Falcon's sensitive radar nose tells it the range is right, it launches itself automatically. A self-contained guidance system causes it to home on the target at tremendous speeds and to follow any evasive action taken by the enemy. Even without a warhead it has brought down target drones in practice.

Men, Missiles and Maintenance

These are the missiles in being or projected for the immediate future. What kind of people will we need to handle them? The Honorable Donald A. Quarles answered that question this way at a press conference.

"You can take the average high school graduate recruited by the Air Force and train him to do the job," he said. "However, when you have to dig into the innards of the thing, you need experts, highly trained sergeants or contractor's representatives serving in the area as maintenance experts. You and I can drive an automobile, but if we want it serviced we take it to the experts who know the systems and have the practical experience needed to fix it."

Guidance systems, for instance, are varied and complicated. Some are simple radar systems. Others, called infrared systems, are heat-seekers, controlled by the heat radiated from the target. There are star-tracking systems that automatically pick up successive pairs of stars and triangulate the missile's position. Then there are inertial systems that can not be jammed. Their gyros maintain a stable platform for reference. Any course deflection is sensed by instruments that signal the control mechanism to make corrections.

These inertial systems are so delicate that the sealed units are built in hospital-like rooms under rigid temperature and humidity controls. Workmen enter through a vestibule where revolving brushes scrape mud and dirt from their shoes. They wear nylon smocks like doctors do, and huge exhaust fans literally vacuum lint and dust from their clothes. Filtered air in the room is kept under pressure to keep dust from sneaking in through opened doors. Airmen out in the boondocks at a missile site can't make repairs on equipment built under such conditions.

Even if the Air Force could afford to build such shops at each site and equip it with the special tools needed, airmen just don't have enough practical experience yet on missile maintenance. To train technicians would cost up to $20,000 per airman and take from two to three years of constant schooling. If airmen were trained to trouble-shoot and repair every one of the 300,000 parts in a missile, they'd still be in tech school at the end of their third hitch.

The Directorate of Maintenance Engineering at Headquarters, USAF, has found an answer. They develop missile maintenance policies and their philosophy is based on the replacement of plug-in type units. Sealed components, from a complete guidance system to a signal amplifier the size of a matchbox, are built as a unit to plug into the missile. When something goes haywire the airman "removes and replaces" the whole unit, and sends the old one back to the factory for repair.

This "black box concept," as it is called by most airmen, isn't popular in many circles. A lot of missile people resent it because they haven't understood why it's necessary. A missile must be ready to go instantly. That means continuous maintenance. A bird expected to do the job of a whole fleet of manned bombers can't be grounded because some tiny component is out for repair. It could make the difference in who wins a war.

So a man is trained first to isolate, remove and replace a faulty unit. After he's gained practical experience on the job working with more experienced airmen and technical representatives, he can be given advanced training. Meanwhile he is a valuable man, still learning, and soon becomes a seasoned maintenance expert of the type Mr. Quarles mentioned.

Airmen With A Future

Training often starts early in the development stage. A lot of airmen now getting their fundamental missile training will likely end up with one of the more advanced birds. Many of the key men selected to set up the new organizations will be graduates of Matador squadrons. They already have the experience and basic skills needed, so they'll need only a little specialized training.

The first airmen assigned to the new advanced missiles, Atlas, Titan and Thor, will be trained by the manufacturers and contractors, who are already at work on their training programs. Operations people get short but comprehensive courses on the whole bird. Maintenance men get longer, more intensive, specialized courses on the mechanical principles.

After this initial phase, airmen will be assigned to a unit where they'll train as a team while the bird is undergoing its final testing. Training has been timed so that qualified people will be ready to move out with the bird as soon as it becomes operational. Camp Cooke, just north of Los Angeles, is being activated to train teams for strategic missiles. Later, Air Training Command will set up tech schools; just as they have for the Matador.

Besides the few airmen already trained in missiles, the initial input will be NCOs from aircraft mechanics (43) and electronic (31) career fields. Basic airmen will need fairly high scores in the Airmen's Qualifying Examination to indicate that they have an aptitude for learning the skills. Backgrounds in high school and college mathematics and science will help a lot. Airmen with support AFSCs, supply, transportation, security and installations, will be selected just as critically.

The same thing is generally true of officers. Many of them are probably now in research and development jobs, but they must have had some administrative and operational backgrounds. Electronic experience is mandatory for Control and Guidance officers and engineering is mandatory for Operations officers.

Classification manuals have been changed to include the necessary skills and to make sure that both airmen and officers once trained in missiles aren't shifted to other fields.

This is just the beginning of the missile era.

Each service is developing guided missiles to fill its own requirements. This year the Air Force is spending 35 percent of its procurement money on missiles. By 1960 half or our money will go for missiles. Air Force will operate strategic missiles designed to destroy the enemy's capability to make war and those that help provide tactical support of ground forces. It is responsible for long range interception of enemy air attacks, and of course, for the deadly air-to-air missiles that put the sting in our fighter-interceptor weapons systems.

The guided missile field is wide open and the sky's no limit. Science and industry are going full throttle in missile development. Now it is up to the airmen, the specialists in air power to deter war and keep the peace by being prepared to use them effectively.

Results of typical winter at Minot, clearing access roads to missile sites.

Ready for launch at Cape Canaveral, 1956.

One Helluva Fight

Submitted by Charles W. Shaw

The coming of winter on the Northern Plains has a special significance for missilemen. They know the bitter cold of the Arctic winds is coming as well as the impassable roads and the blinding snowstorms that make alert crews, security troops, and maintenance teams curse under their breath all the way out to the far-flung sites.

During the winter of 1968-69, Minot AFB experienced the largest accumulation of snow, 46.9 inches, since the installation of the Minuteman complex in 1963. As the snow continued to settle and drift over the North Dakota prairie, veteran missilemen on base also apprehensively recalled the winter of 1964-65, when two missile sites were lost to flooding that resulted from 42.5 inches of snow.

By the end of February that record snowfall had already been surpassed, and the worsening situation was apparent to alert crews and support personnel on duty in the field. Throughout the 10,000 square mile missile complex, giant drifts had crept up over the windows of Launch Control Facility support buildings. At the isolated unmanned launchers, shifting drift patterns played havoc with security monitoring systems. Maintenance teams encountered lengthy delays awaiting site clearance; virtually every site was in need of snow removal.

On 6 March, 10 days before the first major thaw, the missile wing commander wrote the base commander to express his concern about the situation. Flood damage, he said, was imminent unless "more drastic action is taken to improve our posture through snow removal and facilitation of site clearance and drainage." Prepared plans were implemented for an all-out effort of emergency proportions. The civil engineering squadron devoted itself to snow removal at the missile sites. All support group and missile wing personnel were made available for any type of flood control duty. The magnitude of their task showed clearly in the findings of survey teams reporting from the field: 50 missile sites, one third of the Minot force, had to be designated priority "1" for snow removal; 18 of these were especially singled out as problem sites.

After the last major snowstorm ended on 9 March, the snow removal effort proceeded unabated at high priority sites. It was a slow and arduous job. The winter's snows, swept by winds of 25 to 30 knots, had drifted many times higher than the official accumulation. according to a standing contract, the State Highway Department was called in to assist with exterior clearing at the acre-sized missile sites while civil engineering teams concentrated on the site surface.

But now a warming trend moved into the Minot area, and temperatures rose steadily above freezing on all but four days of the arriving spring. Beneath its surface crust, the endless snowscape was transforming into the peril feared by the military and civilian community alike: flood waters. By the beginning of April, flood control personnel flying by helicopter to threatened sites surveyed a massive swamp where only a few months earlier grain fields had stretched to the horizon. Within three weeks the city of Minot was inundated.

As the warming trend became evident, preparations for the worst moved into high gear on the base. Diaphragm pumps and high volume

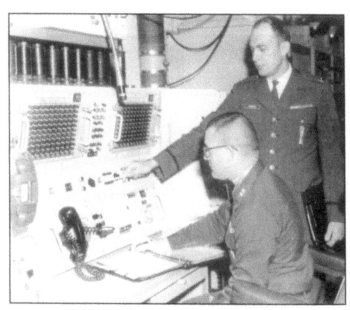

Captains Simpson and Downing, 321 SMW.

STC Alteration Task Force. Ellsworth AFB, SD. March 1973.

1969 Olympic Arena team members, 321 SMW.

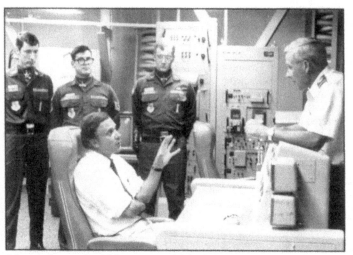

Arkansas Senator David Pryor is Briefed by Col. Gary E. Marsh, 380th SMW commander. Duty crew members Capt. Joseph Mulcany, MCCC; T/Sgt Jack Kingsley BMAT; and Robert Hartzell, DMCC. 9 July 1979.

Capt. Brian "Chappie" Chappell and Capt. Darcie Heth.

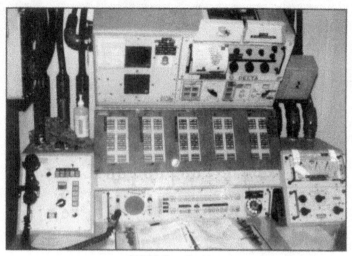

MCCC Crew Position.

Transport-Erector launcher in firing position.

Preparing for flight ops during training at Montabawr training site. September 1988.

Navy pumps were distributed to distant Launch Control Facilities, while 40,000 sandbags were filled on base. Officers and NCOs were dispatched to each LCF to keep watch around the clock on water depth sticks implanted throughout their flights. To support flood duty personnel, extra rations were sent to each LCF and cooks prepared meals around the clock. Meanwhile, civil engineering and missile maintenance engineers conducted land surveys, emphasizing culverts and drainage profiles. Where the need was critical, natural melting of snow "dams" received a substantial assist from steam generators.

By far the most feverish activity took place at Foxtrot 6, one of the southern sites in the missile complex. It was located near the bottom of a land depression that proved to have little natural drainage capacity, but a remarkable ability to trap snow. After the site had been cleared twice, it became apparent that the real danger lay on the hills surrounding the launcher and consequently a four-foot dam of sandbags and polyethylene plastic was constructed around the entire site. A Navy pump put into operation on 22 March seemed to be holding its own against the slowly encroaching runoff waters.

The critical hours occurred through the night of 7 April. In mid-afternoon, two four-man flood control teams working on site reported that water trapped in the snow on a hill southeast of the site was beginning to break loose. Shortly thereafter, a torrent of water six feet wide began streaming down the hill into the already full ditches and culverts protecting the site. By 1800 hours, eight pumps, 3,000 sandbags, and 69 persons to man them were amassed on site. A maintenance team with sump pumps was also standing by for possible launcher penetration and missile shutdown.

When the missile wing vice commander arrived on the scene to take charge of operations at 1850 hours, a trench six feet wide and one and one-half feet deep had been dug across the nearby county road to facilitate natural drainage; all pumps were working at full capacity to hasten the process. The water, however, continued to rise. Fifty people and additional sandbags departed the base for Fox 6, while 60 more men on base continued to fill sandbags. By 2200 hours, there were 119 people and 4500 sandbags on site. Construction of additional dams around the launcher and its underground support began immediately.

Events took a critical turn about a half hour before midnight, when a flatbed truck loaded with sandbags broke down on the road to the site, hampering the movement of personnel and equipment. However, terms on site had by that time widened the ditch in the county road to 25 feet, and by midnight the water level began, barely perceptibly, to go down. As dawn approached, it was apparent the flood was definitely receding.

Fox 6 had survived the emergency; it was not alone. At the height of the flood threat, 56 dams had been built around missile sites and 67 pumps were in operation, all involving the continuous work of hundreds of men. On the peak day of activity, 176 men had been diverted from normal duties for snow removal and flood control work. Even so, these statistics were not so impressive as those accumulated during the thaw in 1965; then, although 1,000 men and 110,000 sandbags were utilized, two sites were lost.

This time, the lesson of preventive measures had been learned, and put to use. Fewer men, equipped with superior plans, had saved all missile sites from severe flood damage. Now another winter is upon us, and the fight will soon begin again.

Reminiscing
by Gene Slegel

Reminiscing about the "Good old days," December 1949 to April 1951, when the MX 771 (Matador) was a secret weapon.

To explain what we were trying to do. We were to launch the missile on a NNW heading from a place in the desert at Holloman AFB, NM.

It was to fly up range to the northwest corner of the range, under the control of the Chase Pilot then turn 90° right and continue for a short distance, turning another 90° or so heading south. The Guidance System would be turned on and the missile was to fly down range to impact at White Sands Test Range. (The Guidance System was later called Schanicle.) All that was easier said than done.

One of the early problems was to get around the first turn, sometimes referred to as coffin corner, the missile would suddenly loose control and come to its end. It was found that the hydraulic lines that closed the JATO doors were getting damaged during JATO burn so when the JATO unit dropped off the doors were energized to close allowing loss of hydraulic fluid and consequent loss of the missile. Solution, elementary, don't close the doors. We made it around the first turn.

The next big problem to show was instability at cruise altitude (40,000 feet). Tests in the Climatic Hangar at Eglin Field revealed an ordinary resistor in the roll circuit would become unstable at -25° F. Generally the temperature at 40,000 feet, altitude is -40° F. After replacing these resistors we continued testing the missile.

On one flight everything was good, so good it was unbelievable. Both turns at the north end of the range were perfect, missile guidance was turned on and the missile responded beautifully. At the south end of the range, just prior to dump (the automatic command to dive at the target), a quick conference between the chase pilot and the governing body at central control agreed everything was going so well that "let's turn the missile around and head up range for a second run." (This was before the expression "more bang for the buck.") The missile was turned in proper sequence and headed up range; everyone was pleased and all was well.

Suddenly the missile went into a dive; not to worry, it was on range. Whoa, it pulled out straight and level about 500 feet off the ground headed for the White Sands Monument Park. A quick and timely decision by the chase pilot to initiate destruct accomplished impact a few miles south of the road between Las Cruces and Alamogordo. A lady traveling that road saw the crash and inquired about the pilot's welfare. A member of the recovery team assured her he was allright. The fact of the matter is he had done right well, destroying the missile before it did any damage, yet allowing valuable information to be collected.

The crew participating in this activity was made up of Martin Company engineers and technicians and USAF officers and enlisted men, in a joint effort that ultimately paid off.

Conrad, Montana
by Mike L. Veres

An hour's drive north of Great Falls along I-15 will bring you to Conrad, right in the heart of Montana's Golden Triangle. In a region where the typical prairie town has but a few hundred people; Conrad, MT is a big place with a population of 3,000. Still, it is a nice little town. It represents the three main cash crops of Montana: wheat, cattle and missiles. Papa-Zero Launch Control Facility is just three miles northeast of town, smack-dab in the middle of the 564th's missile complex. I pulled my very first alert at Papa in July 1976. But Conrad is special to me for other reasons.

Conrad is a surprisingly rich town for being in such a remote location, with a lot of really nice homes on the outskirts of town. It has an unbelievable number of banks to service the local wheat farmers and cattle ranchers. Most of these farmers and ranchers owned thousands of acres (they often spoke in terms of square miles, rather than how many acres they owned), and were "land rich millionaires," at least on paper. This was largely due to the high inflation rates that prevailed throughout the 1970s. In reality, they could not sell their land and make much of a profit. Most of them had serious cash flow problems, usually needing to take out annual loans to plant crops and maintain equipment.

So some of the local people tried other ways to make a living. The "Target Bar" was built a few miles east of Conrad by an enterprising businessman when construction was begun on the Anti-Ballistic Missile (ABM) system. The bar was built right across the road from the ABM site. Although much work was done, the site was never completed due to the ABM Treaty being signed. The Target Bar itself was completed, but when the ABM program was canceled, the owner moved the building back into town where it is now the local Moose Hall. All that remains of the bar at its original location is a weed-infested parking lot, a concrete slab, and the sign. The sign, naturally enough, consisted of an archery style target on a white background, with a black arrow piercing the center of the red bulls eye. Today it is only a landmark.

Conrad also is home to the wooden mold used to build the concrete shell of Papa Launch Control Center. The mold is a huge oval shaped structure painted sort of a silvery color. It sits in the yard of the Cargill grain elevators and railroad siding, fading and peeling in the sun. It is now used to store some of their equipment. Coincidentally, back when we lived in Akron, OH, my wife Susan had worked for an insurance company which was a subsidiary of Cargill. It was really little more than a tax write-off for the giant corporation.

Conrad also has the Home Cafe," where an excellent big, farm-type breakfast could be had for three dollars. Their hamburgers and fries were good, too. There was also a small bakery which did all of their own work. A dozen fresh baked oatmeal raisin cookies could be had for a dollar. I often stopped there on my way out to alert to buy cookies.

Conrad is near the western side of the Golden Triangle, so called because wheat farming dominated both the landscape and the economy. This area had been sculpted by the ancient work of glaciers. It featured land that was generally flat, but with some low, gently rolling hills punctuated by sudden cuts harboring narrow riverbeds and thick stands of trees. About halfway between Conrad and Tango (our squadron command post near Valier), along a back road sometimes used by crews bored with the routine of always taking the same old route, there was an unusual sandstone rock formation. Several massive chunks of sandstone the size of small houses stood in the shallow depression of a field. These rocks were a little harder than the surrounding rocks and land, and were carved out by glacial water as it melted after the last ice age some 14,000 years ago. Fourteen thousand years and still standing. As I drove past those rocks, I often wondered how long our missile sites would be there, and what some future archaeologist would think of them.

The Golden Triangles' southern corner was at Great Falls. It stretched north to the Canadian border, with Shelby on the west and Havre on the east. Shelby was the scene of the famous 1923 prize fight between Jack Dempsey and John Gibbons. Some of the old-timers still talked about that 15 round fight. This northern part of the Golden Triangle is known as "The Highline" because of the Amtrak rail line and its proximity to Canada. Amtrak runs a train called The Empire Builder through Shelby and Havre, connecting the distant cities of Minneapolis and Chicago to Glacier National Park and points west.

For my first year on the crew force, we operated under a schedule that had two crews on site most of the time. Back then, both men were required to be awake at all times, so the two crews would trade off shifts that ran from six to twelve hours underground. This time varied with the seasons, because they were not allowed to depart the LCF until sunrise. The three shifts were performed over a 40 hour period for a total time in the LCC of 24 hours. During a crews' "time off" topside, they would have breakfast or diner; play pool, or just shoot the breeze with the other people on site. You could walk around outside or shoot some baskets, maybe even take a shower before going to bed. You would see the facility manager (a middle-grade sergeant) mowing the lawn, and the Security Alert Team troops getting saddled up to investigate a security situation on a remote Launch Facility (LF). You might even meet the maintenance team you talked to over the hard phone line from one of your LFs, dog tired from a long day's work, in for supper and a well deserved night's rest before going on to another LF in the morning. In short, you could get to know some of the other people pulling duty at these remote sites.

This schedule made it possible to see some gorgeous sunrises, especially from Papa. I remember coming up out of the hole on many cold, clear winter morning. The snow on the ground crunched under my boots as I loaded my gear into the big blue Chevy Suburbans that us crew dogs used. A look to the west was a real treat. Beyond the nearby wheatfields was the front range of the great Rocky Mountains only 40 miles away, all blue and capped with snow; their peaks glowing pink with the rising sun. Katherine Lee Bates' immortal song *America the Beautiful* leapt to the front of my mind every time:

"...O beautiful for Spacious skies, for amber waves of grain; for purple mountain majesties, above the fruited plain..."

Bates wrote this song following an 1893 trip to the summit of Pike's Peak in Colorado, but I'll bet she would have also wrote it if she visited Conrad at sunrise on a clear winter day. This spectacular view made me proud to be an American, and proud to be a part of protecting this great country.

Convoy Ops from Wueschheim AS, Germany to training site.

Training GLCM "All Up Round" Bravo Flight Asset. August 87.

Recovery vehicle assisting in on site maintenance of LCC.

In a cost-saving effort that reduced the crew force by about one-third, the schedule was changed such that only one crew was normally on site, and spent 24 continuous hours in the LCC. This was made possible by placing special seals on certain critical hardware items to prevent and reveal any attempt at tampering, thus allowing the two crewmen to take turns sleeping in the LCC.

When the schedule changed, SAC lost more than some crewmen. The crews lost some sunrises. They lost a rapport with the people topside. It made the job colder, more mechanical and a little less human.

My Encounter With The Beast
by Mike L. Veres

"Just one more question," asked the deputy, "How did they come to call the first test site Trinity?"

The Commander answered, "Oppenheimer chose that name himself. According to Lansing Lamont's book, *Day of Trinity*, it so happens that when he got the phone call that Gen Groves' deputy had finally chosen the place, he was reading a book of John Donnes's poems. He recalled the opening of a holy sonnet that he had read:

'Batter my heart, three-person'd God; for, you as yet but knock, breathe, shine, and seek to mend...'

The deputy said, "It's strange that the creator of the atomic bomb would choose a reference to God as the name of the place to test the bomb."

"Is it really?" countered the commander. "These men knew what they were doing. They knew that if the bomb worked, it would be a terrible, unholy thing. It would change the world forever, maybe even destroy it; certainly at the very least, cause death and destruction on an unprecedented scale. But they also knew the bomb was necessary to stop the great evil of Hitler and an entire world at war with hundreds of Americans dying each day it went on. They wrestled with their conscience all the time."

"Just like us," said the deputy.

"Yes," the commander answered, "just like us." He paused for a moment, looking thoughtful as he puffed his pipe. Taking it from his mouth, he added, "It's an odd deal that God and the Devil have made. But it works."

The events related above are factual, although the conversation is an excerpt from my novel based on my experiences on the Minuteman crew force. What follows actually did happen, and my impressions of those events are real. See if you feel the same way.

It was a cold and cloudy day in the Spring of 1977. There was a light dusting of snow on the ground, and intermittent flurries were coming down. This was my first assignment as code officer (I was still a second lieutenant). I was sent to a Launch Facility (LF) in the 12th SMS, to the southwest of Great Falls, MT. This missile, an older Minuteman - II with its one big bomb, was undergoing a 'can change,' meaning that the guidance computer had failed and needed replacing. When I got there, the maintenance team already had the launcher lid open, but it would be a while before they would have the can pulled. So, I must await the team chief's call to duty, and try to stay warm in a drafty old pickup truck.

The whole reason I was out there today was to be part of a "Two-man/One officer" team to carry the Permutation Plug from this missiles' computer back to the Codes Vault at Malmstrom. The P-plug is one of several links in the two-sided chain of launch and enable codes used to ensure that no unauthorized launch command reaches a missile. Since this can had malfunctioned, there was no way to be sure that its codes have been overwritten, so it and the P-plug must be returned to base separately. Otherwise, the complete launch code may be together in a vulnerable environment, and that is a very big No-No.

Both the enlisted airmen on the team and myself carry sidearms for this duty. He waits in the truck with me, studying for a biology course he is taking at the College of Great Falls. He wants to be a teacher when his hitch is up. We also carry a letter from the wing deputy commander for Operations authorizing us to be armed, and stating that the strongbox we are carrying is a special federal shipment not subject to search by civilian authorities. The box is locked with two heavy duty combination padlocks. We each know the combination to only one of the locks.

In order to do a can change, it is necessary to first remove the warhead, since the can sits atop the third stage immediately below the re-entry vehicle (RV) containing the warhead. The can spans the full diameter of the third stage. To change the can, a special tractor-trailer rig, a big 18-wheeler known as a payload transporter (or "PT van" for short), is parked over the launcher lid with the wheels of the trailer straddling the lid. The maintenance team penetrates the LF and opens the lid. Then a crane is rigged from the trailer. The RV is pulled up into the truck through trap doors built into the floor of the trailer, and is then placed upon a special mount and bolted down while a new can is swapped out for the old one.

This was my first chance to see a real live missile on alert, so I badgered the team chief, an old tech sergeant who knew the ropes, to let me into the PT van for a look around. The RV was still attached to the missile. I looked down into the silo, seeing one team member on the "diving board" cantilevered out from the Launch Equipment Room (LER) which surrounds the missile. He was busily loosening the bolts attaching the RV to the can. Another man was in the LER behind him taking the bolts. Their task was lit by incandescent worklights shining down from the trailer.

And there it was! The beast was big and ugly, neon green in color. It was not very aerodynamic in appearance. These monsters did not rely much on finesse to reach their targets. Rather, they bulled their way down the fiery re-entry path through the atmosphere, hitting their targets with the brute force of some 100 Hiroshimas. I couldn't help staring at the thing for at least a full minute in silence, not really knowing what to think. I ignored the men just a step away from an 80 foot fall to the bottom of the silo.

After a while my curiosity was satisfied, so I naively offered to lend a hand at the work. The team chief politely refused my offers of help,

Yes it happened — 1970

Me? No — I thought you chocked it.

I really wouldn't have been much help anyway, and shooed me back to the Air Force pickup truck that I "own" for the day. So I sat in the truck drinking coffee from my thermos, smoking my pipe, and trying to work on my Squadron Officers' School as I pondered the beast.

It is a very eerie, haunting feeling to stand an arm's length away from such terrible destructive power. You never forget the first couple of times you get that close to a warhead containing more power than all of the explosive force released during the entire Second World War. It is at once seductive and frightening. I wondered if this feeling was something like what

the builders of the first atomic bomb felt as they were experimenting with the nuclear core of the bomb, trying to determine how much fissionable material was needed to produce a nuclear explosion; and the corollary problem of how big a blast would result. These men referred to the process as "tickling the dragon's tail," an apt description if ever I heard one. Some of them died doing it.

I was on crew in the 564th, which had the newer Minuteman III missile. Our warheads were smaller than the one I had just seen, but were still roughly 10-20 times more powerful than the early atomic bombs. The first atomic bomb, which was exploded at Trinity Site, NM (on what is now the White Sands Missile Range) on 16 July 1945 yielded the equivalent explosive power of 20,000 tons of TNT.

Twenty thousand tons of TNT does indeed make a very big explosion, but compared even to the medium sized warheads on MM-III, those early bombs were mere firecrackers. It only took a few years to develop the so-called superbombs, otherwise known as fusion bombs, hydrogen bombs, or more simply: The H-bomb. The smallest of them are about an order of magnitude more powerful than those early atomic bombs. Some of them have yields as large as 20 megatons, but most of today's warheads are much smaller, usually falling in the range of 100 to 400 kilotons.

Even though the theoretical possibility of the H-bomb was known before the first atomic bomb was completed, the United States did not proceed to build these superbombs until after the Soviet Union exploded their own A-bomb in 1949. This event came as a real shock to the American atomic weapons community because they knew what a colossal effort the Manhattan Project was, and that only the unmatched industrial capacity of the United States, along with the brilliant scientists who fled from the Nazi terror in Europe, could have accomplished this task in so short a time. How could the Soviet Union, a country that was only beginning to become industrially competitive before the outbreak of World War II, now prostrate from the combined ravages Hitler and Stalin, have built their own atomic bomb in only four years? The obvious reason for this was the treachery of the early "atomic spies" like Klaus Fuchs and Julius and Ethel Rosenberg, among others. It takes more than theoretical scientific knowledge to build atomic weapons, it takes advanced engineering and industrial methods that the Soviets simply could not have possessed in the late 1940s without the crucial help of well place spies. The information thus gained was of immense help in avoiding the numerous blind alleys and parallel development efforts that were a key part of the Manhattan Project. The United States had the money and industrial capacity enabling it to take this approach to speed the development and ensure success; the Soviets did not.

We can only guess what direction the world would have taken had the Soviets not gotten the bomb for another 10 or 20 years, as most American atomic weapons experts of the day had estimated. The Soviets would not have been in a position to threaten us and Europe, and we would not have had to spend such sums of money and other national resources on an atomic arms race. It is conceivable that the numerous episodes of Communist insurgencies and Soviet-supported coup de etat's around the world may not have happened, or at least there may have been fewer of them. Lacking a nuclear arsenal, the Soviets would have been no real competition for the United States, and they would have been far less attractive as a role model to the developing countries. It is even possible that the movement for international control of the bomb would have succeeded, and that by the time the Soviets did manage to develop nuclear weapons on their own, the international community could have taken steps to stop them at an early point.

This, then, may be the true legacy of Hiroshima: A warning to humanity of the power we are capable of unleashing upon each other, and the ultimate stakes of this half-century long poker game we have been playing. Had those two Japanese cities not died, the power of nuclear weapons would have remained effectively theoretical. It is one thing to go the New Mexico desert and vaporize steel, turn sand to green glass, and blast a big crater in the ground. It is quite another thing to use one airplane and one bomb to obliterate one city in one second. I am convinced that without Hiroshima, the world would never have understood the situation, and we would eventually have suffered a real nuclear war. Then, we would not be debating over the loss of two cities and a couple of hundred thousand people; in fact, there may not be anyone left to debate anything.

A couple of months later, I was at an LF in the 564th's Tango flight for the same reason. I again climbed into the PT van to see the warhead. Probably because I knew that it was physically different. Maybe because it still held a weird fascination for me. You never quite get over that reaction. Maybe the guys who routinely maintain the bombs do, but not me. I had to be ready to launch them at any time, and that is anything but routine.

Coincidentally, at the time of my first code officer assignment, my wife Susan and I were planning to brighten up our house on base by painting the doors in each room. She agreed to shop for the paint while I was out on code officer duty that day. When I returned home that evening, I found she had fulfilled my color request only too well. When I saw that the paint she bought perfectly matched the green of the warhead, I let out a loud groan that at first mystified her. When I explained where else I had seen that color on that same day, we both thought it was funny. Later, while painting the doors, a bit of the paint splashed on the side of my desk. To this day, whenever I notice that paint on my desk, I am reminded of my first encounter with the beast on that cold and snowy day on the Montana prairie so many years ago.

Sparky's Story
by Mike L. Veres

This story happened in late 1979. I had upgraded to missile combat crew commander in the 564th SMS at Malmstrom AFB, MT that summer. Sparky was my first deputy, fresh out of Vandenberg. I was lucky to get him, he was a real fine crew dog and a nice guy. Here's how he became one of those classic legends of crew life, a real piece of missile lore, and every word of it is true. This one's for you, Sparky, wherever you are!

Sparky was a typical cool California kid from San Diego. Despite being a bit laid back, he was a solid, straight guy who had done well at Vandenberg and was shaping up into a good crewman. He got along well with people and everybody liked him. He even bore an uncanny resemblance to the actor Anthony Edwards who played "Goose" in the movie *Top Gun*.

Sparky and I were at Papa one time when he got bored with the routine in the wee hours of the morning and kept dozing off while I was on my sleep shift. To help stay awake, he decided

to study the "supplemental training materials," actually a fine collection of *Playboy, Penthouse* and *Oui* magazines, that crews kept hidden in a blank drawer of the medium frequency radio rack. Sparky figured that one way or another, he would stay up. But in his half-awake state, he grabbed the wrong set of handles on the rack. Instead of opening the blank drawer, he actually opened the radio frequency amplifier drawer. And as luck would have it, he picked the worst possible moment. Just as the drawer separated from it's connections at the rear, the radio hit its transmit cycle, sending a powerful surge of high voltage electricity through the system. The arcing caused a small fire and some smoke.

Sparky jumped back and yelped, although I did not hear it. The first I knew of the incident was when Sparky ran around to the bunk and started slapping the curtain, yelling something like "Wake up Mike, I really screwed up this time."

I was instantly awake. I threw open the curtain and pulled on my pants, then hurried around to the Deputy's Console, where Sparky had regained enough composure to open his T.O. to the "LCC Electrical Fire or Overheat" checklist. There was a smell of burnt wiring in the air, but the smoke itself had already dissipated in the rapid airflow.

Already, transmission failure printouts were starting to come in from some of Papa's LFs. I knew that in order to prevent further arcing I would have to open the big MF radio transmitter circuit breaker located on the 32 volt battery charger near the bunk. But I also knew that Sparky was still a little rattled, and I myself had been sound asleep less than a minute ago. I saw that there were no open flames or large volumes of smoke, so this did not appear to be an immediate life threatening situation. I elected to follow the checklist step by step, knowing that good checklist discipline would save us from mistakes caused by foggy, sleepy brains. I figured that there had already been enough of that for one night.

We ran the checklist together, performing each step as applicable. It directed us to Table 4-2, which was several pages long and listed every circuit breaker in the LCC. It did indeed tell us to open the MF radio transmitter circuit breaker, as I knew it would. I briskly went around to the 32 volt battery charger and opened the breaker. This was actually one of the biggest breakers in the capsule, being a rotating lever almost a foot long. It looked like the speed controls on the transformer of the electric train set I had as a boy. That done, I rejoined Sparky at the deputy's console to complete the checklist.

The only things left to do on the fire checklist were to see if the fire was out (it was), close cooling air dampers if required (only if the fire was still burning), apply fire extinguishers if needed (they weren't), realign the squadron, and make the reports to Job Control and WCP. But now that the circuit breaker was open, there were some more fault indications to contend with. We finished running the fault procedures in a couple of minutes, and satisfied ourselves that all indications were as expected.

It was time to call the crew at our squadron command post, Tango. I explained that the MF radio had apparently shorted out and we had to open the circuit breaker. Tango got the rest of the squadron up on the Hardened Voice Channel (HVC) and realigned the squadron to make sure that all of the missiles were properly monitored. Since HVC was like an old-style telephone party line, I didn't tell the whole story just yet.

I knew I had to get to the bottom of this fast, so I went to the rear of the capsule and poured a cup of coffee. Besides, I was too keyed up to go back to sleep now. I got my cigarettes and stepped over to the deputy's console where Sparky was still sitting, staring straight ahead with a look of disbelief and disappointment on his face. I shook the pack towards him in the old sociable gesture of offering a smoke to a buddy.

"Thanks, Mike," Sparky said as he accepted the offer. I took out a cigarette for myself, tapping the end smartly against my Zippo a couple of times before putting it in my mouth and lighting it. Then I leaned over to light Sparky's cigarette. Both of us inhaled deeply, letting the rich smoke of the unfiltered Camel have its calming effect.

I broke the silence, "Wanna tell me what happened?"

"Not much to tell, Mike" Sparky began. "I was reading the monthly weapon system training package and kept nodding off. I figured I needed a break, and thought glancing through a *Playboy* would help. I reached to open the drawer, but I must have been more tired than I thought. I guess I grabbed the wrong drawer and opened it. Damn, I'm real sorry about that. It was really stupid."

"Anything else?"

"No, that's it," answered Sparky. Then he added quietly, "What do you think they'll do to me?" The enormity of his screw-up was beginning to sink in.

"I don't know."

I was beginning to wonder myself, and about how badly we both might get hurt by this. Missile duty made you paranoid about things like that. But one thing I believed is that we were a crew, and we would face things together. I knew that I would do everything I could to help Sparky.

"You have a couple of things going for you," I told him. "First, you have a good record and a reputation for being a smart, competent crewman. I'll vouch for that. Second, people like you. Third, you're just a second lieutenant, so you have plenty of time to recover."

Sparky managed a laugh, "Gee, at last I've found something good about being a butterbar."

"Right," I said, "Look, people are going to ask what I did to you for this. You know what you did was wrong. If you have trouble staying awake, get me up. I don't think you'll do anything like this again, and you're likely to be very careful in the future. So if anybody asks, you've been counseled on the matter by me. Got it?"

"Got it, Mike."

"OK, I'm going to tell Tango what happened. They and the squadron commander will have to hear the full story. But I'll try to keep my reports to WCP and Job Control to a minimum."

"Thanks, Mike."

I sent Sparky to bed, then called Tango on the phone so that I could relay the truth in private. After that I called WCP and Job, and made the log entries.

Inevitably, word of the incident got around. There was a lot of kidding about it, but Sparky took it in good humor. After a while, he didn't

even mind being called "Sparky." Eventually, he made to the Standboard shop. They didn't even dock his pay for the quarter million dollars in damages.

Elegy To A Silo Queen
by Maj Robert A. Wyckoff

In nineteen hundred and sixty-three,
The queen was born to shouts of glee.
In the season of holly and the red poinsettia,
Father was Martin, Mom: Marietta.
Her full name was 63-77-2-4.
Oh, she was a Titan II to the core.
Her seams were straight,
her skin did shine,
And she went by the nickname of B-69.
She did her bit in the Cold War fight,
Pulled her alerts at a Little Rock site.
Came to V'berg and found life so gay,
She scuttled the launch and
decided to stay.
She made the grand tour of
Bravo and Charlie,
Haunted those sites,
like the ghost of Marley.
She became in time, a witch of sorts,
What with nine emplacements and
two aborts.
She was enough to try a good man's soul,

Spent more time at depot than she did
in the hole.
The troubles she caused put two
colonels in traction,
She was one of those broads that's all
tease and no action.
Is it not enough to turn any man gray,
To try to launch a static display?
Her life was harsh, she no longer enchants,
But what do you want after
two engine transplants?
Like a matron fighting what time incurs,
There was not much left that was
originally hers.
Though her bulkheads reversed
and her seams did corrode,
She had the good grace not to explode.
What else could she pull from her
bag of tricks,
To continue to homestead in the
Vandenberg sticks?
Her batteries then at T-Zero's approach,
Wouldn't even respond to

the brogan approach.
But how the hell did she
work the last dodge,
And run a freighter aground at Kwaj?
I can still hear the whispered DOV prayer:
"Lift-off you mother...get out of our hair.
Hear this plea from Ted and Keich,
Please lift your skirts and just
clear the damn beach."
This time she flew
(though she was somewhat late),
For her paperwork equaled her
max, gross weight.
Let's lift our glasses on one final accord,
Surely one kind thought we each can afford,
Now that B-69's gone to her reward.

The Origin Of The Poem Missileer

The poem *Missileer* was written during February and March of 1972 at Vandenberg AFB by then Capt Robert A. Wyckoff. During his five year assignment at Malmstrom AFB, as a crew

member/commander in the 564th SMS and as a 341st SMW EWO instructor, he became known for his ribald farewell "friars roast" poems for departing squadron members. After reassignment to Vandenberg he and Col Tom Morris, chief of the Ballistic Missile Staff Course (BMSC) were participating in the weekly observation of "Happy Hour" at the Officers Club. Col Morris had the enviable assignment as the Olympic Arena (annual missile competition) banquet project officer. Knowing of Capt Wyckoff's writing exploits, he suggested that Wyckoff prepare a poem for the closing event of the '72 banquet. Wycoff agreed, but Morris observed that, considering the nature of the audience, it should be a serious work, rather than Wyckoff's accustomed humorous (usually) poems. After two months of spare time effort and substantial rehearsal, the Commander, 1st Strategic Aerospace Division determined that the banquet program was too long and cut the presentation of *Missileer*. Col Morris took the pains to have the poem published in the Poets corner feature in the *Air Force Times* as a consolation prize and subsequently succeeded in including the author's oral presentation of *Missileer* in the '73 Olympic Arena banquet ceremonies.

The poem had an immediate impact and has since been published by the Air Force in five editions. At one time copies were displayed in all active Missile Launch Control Centers in the Strategic Air Command and was found in such unlikely places as the entry foyer of the FPS-16 radar building atop Tranquillion Peak on South Vandenberg AFB.

While the poem was written at Vandenberg (at 1104 Cypress, Vandenberg married officers quarters to be precise), it was based on Wyckoff's experiences at Malmstrom and was intended to depict the professional lives of all missileers. The consistent reference to "men" in the text is not intended to slight the accomplishments of female missileers, but at that time the poem was written, and for many years thereafter, there were no women on the crew force.

Wyckoff is married to the former Eileen Hansen with whom he has raised a son, Steven, and a daughter, Lynne. They reside in Vandenberg Village, near Lompoc, CA where both are active in private aviation. Wyckoff retired from active duty in 1982 and is now a program director for the Western Commercial Space Center.

Anecdote
Submitted by LtCol Charles Wynne

NOTE: *LtCol Charles Wynne spent the first eight years of his career in missiles before transitioning to a career in public affairs. Ironically, he says the highlight of his career came when the two disciples merged shortly after he came off crew as a ground launch cruise missile launch control officer to become the public affairs officer at RAF Greenham Common, UK in 1988. The following is excerpted from a newspaper article that appeared in the April 3, 1991 edition of his hometown newspaper, the "Rantoul Press."*

"If I had to choose one event that stands out in my career, it would be my experience when the Soviets came to RAF Greenham Common," he says.

As the public affairs officer for the 501st Tactical Missile Wing at the cruise missile base 45 miles southwest of London, Wynne was in charge of all press coverage when the Soviets made their first trip to England as part of the Intermediate Nuclear Forces (INF) treaty verification process.

"We had more than 400 media from 13 different countries. We even had correspondents from the East Bloc and Soviet Union there," he said.

Signed in December 1987, the treaty called for the removal of all American and Soviet intermediate range nuclear missiles from Europe. One of the most significant aspects of the treaty called for on-site verification by inspectors from both countries as a means of determining treaty compliance.

"I was fortunate to be in the right place at the right time. It was a truly historical event. If the on-site verification clause had never been hammered out by the negotiators in Vienna, none

of this would have ever happened," said Wynne

Several aspects of the Soviet visits still stick out for the former Titan II and ground launch cruise missile launch control officer.

"For one, the Soviets landed an hour early. Fortunately, we were in the 'hurry up and wait' mode, so everybody was in place except for the last-minute arrivals," he said.

The media who got there late actually got the best video footage since we stopped the bus directly beside the runway for them to get pictures of the plane landing. The ones who had arrived on time were already waiting at the arrival area and didn't get anything but boring pictures of the plane taxing to a halt, he said.

Wynne was also responsible for writing the speech his commander gave in welcoming the Soviet inspectors.

"After about 20 attempts, I came up with something that was acceptable to my boss and the British base commander. It was short and sweet and nowhere near as verbose as what the Soviet team chief had to say," he added with a laugh.

One other telling incident occurred during that first inspection. All sides had agreed to allow the media to cover the Soviet arrival and departure. In preparation for the team's departure, Wynne instructed the bus drivers and team chiefs where to park the bus to maximize the media's photo opportunities as the Soviets boarded their plane to return home following the historic visit.

"When it came time for the bus to stop, it kept right on going until it was parked directly next to the steps leading up to the aircraft," he said.

Worse yet, it was parked directly between the media and the airplane meaning no photographs were taken of the Soviets boarding the plane.

"We all just stood there. Nobody knew what was going on and the media was getting antsy. Fortunately, a few of the Soviet inspectors came back out and gave the obligatory wave goodbye from the top of the stairs," said Wynne.

He later found out the Soviets changed the plan to avoid being photographed boarding the plane. Just before leaving, the inspectors had been given permission to shop at the base exchange and commissary.

"When it came time to board the plane for the trip home, they each had bags of Western contraband under their arms. They literally bought the exchange out of things like ballpoint pens, pads of paper, batteries, mops and undergarments. But they didn't want anybody to know they were taking the stuff back with them," said Wynne.

The public affairs staff at RAF Greenham Common later won one of the most prestigious public affairs awards given by the USAFs in Europe for their work during the first Soviet inspection in the United Kingdom in conjunction with the INF treaty.

When the Soviets returned for a second inspection, Wynne was selected to escort the local minister of Parliament, Sir Michael McNair, as he met with the Soviet inspection team chief, a major general from the Soviet Rocket Forces.

"That was a very exciting experience as

well. I was the only American in the room and was surrounded by KGB agents, who made up the majority of inspectors. It was kind of spooky." he added.

"The whole INF experience was just fabulous for me. It was a lot of hard work but I saw history being made," he said in reflection.

As for other highlights in his career, the 1979 University of Illinois journalism graduate mentioned just two.

First, he was fortunate enough to eat dinner with members of the British unit whose predecessors sacked the White House in 1814.

"We actually ate dinner in their mess and we all got a laugh out of the White House thing," he said fondly.

The second thing he will always remember is more personal.

"My son, Nicklaus, was born just outside of Oxford and life hasn't been the same since."

There I Was
Submitted by Alonzo Hall III

(1) I think it was back in the late 70s while serving as an ICBM maintenance team chief, my team member and I were performing maintenance in a Minuteman II silo when we received a call from capsule (Launch Control Center) to prepare for EWO (Emergency War Order) launch. Needless to say we performed the required tasks very quickly and departed the site. The old saying "Train like you fight; fight like you train" was very evident, because we didn't even stop to think about what we were doing or its implications until we were sitting on a hilltop watching the site and waiting for the end to come. Fortunately the warning system problem was corrected and we were around to deactivate the site Hotel 8, near Union Center, SD in 1994.

(2) As a member of the On-Site Inspection Agency (OSIA) we performed numerous inspections and verifications in Russia (former Union of Soviet Socialist Republic) in support of the Intermediate-Range Nuclear Forces Treaty. At the completion of one of our longer elimination inspections, we were all looking forward to departing Moscow and heading back to the west. Myself, I was looking forward to the taste of Coca-Cola again, because the only pop we could get in country was Pepsi and Fanta.

Our team was informed that we would do a consecutive inspection so we spent another night in Moscow before making the declaration for the next inspection site. That night while in the room with my fellow team member, I commented how nice it would be to get a coke to drink instead of a Pepsi. Upon arrival at the new inspection site and after the formalities, we were directed to our quarters. In each of our rooms there was a small refrigerator and will wonders never cease, only mine was filled with coke, and it tasted great. Do you think they were listening in our hotel room in Moscow?

MX - What Does It Mean?
by Col Jack G. Hilden

Prior to being renamed by President Reagan, the Peacekeeper Program was called MX. Many people have wondered what MX means. Was it meant to mean missile

experimental? Was it the unknown missile Was it the unnamed follow-on to Minuteman? I've heard many people speculate and several claim to know.

In order to understand the meaning of MX, you must understand where the name came from and why it was chosen. In 1971, the program was called the Advanced ICBM Technology Program. The Navy had their ULMS program. The Army was pushing safeguard and the Air Force didn't want anything to stand in the way of B-1. Other strategic development programs were not well received. ICBM development funds were almost non-existent.

RDQSM, the small Air Staff ICBM requirements office, was almost the only voice in Washington pushing for ICBM development funds. By the time we explained that we wanted to discuss the Advanced ICBM Technology Program with Pentagon powers-that-be or congressional staffers, they said lets go on to the next subject. Out of frustration we decided we had to have a sexy name like the other strategic programs before anyone would listen to us.

One afternoon, the five of us sat down by the chalk board and started to list possible names. Many were suggested, MX among them, but none seemed to catch our fancy. We left them on the board that night and decided to go at it again the next morning. The next morning someone asked, what does MX mean? We went through all the possibilities and concluded that it could mean all of them or none of them. We felt that the beauty of MX was that it sounded like an ICBM development program and that it might be able to get attention for funds in a very adverse environment.

I immediately went downstairs and told my boss, BrigGen Chic Adams, that unless he stopped us, he would start hearing about "MX" and would never hear "Advanced ICBM Technology Program" again. Whether the name change was good or had any effect is open to conjecture, but we were able to get increased future fiscal year funding for MX shortly after the change. You can love or hate the name MX and give it any of several meanings as you wish but, in my judgment, it served an important function in getting funds budgeted for a critical ICBM program.

Penelope Jackalope, Wiley Coyote and The Wrought Iron 99
by Col Jack G. Hilden

I received orders in May of 1974 to report to Ellsworth AFB for assignment as vice commander of the 44th Strategic Missile Wing. One of my last duties as deputy commander for maintenance at Malmstrom AFB prior to departure was to welcome the 341st SMW team back from a successful Olympic Arena missile competition. Col Ralph Scott, the wing commander, came down the stairs from the KC-135 first. He was carrying a beautiful jackalope. It had the head and forebody of a jack rabbit, horns of an antelope and the delicate wings and tail of a pheasant.

To my surprise he presented the jackalope to me as the grinning team looked on. He explained that the jackalope had recently been the centerpiece of the 44th SMWs display at Olympic Arena. I was to deliver it to my new boss, Col William Bush, the commander of the 44th. As the wind on the ramp blew the feathers of the jackalope around, I knew that I had to protect that beast with my life or my name would be mud at the 44th. A message was sent to Col Bush announcing that the jackalope was safe and that their new vice commander would deliver it when he reported for duty.

The awards banquet was held a few nights later at the Officer's Club. In order to show that I was a good sport, I brought the jackalope and placed it on the table with the rest of the awards. A critical error! Half-way through dessert, I looked up and saw that the jackalope was missing. I immediately cornered Sgt Kenny Bunker and told him that the joke had gone far enough. I was due leave for Ellsworth early the next morning. Of course, he claimed complete innocence. Nobody at the banquet, including Col Scott, would help me recover the jackalope.

I didn't sleep that night. What a way to report to a new assignment! What would I tell Col Bush? Wall Drug Store had loaned the jackalope to the 44th and was anxious to get it back. About 1:00 a.m. the doorbell rang. I rushed to the door to find a large dishpack carton on the step. It was well sealed and was addressed to Col Bush. "Very fragile," "This end up," "Critical cargo," and "To be opened only by the addressee" were very prominently printed on the carton. I took the carton into the house and went back to bed relieved. I had the jackalope back. Then it occurred to me that the carton felt too light to contain the jackalope. I had to find out what was in that carton. I carefully slit the tape so I could properly reseal the carton. Inside were pheasant feathers, rabbit fur, a stuffed coyote head and a note saying, "I can't believe I ate the whole thing." A corner of a 20 dollar bill was stuck on one of the coyote's fangs. What now? I had no choice but to reseal the carton and deliver it to Col Bush. No more sleep that night!

Early the next morning, I started out on the longest 540 mile drive in history. Col Bush had invited me to dinner at his quarters. The closer to Ellsworth I got, the more I became convinced it would be my last supper. I finally arrived at his quarters and met my new boss and his wife. I stammered that I had a package to deliver to him. We took it inside and he carefully opened it. He looked inside for a full minute and then burst out laughing. The dinner was delicious and it wasn't to be my last one after all.

Several exchanges of messages followed discussing the lineage of Wiley Coyote, his criminal behavior toward Penelope Jackalope and how an exchange of prisoners could be arranged. The exchange finally took place in neutral territory and Penelope was returned to Wall Drug Store partially restoring the 44th to the good graces of the community.

We at the 44th couldn't allow this humiliating defeat to go unchallenged. How could we strike back? The pride and joy of the 341st was a wrought iron 99 sitting in a place of honor behind Col Scott's desk. It had been presented to him on a red carpet in front of the full missile maintenance squadron after we had held 99% effectiveness for a full month. That would be our target. One of our best scouts volunteered to penetrate the enemy lines, distract the command post long enough to get the keys to Col Scott's office and make off with the hostage in the middle of the night. His mission was a total success. They never figured out what happened. The new hostage was placed in job control at the 44th and was not returned to Col Scott until after he retired.

The Night The Balloon Burst
Submitted by Wells E. Hunt Jr.

The Atlas F contractor, General Dynamics, performed the initial installation and checkout of the missiles before turning them over to the Air Force. It was during installation of one of

Minuteman Launch.

the Atlas F's onto the launch platform that the balloon burst.

The missile on its transporter was backed onto one side of the launch platform and attached to fittings on the launch platform. On the opposite side, a crane-like device was extended over the silo opening and attached to a fitting at the top of the thrust section. The erection device was essentially a long threaded shaft thru a fixed nut. As the shaft was turned, it pulled the Atlas to an erect position on the launch platform. All was proceeding normally one night and the missile was at about 60 degrees up from horizontal. Suddenly, the nut on the erector failed. This giant stainless steel balloon headed back down. Knowing that its impact with the transporter would be catastrophic, workers started scattering, some of them easily clearing the silo opening. One even went over the perimeter fence in two steps! The missile hit bottom and exploded into shrapnel, pieces of stainless steel less than the thickness of a dime went sailing everywhere like lethal Frisbees! Luckily, no one was injured, probably due to quick reflexes.

Hold That Button
Submitted by Wells E. Hunt Jr.

The Atlas F system had a less than ideal safety record, four of them detonating in their silo during dual propellant loading exercises (PLX). It was during a PLX that I endured my longest day. I was Deputy MCCC on the crew designated to perform a PLX during one of the initial Operational Readiness Inspections at Lincoln AFB, NE. At the time, we were still using liquid oxygen (LOX) and RP-1 for the PLX (later, the LOX would be removed and liquid nitrogen put into the tanks for a PLX). All went well with the propellant loading and countdown until the missile was raised to the surface. At that time, events took a turn for the worse. Due to automatic propellant problems, we had to perform a manual lowering of the missile and a manual offload of the LOX, all of this took a long time.

A primary responsibility of the DMCCC on the crew was to monitor the tank pressures to ensure that a positive differential was maintained between the fuel and LOX tanks to prevent reversal of the intermediate bulkhead and a resulting explosion. During the manual offload of LOX, I observed that the pressure in the fuel tank kept decaying. As a result, I had to continuously hold down the fuel tank pressurization button, a small button about 1/4 inch in diameter. Once the missile was lowered into the silo, the LOX offloaded, and the missile placed into stretch, I could take my thumb off the button. One of the enlisted crew members and I then went to the pressurization unit located on the launch platform and found the relief valve for the fuel tank stuck in the open position. We were on the brink of disaster without the constant addition of pressure to the fuel tank. Thank goodness the helium supply held out. To this day, I have an indentation in the pad of my right thumb as a reminder of my Atlas duty.

Bi-Lingual Curtain Raiser
Submitted by John W. Haley III

My Titan II crew, S-117 was reformed to include Capt Stayton, deputy missile crew commander-DMCC; MSgt Miguel de Zarraga de de Zarrage, ballistic missile analyst technician (BMAT); and SSgt Michael Santistevan, missile facilities technician (MFT). We won a local competition in order to be one of two crews to represent the 390th SMW at the first Missile Combat Competition Curtain Raiser in the summer of 1967. We were trained by Lt Kelley and his crew. During our last scored exercise in the Titan II launch control trainer, we received a simulated launch message, and almost simultaneously, an uncontrollable fire in the silo. Launch or fight the fire? The message was "Good" and we decided to launch. MSgt de Zarraga whirled around (to attend his guidance console) and ran head on into a steel support pole, I continued the launch checklist but de Zarraga responded in Spanish (Cuban style), Santistevan answered in Spanish (Portuguese style), I read the checklist in Spanish (Gringo style) and we launched (simulated) SAC style. The judging crew said we only had one error, failure to launch in English and they flunked us. We protested. The competition review board ruled that there was no stated or written requirement to launch in English (There was in the next competition) so our perfect score was re-instated. We came out number 3 of 6 Titan II wings and number 7 of 18 total competing missile wings. Not bad for a crew that had only been together 60 days. We proved that with SAC missile training we were truly interchangeable and could launch in any language.

NB: Two hours after this exercise, Col R.R. Scott, 390th SMW Commander, handed me my orders to fly C-130s as a navigator in Vietnam. To err is human, to forgive is not SAC policy.

Memoirs
Submitted by John C. Leber Jr.

Since LtCol John C. Leber Jr. took command in November 1985, the 90th Services Squadron has gone through many changes. As a result of hard work and dedicated effort, significant improvements in the quality of service and facilities for personnel assigned to Francis E. Warren AFB, has steadily improved. The Chadwell Enlisted Dining Facility has been renovated and improved to the point where it is the best in the command. The Chadwell Dining Facility is presently one of the three finalist for the SAC R.T. Rhiney Award. The winner of the award will go forward as a competitor for the Air Force Hennessy Trophy Competition. The SAC Inspector General's visit to F.E. Warren AFB in June 1987 resulted in an overall excellent rating for the Services Squadron. This was the first excellent for a Services Squadron in command this year. All food service sections were rated excellent or better. The food service accounting section went from 18th in the command in 1985 to 3rd in 1986 and now has zero errors for the first three quarters of 1987. Furnishings Management achieved 100% accountability in all furnishings and furniture accounts for the 13 dormitories, three transient lodging facilities (TLF) and seven visiting officers/junior quarters. Each of these facilities was rated excellent by the IG.

The standard of excellence instilled by LtCol Leber, inspired many personnel to achieve honors in their on and off duty endeavors. These awards included Noncommissioned Officer Leadership School Honor Graduate, a Wing Airman of the Quarter, two base level Noncommissioned Officer Preparatory Course Honor Graduates Awards, an Honor Graduate at the First Sergeant's Academy, a Quality Assurance Evaluator (QAE) of the Quarter and four Below-the-Zone promotions.

The Prime Ribs program was rated excellent during both IG inspections. Mobility training and equipment are top quality. The Harvest Eagle Mobile Kitchen Trailer (MKT) was received and fully field tested. Prime Rib deployments supporting two Medical Red Flag exercises, a Civil Engineering Prime Beef Deployment during the IG inspection; and a CES Prime Beef Deployment to Chimney Park, WY, completed "The Best Deployment" of Prime Ribs observed by the IG.

The Billeting Facility Renovation program and upgrade of facilities will be complete in 1989. All Transient Lodging Facilities (TLFs) have been renovated and will be completely furnished with new furniture. A new billeting office will be renovated in November 1987. Distinguished visitors facilities are currently under renovation and will be complete in mid-1988.

New furniture for five enlisted dormitories has been purchased. Seven dormitories will enter renovation in the next fiscal year. An additional $459,000 worth of new dormitory furniture is programmed for the next fiscal year. Vastly improved washer and dryer service to all billeting facilities was contracted in FY87. The most significant improvement in this area was the billeting office going from a loss to a profitable operation, giving vastly improved accounting, inventory control and billeting reservation services.

Self-Help projects which greatly enhanced the quality of both living and work environments are present throughout the 40 plus buildings in the services' area of responsibility. Services Squadron personnel have deployed on several occasions to the Middle East in support of the Air Force requirements. The foil pack kitchen renovation will reopen in mid-FY88. The Linen Exchange Facility was called the "Best such facility in the command," during the CINSAC visit.

Quality customer service in all services facilities is the rule. During the past 18 months a new Air Force Commissary and Army Air Force Exchange Facilities were opened. A newly renovated Base Theater also opened.

Significant quality of life improvements have been made for all F.E. Warren AFB personnel.

Base Missileman Launch Titan II From Vandenberg
Submitted by Walter Kundis

A Titan II Intercontinental Ballistic Missile was successfully launched from Vandenberg AFB, CA, Friday by Crew S-116, composed of Capt John H. Womack Jr., missile combat crew commander; 1stLt James W. Harshbargar, deputy crew commander; SMSgt Walter Kundis, missile systems analyst technician; and TSgt James W. Meddress, missile facilities technician.

The missile was fired from Vandenberg at 3:56 a.m. Preliminary reports indicated that the missile was on course for a selected target area more than 5,000 miles down the Air Force's western test range.

Operations and maintenance personnel participating in the launch were under the supervision of LtCol Dave Counts, task force commander. Personnel participating were LtCol Rhuena B. Pfeiffer, 1stLt Myron M. Langhoff, SMSgt Richard L. Ott, MSgt Walter L. Harrel, TSgt Donald E. Jagger, TSgt Hamp Parker, TSgt Richard L. Stults, SSgt John O. Hannon, SSgt Richard P. Marzyn, SSgt Philip T. Hutchinson, SSgt Bobby L. Crossley, SSgt Delbert L. Hancock and SSgt Jerry J. Gloston.

Other members participating in the firing were SSgt Cassel Davis, SSgt Harry L. Dorsett, SSgt Herbert G. Roper, SSgt Marvin D. Taylor, SSgt Samuel Walker, SSgt Gary L. Burnett and SSgt James D. Taylor.

Airmen participating in the launching were A1/c Larry E. Jewell, A1/c James D. Evans, A1/c James A. Ott, A1/c Roger L. Richie, A1/c Jerry W. White and A1/c Harry J. Dee.

Other airmen involved were A2/c Loyal E. Stricklin, A2/c Robert D. Brixey, A2/c William M. Palmer, A2/c Stephen C.M. Gill, A2/c James E. Lloyd, A2/c Jan L. Hepworth, A2/c Michael T. Jankowski, A2/c Gerald F. Pyne, A2/c William J. Fox, A2/c Paul M. Flores, A2/c Jeffrey D. Guyer and A2/c Lynn B. Daniel.

Titan II Launch
by Capt. Joseph Bacik

The 374th Strategic Missile Squadron, with a complement of 98 officers and 111 enlisted personnel, was formally assigned to Little Rock AFB in September 1962. Initially, personnel arrived from various commands within the Air Force.

This diverse group of men had to be assigned to crews and molded and trained to operate as highly efficient and professional combat teams. To accomplish this formidable task, an intensive training program was initiated, and in the end 40 combat crews functioned as one. By December 1963 the 374th became fully operational and the mission of the squadron became reality - to destroy enemy targets when so directed.

Since becoming operational the 374th has compiled an admirable record. An unparalleled rating of 100 percent in written tests, as well as crew performance demonstrations, was achieved by the squadron during the first Operational Readiness Inspection (ORI) conducted by Strategic Air Command's Inspector General.

This squadron has given enthusiastic support to SAC's Big "R" program with 253 suggestions contributed. Fourteen have been approved and 100 are still in various stages of evaluation. No less than 70% of the crews hold the coveted title of "Senior Crew." The families of the officers and airmen of the 374th are understandably proud of the job done by their men.

Titan II Launch.

308th Bombardment Wing
Submitted by Walter Kundis

The 308th Strategic Missile Wing was originally constituted as the 308th Bombardment Group (Heavy), 28 January 1942. The Group, then composed of the 373rd, 374th, 375th and 425th Bombardment Squadrons, was activated at Gowen Field, ID, 15 April 1942.

The 308th, one of original units of 14th Air Force, moved to Kunming, China, early in 1943, with the air echelon flying by way of Africa, and the ground echelon traveling by ship across the Pacific.

The Group made many trips over the Hump to India to obtain gasoline, oil, bombs, spare parts, and other items needed to prepare for and then to sustain its combat operations. In order to remain operational while stationed in China, the 308th found it necessary to fly three supply missions to India for every one combat mission flown in China.

The 308th Group supported Chinese ground forces; attacked airfields, coal yards, docks, oil refineries and fuel dumps in French Indochina, mined rivers and ports, bombed shops and docks at Rangoon, attacked Japanese shipping in the East China Sea, Formosa Strait, South China Sea and the Gulf of Tonkin.

The Group received a Distinguished Unit Citation for an unescorted bombing attack, conducted through antiaircraft fire and fighter defenses, against docks and warehouses at Hankow 21 August 1943. A second Distinguished Unit Citation was received for interdiction of Japanese shipping during 1944-45.

One member of the 308th, Maj Horace S. Carswell Jr. (Carswell AFB, TX) was posthumously awarded the Medal of Honor for staying with his battle damaged aircraft in an attempt to save a wounded crew member 26 October 1944.

In June 1945, the 308th was moved to Rupsi, India, where it flew gasoline over the Hump until 7 December 1945, when the unit returned to the United States. The Group was deactivated 6 January 1946.

Approximately 10 months later, 17 October 1946, the unit was reactivated at Morrison Field, FL, having been redesignated the 308th Reconnaissance Group (Weather). Following five years service in weather reconnaissance and weather reconnaissance training, the Group was once more deactivated 5 January 1951.

Redesignated the 308th Bombardment Group (Medium), the unit was activated for the third time 10 October 1951. Equipped with B-29s, the 308th Bombardment Group was assigned to the 308th Bombardment Wing in Strategic Air Command. The Group was deactivated on 16 June 1952. However, the 373rd, 374th, 375th and 425th Bombardment Squadrons remained assigned to the bomb wing until they were deactivated in June 1959. The wing was deactivated 25 June 1961, and was redesignated the 308th Strategic Missile Wing 29 November 1961.

On 1 April 1962 the 308th was activated and redesignated the 308th Strategic Missile Wing, marking its entrance into the missile age. On 31 December 1963 all 18 missile complexes, which are strategically located throughout northern Arkansas, were fully operational and were turned over to Strategic Air Command and the 308th SMW by Ballistics Systems Division, Air Force Systems Command.

On its reactivation in 1962, the 308th became the last of only three units assigned Titan II Intercontinental Ballistic Missiles. These missiles are 103 feet long, 10 feet in diameter, weighing over 150 tons. The missile engines produce a 430,000 pound first stage thrust and a 100,000 pound second stage thrust capable of powering a payload to targets over 6,000 miles away.

A hallmark in the operational development of the Titan II was made 2 October 1964, when a crew from the 308th SMW launched a Titan II ICBM down the Pacific Missile Test Range from Vandenberg AFB, CA. This marked the first time that a combat ready crew from an operational Titan II missile wing performed such a feat. After October 1964, eight more launches were conducted by crews from the 308th SMW at Vandenberg AFB. These launches occurred 16

Heavyweight of SAC's Large ICBM Force, Titan II launched directly from its underground storage silo. Sequence, starting at top left, shows smoke and flames emerging from the exhaust escape vents, then the missile appearing on the start of its intercontinental range flight from Vandenberg AFB.

August 1965, 26 August 1965, 20 October 1965, 27 November 1965, 22 December 1965, 5 April 1966, 22 July 1966 and 12 April 1967.

In April 1967 the 308th SMW participated in the first annual SAC Missile Combat Competition, designed to select the best missile unit, best missile combat crew and best targeting team in SAC. The wing finished third in the overall competition, but its targeting team was selected as the best in SAC.

Presently, the 308th SMW is commanded by Col Edward A. Vivian. The wing vice commander is Col Leo C. Brooks. The deputy commander for Operations is Col John P. Foster while Col William B. Cofield is deputy commander for Maintenance.

Today, as in World War II, the 308th SMW is living up to its motto of "Non Sibi Sed Aliis" (not for self, but for others), meaning that the 308th has progressed from support of Chinese citizens seeking to rid their country of an enemy to an era of devoted duty by 24-hour a day readiness in safe guarding the well being of the American people.

On My Mark...
Submitted by Lawrence Hasbrouck

"Looking Glass, this is the Vandenberg launch director," a sober voice echoed. A silent hush fell over the EC-135 crew as they looked anxiously to one, then the other. Nothing could be heard save the laboring engines and rushing air.

"Is this it?"

"Is this the time?"

The suspense is quickly broken by the return of the director.

"Initiate an execute launch command on my mark. Advise when transmission has ended," the voice demanded.

"5-4-3-2-1-mark."

Instinctively, the launch keys are rotated by LtCol Jerry Chamberlen and Capt Kenneth Fox, and launch data is transmitted. Within seconds the directors computers are programmed for launch.

"Vandenberg launch control, this is Looking Glass, transmissions have ended," airborne launch director, LtCol Larry Hasbrouck advises.

"Looking Glass, this is Vandenberg launch director; we have launch in progress; we have first stage ignition and lift-off."

Only a very few minutes have expired in the simulated airborne Minuteman launch exercise since the ground launch director first contacted the Looking Glass airborne controller for launch instructions.

The simulated exercise was held Friday to insure continued success of the missile launch program. The system has been in effect since 1967.

A radio command can be initiated from modified EC-135 aircraft with the aid of several control panels. A computer, missile crew, flight crew and radio operators can initiate a radio command which will cause the desired missile or missiles to strike pre-selected targets.

If ground communications between launch control centers and missiles has been lost, the missiles can be launched by an airborne launch control center installed on post attack command and control system aircraft or in SAC airborne command post aircraft.

"The Minuteman Intercontinental Ballistic Missile program is the most important single element of deterrence we have. It represents the largest part of the triad of United States strategic offensive weapons and is virtually 100% on the alert all of the time," Gen Bruce K. Holloway, Strategic Air Command commander in chief said on the occasion of the 10th anniversary of the Minuteman Missile.

"Land-launched Minuteman Missiles are the nation's fastest reacting and most cost-effective strategic system," he said.

Aircraft from Offutt and Ellsworth, SD have been specifically modified to provide a secure secondary Minuteman launch system. Each EC-135 aircraft can launch any or all of the Minuteman strike force.

Twenty-three Minutemen have been launched from Vandenberg firing range to insure the systems capability.

The Gladiators
by Richard J. Biggs

"Why are there so few missile wing commanders whose background is missile maintenance?"

I was flying on an executive jet on the way to Grand Forks, when I was asked the above question by a two-star general who was also on his way to Grand Forks. He was the vice-commander of the 15th Air Force and I was the 15th Air Force Director of missile maintenance. I believed he had asked a serious question, so I gave him a serious answer; "In order to be considered for the wing commander's slot one must be a colonel. There aren't too many young missile maintenance officers who live long enough to become colonels. We have a way of eating our young."

My first experience in maintenance was at Ellsworth Air Force Base, SD. I was assigned as OIC of the Technical Analysis Engineering Teams. We worked for the deputy commander for maintenance, therefore we were maintenance troops. In reality we were a hybrid of engineering and maintenance. Each team was made up of a graduate engineer and an experienced maintenance NCO. It was a marriage of convenience. The graduate engineer was there to add mystery and sophistication to what we did, and the NCO was there to make sure that we didn't do anything dumb. The NCOs were the professionals on the team. They knew the missile system as no one else in the field did; they did most of the work, and they kept a wary eye on the young 2nd lieutenant or 1st lieutenant assigned to their care. They were good, but they were not the "gladiators" I met when assigned to a "real" missile maintenance job.

In January 1972 I was assigned as the MIMS Maintenance Supervisor at Minot AFB. There I met the young officers and non-commissioned officers and airmen whom I call Gladiators. They were truly young, mostly lieutenants, captains, staff sergeants and below, trying to cope with job assignments that called for majors, lieutenant colonels, technical sergeants and above. They could do the job but the pressure was overbearing and the attrition rate was high.

These were the young people we entrusted to keep our fleet of missiles on strategic alert. Supposedly the "best of the best." Yet, we did not trust them to make a decision. The team chiefs were not allowed to manage their team members; branch supervisors were not allowed to manage their teams; and even the maintenance supervisor was not allowed to manage assigned resources. Mistakes, any mistake, was not tolerated.

That is why I call them Gladiators, these young missile maintainers I met at Minot. Every day was a battle, a battle against the elements, the schedule, supervision and even against the system. If they did everything correctly, they were rewarded by getting to do it again the next day. If some how they made a mistake, they were dead. Today's performance was the measure of existence. Tomorrow, or longevity, was not a major concern of the Gladiator. He fought and won today, or there was no tomorrow.

We truly ate our young.

Looking Back
by BG Elmer T. Brooks

In retrospect I guess I can count myself as fortunate to have survived a career associated with liquid-fueled missiles. These cantankerous "beasts" have left me with many memories, good and bad.

Most memorable to me was the 48 hours straight of Atlas F missile alert duty our crew performed at the height of the Cuban Missile Crisis in October 1963. I was commander of a tight knit crew of five young men trying mightily to be brave and focused in the face of impending nuclear holocaust. We worried about the fate of the world, and especially about our loved ones. Several crew members wanted to call and warn their families to evacuate the area, even to come and take shelter in our Launch Control Center. As much as I empathized, I had to turn down such notions. After hearing (CincSAC) Gen Power's unprecedented "call to arms" come over the primary alerting system, I turned our attention to what it would take to ready our off-alert missile for launch. We determined that a simple "kluge" could circumvent the malfunction and allow us to launch safely. I passed it on to maintenance for authorization - which never came. Authorization or not, we knew what we had to do if a launch message was received. Thank God, that message also never came.

Another instance I'll never forget is a dual propellant loading exercise my Atlas F crew conducted while undergoing a Stanboard check in 1963. This occurred before SAC became sadder and wiser about using RP-1 and LOX (instead of LN2) for exercises, and before the tech data emergency procedures (especially those pertaining to AGE) did more than scratch the surface of the catastrophic things that could happen and how to recover from them. After loading LOX, and in the middle of the launcher platform ascent sequence, we experienced a major malfunction. The primary diesel power generator shut-down automatically. We switched to the standby generator. Shortly, we noticed that it too was overheating rapidly and we had to shut it down. That caused the launcher platform to stop rising, before the liquid oxygen boil-off valve on the missile had cleared the silo cap. The super-cooled gaseous oxygen began to boil-off and vent through the valve, impinging on the steel launcher platform cables. Similar experiences later at Rosewell would prove that this can be a disastrous situation, the launcher platform suspension cables could snap, dropping the launcher platform and loaded missile into the silo. Explosive! Of course, our trusty "Dash One" didn't cover this predicament. And, when I turned to the knowledgeable Stanboard Crew Commander to rescue us, I could tell by the pained expression on his face that he didn't have the answer either. Fortunately, I had an outstanding crew. We had spent many hours of alert duty studying the cranky aerospace ground equipment (AGE) systems and how they interacted with one another and with the missile systems. We knew that this complex had been experiencing problems with the automatic water valves serving the cooling tower. And, when we panned the topside TV camera to the water cooling tower, we could see that there was no hot water circulating through it because there was no steam cloud. Necessity being the mother of invention, I decided to improvise. After a quick review of the plumbing schematics, we devised a recovery plan that the Stanboard Commander bought off. I sent the missile facilities technician and the ballistic missile analyst technician into the silo to reposition several water valves manually in order to reroute the cooling water for the diesels. Thank God, it worked! We restarted the generators and successfully backed out of the PLX.

My tenure as commander of the 381st SMW, McConnell AFB, KS, provided lasting memories as well (each one a story in itself). Among them were:

•Dealing with the intense scrutiny from all quarters, especially the media and the politicians, in the aftermath of the Titan II catastrophe at Rock, KS, when I assumed command of the wing.

•Personally overseeing the complete replacement of the ineffective butyl seals with polymer seals in the missile oxidizer system of every missile in our fleet. The extended period of leak-free status which followed was pure bliss.

•Seeing the first female missile combat crew member stand alert and the first female law enforcement security policeman on duty.

•The positive response to the town forums that I initiated to enlighten the residents and elected officials of the communities near our missile complexes about our weapon system and its mission.

•Discovery in 1981 of an apparent traitor in our midst. We were shocked beyond belief to receive a report that a deputy missile combat crew commander had been observed delivering what was believed to be extremely sensitive national security information to the Soviet Embassy in Washington, DC. To add insult to injury, he was able to avoid criminal prosecution on a technicality raised by his defense, Attorney F. Lee Bailey.

•On the other hand, there were many proud moments. In 1980, we won the Air Force Missile Safety Award, the Air Force Outstanding Ground Launch Missile Maintenance Award, the Chadwell Trophy (Best Missile Maintenance in SAC), and the coveted Blanchard Trophy (Best Wing in the SAC Missile Combat Competition).

The Day Almost Everything Went Wrong At Juliet-Five
by Richard J. Biggs

This really is not a funny story, nor is it a short story. However, the passage of time somehow allows me to smile a little when I think of everything that went wrong. It happened some 27 years ago. Time and age have blurred some of the details and soften some of the frustration, but I remember well that cold Thursday in 1972 when almost everything went wrong at Launch Facility Juliet-Five, 91st Strategic Missile Wing, Minot AFB, ND. That this happened at Minot will come as no surprise to those who have been to Minot and certainly not as a surprise to those who have been associated with missile maintenance.

First, I must provide a word of explanation. This is the way I remember the story. All of the accounts, save one, are first hand and witnessed by me. The one exception is my version of how the missile ended up on the launcher lid with the lid halfway open. This exception is a matter of conjecture based on second-hand and third-hand accounts, tempered by my knowledge of procedures and the layout of the Launch Facility. If anyone is offended by my telling of this story, I sincerely apologize.

Ronald J. Bishop Jr. was, and still is, a friend of mine. He and I suffered through one and half years of pure hell from January 1972 to July 1973. He remained at Minot for another year after I departed. On a brighter note, we both were on the list of those promoted to colonel released in late 1974.

I was the MIMS maintenance supervisor and Ron Bishop was the Missile Installation and Maintenance Squadron (MIMS) Commander. We were two miserable lieutenant colonels just waiting for fame or misfortune to move us past what was probably the worst assignment we had ever experienced. I don't know about Ron, but I fully expected misfortune, and most probably infamy, would be my destiny. Although I had been assigned as the chief of the Technical Engineering Analysis Teams for five years at Ellsworth AFB, this was my first venture into "real" missile maintenance. After one year at Minot, I knew there would be no fame in this assignment.

It was a cold Thursday morning (only three more workdays until Monday). I had just walked in the door of the dreary building that housed most of the MIMS functions when I was confronted by the officer-in-charge of the Organizational Maintenance Branch. I could see that he was deeply troubled even at that early hour of the day (it was just past 6:30 in the morning). As we walked down the hall to my office, he kept saying "We are in big trouble." After three or four of these pronouncements, I finally bit and asked, "What the hell are you saying." He replied with seven words that really meant "big trouble," "We set the missile on the door." Of course I knew the procedures and therefore knew that it was impossible to set the

missile on the door of the Launch Facility. The more I insisted that we could not have done that, the more he insisted that we had done exactly that. By this time I was in my office and the hot line from Job Control was fairly jumping off the table with its insistent ring. I picked up the phone and got the official word. We were indeed in big trouble.

Before I continue this sad tale, let me set the stage. On the previous day, Wednesday, we had begun a missile removal and replacement at Launch Facility Juliet-Five. A Missile Maintenance Team (MMT) with an empty payload transporter (PT) had dispatched early that day. This PT would be used to store, on-site, the re-entry system (RS), propulsion system rocket engine (PSRE) and the guidance set. All three of these components had to be removed prior to removing the old missile. A Missile Handling Team (MHT) in an empty transporter erector (TE) was dispatched some two or three hours later to be on site when the missile was ready to be removed. A second MHT in a TE loaded with the replacement missile was dispatched approximately four hours after the first TE. This TE was scheduled so as to be on site right after the first TE had departed the site with the missile being removed. This bit of musical chairs allowed us to remove and replace a missile and have the new missile up and in "Strategic Alert" (ready to go) with less than 26 hours of downtime. We had done this many times and it had worked like a dream. This was not going to be one of those times.

When I went home that Wednesday evening, the defective missile had been removed and was on its way back to the base. The maintenance teams and two field supervisors were on site awaiting the arrival of the replacement missile. Everything was going according to plan.

Well back to that Thursday morning... Ron Bishop arrived at my office about the same time that I did. It was decided that I would go out to the site, find out what had happen and report that information back to Job Control. I called Transportation Control Center (TCC), filed a trip ticket with my permanently assigned trip number and made preparations to depart. By this time, bigger wheels in the wing had made the decision that Ron Bishop would also make the trip to Juliet-Five. So, he called TCC, filed his trip ticket and received his permanently assigned trip number. He and I climbed into a station wagon and departed for Juliet-Five at approximately 8:00 a.m. Little did we suspect that we would forever be associated with the continuing comedy of errors unfolding at Juliet-Five.

On the way to the site, we tried to assure each other that there was no way that one of our maintenance teams had set a missile on the launcher door. When we pulled up to the Launch Facility gate, we were met by the security guard. He asked for our trip number. I was driving, so I gave my trip number. After he made a radio call, he asked for a name; Ron Bishop gave him his name. Naturally the name and the trip number did not match. So, we were busted.

By now it was after 9:00 o'clock in the morning. I suppose everyone in the Free World, except this security guard, knew that Ron Bishop and I were driving out to Juliet-Five because we had a potentially hazardous condition on site.

After some 30 minutes of heated discussion at the gate, we were allowed on site. We went into the silo and looked up through the collimator slot (the diving board was closed as it should be when lowering a missile). Though our access was somewhat limited we could see all we had to see. The missile was clearly on the launcher lid. The lid was more than half-way open and the missile was perched on the edge of the lid. The hoist cables were dangling down the side of the missile. The missile was leaning toward and apparently resting on the bottom side of the transporter erector. This was indeed a frightening sight.

Ron dutifully reported what we had seen. By this time the Missile Potential Hazard Net (MPHN) had been activated. That meant that we had representatives from the Wing, 15th Air Force, SAC Headquarters, Boeing, Ogden Air Material Area, and God knows who else listening in to what was going on at Juliet-Five. We, on site, were to do nothing until directed by these representatives, now known as the Missile Potential Hazard Team (MPHT).

We were instructed to assess the condition of the equipment above site. Remember, we had a payload transporter loaded with an RS, a PSRE, and guidance set, a security guard camper, a loaded erected transporter erector, two security guards, a missile maintenance team (five men), a missile handling team (four men), two communication specialists, two field supervisors (an officer and a senior NCO), and two really miserable lieutenant colonels on site. First it was decided that we had to refuel the transporter erector tractor that had been running for many hours since it departed the base the previous day. A contract gasoline tanker arrived on site and preparations were made to refuel the TE tractor. However it was decided to idle the tractor during the refueling. We could not shutdown the tractor engine completely because engine power was required in order to supply hydraulic power to the erector.

You guessed it. When we placed the tractor in idle, the hydraulic power provided was not enough to hold the TE in the erected position. The weight of the missile had shifted and the TE was decidedly off-center.

Just as we began the refueling we heard a popping, groaning sound. Ron asked "What is that?" "The TE is coming down" calmly replied the MMT chief. All hell broke loose.

Ron directed the MHT chief to restore operational RPM to the tractor and re-erect the TE. I ran to open the gates and get some of the people and the big vehicles, especially the payload transporter, off the site. After all, we had all kinds of explosives, hypergolic fuels, poison gasses, and a gasoline tanker plus many people on that one acre missile site.

But it was not to be, the security guard was doing a perimeter check and the gate was locked. By the time I ran the guard down, it was apparent that the actions taken by Ron had stabilized the situation. I did have a little talk with the guard and instructed him that the gate would be unlocked and propped open as long as this situation continued. He protested that he was responsible for security and had to control access to the site. I suggested that he could lie down in

the middle of the road into site if he wanted but that the gates would remain open. And they did.

We completed refueling the TE tractor and the tanker truck departed. One of the most volatile sources of combustion was no longer on-site. Things seemed to be looking better.

However, one of the MMT members stationed in the silo when the TE was re-erected reported that he had heard something fall out of the TE. Now we really had a mess. Not only was the missile cocked in the TE with hoist cables dangling loose, we also had a TE with some unknown damage.

Let me digress a bit and tell you how I surmise that the missile came to be resting on the half-opened launcher lid. This information came to me much later during calm conversations long after the incident. Well, it seems that the teams were ready to lower the missile in the wee hours of that Thursday morning. It was cold, very cold. There were two MMT members in the silo to observe the missile as it was lowered. Two other MMT members were topside opening the launcher lid. The MMT chief was also topside talking to the field supervisors and the MHT chief. An MHT member was in the TE tractor. The tractor windows were closed to ward off the severe cold. The TE console which controlled the TE hoist was positioned outside the tractor cab. The MMT chief was experiencing problems with the intercom system. (It was not known then, but the problem was that the mikes that were exposed to the elements were freezing, making communications erratic.) The MMT chief made the decision that the missile would not be lowered until full intercom had been restored. Communication specialists from the base were requested.

Meanwhile the MMT members topside had already begun to open the launcher lid, a process that could take as long as an hour or more because the heavy door had to be inched open by a mechanical device we called a mule. When the MMT chief decided to delay the lowering of the missile, this message was conveyed to these MMT members and they halted their efforts. The launcher lid was a little more than half open. The MHT chief walked over to the TE tractor and told the team member in the tractor something to the effect that "We are having problems with the intercom, the Comm specialists are on their way. When they get here and fix the problem, we will go ahead and lower the missile." The MHT member only heard "Go ahead and lower the missile." And he did.

So the launcher lid was stopped over halfway open and the missile was being lowered without MMT members actively observing in the silo. The missile settled onto the lid, tipped over until it came to rest on the bottom side of the erector. The hoist cables continued to come off the hoist drums and dangle loosely along the side of the missile. This continued until one of the MMT members in the silo, for some unknown reason, looked up through the collimator slot and saw the bottom of the missile resting on the launcher lid. He raced upstairs, notified his team chief and the lowering of the missile was terminated. That is how I think the missile ended up on the launcher lid.

Meanwhile back at the base and the rest of the Minuteman world, the MPHT had been at

work. A civilian team from Ogden was flying in to inspect the site, provide technical assistance and aid in the recovery from this unprecedented predicament. Ron Bishop had been instructed to return to the base and I was designated as the "On-Site Commander." I guessed the reasoning was there is no need to sacrifice two lieutenant colonels, and Ron was the younger and the more promising. I was not too happy, my day of infamy was at hand.

Actually, I found out much later that a friend of mine stationed at SAC Headquarters, had suggested that I be designated "On-Site Commander." He just thought I was the better choice because of my previous experience as part of the Technical Engineering Analysis Teams at Ellsworth.

Nevertheless, I was now in charge, for better or for worse. Of course, I could not take any action without clearing it first with the MPHT. Soon four large dump trucks filled with dirt arrived at the site and I was informed that the trucks would be placed at the four cardinal directions around the site. Stabilizing cables would be strung from each of the trucks to the top of the TE. There was some fear that a strong wind gust might bring down the unbalanced TE.

A camper also arrived and I was instructed to get some sleep while we awaited the arrival of technical assistance from Ogden. Yeah, I was going to get a lot of sleep in a camper parked beside an unbalanced TE held up by cables attached to four dump trucks. It had never occurred to me or anyone else on site that any kind of wind, short of a tornado, could blow over the TE that was properly anchored at the pylons and supported by fully extended actuators.

Soon a very large crane, I estimate it had a least a hundred foot boom, arrived and began to set up outside the perimeter fence. Some lucky member of the Technical Engineering Analysis Teams had already climbed the ladder attached to the outside of the TE to an access door near the top of the TE. He took a look inside, discovered he could not see much and returned to ground level. Not much progress but the arrival of the crane set the stage for what was to come next.

The members of the Ogden team arrived just as night was falling. We had so many lights, floodlights and spotlights, that it was virtually daylight on site. Earlier we had received permission to move the payload transporter with its hybrid load back to the base, so another potential source of mayhem was removed from the site. After some conversation with the members of the Ogden team, it was decided that two of them would have to go inside the TE through the access door. One at a time, they were raised by the crane and positioned so they could enter the access door. Once in position inside the TE, they reported that there was some minor damage to various hoist components which could be repaired by replacing the components. They also reported over 50 feet of loose hoist cable. The replacement components were made available almost immediately, thanks to the efforts of the MPHT. We were almost ready to begin efforts to raise the missile back into the TE.

The Ogden team had been inside the TE for almost four hours. Here I must add that there were no step ledges or shelves inside the TE. The two team members had been standing on the top of the missile the whole time they had been inside the TE. The loose hoist cables were still dangling alongside the missile. We had to devise a way to feed the cables back onto the hoist drums before we could do anything else. However the MPHT decided that the missile should be tied off to preclude any pendulum motion after it was lifted free of the launcher lid. So we began what to me was a futile attempt to tie off the missile so that it would not swing once free of the launcher lid. After I reported that the missile had been tied off, the wing commander wanted me to guarantee to him that the missile would not swing.

I tried to explain that I could not guarantee this. The missile was in a leaning position. The missile weighed over 50,000 pounds and it was suspended with the very heavy end down. The laws of physic would prevail. It was to no avail. He wanted his guarantee, but I was adamant, there would be no guarantee from me.

At this point, one of the Ogden team members had had enough. He asked to be removed from the TE, so both of the team members were removed and all efforts came to a stop. Some cooler heads on the MPHT net prevailed and it was decided that I, as the on-site commander, would do my best to see that the missile was tied off so as to minimize any pendulum motion, and we would proceed with the recovery efforts as soon as another Odgen team member could be found to replace the one that had quit.

The Ogden team member and I sit down in the camper and he calmly told me what had to be done to recover the missile. First the loose hoist cables had to be fed slowly and properly onto the hoist drums; slowly, because he would be hand feeding the cables onto the drums so they rewind properly and there would be no kinks in the cables. He also told me that after the tension was restored to the hoist cables, we would raise the missile slowly and gently so that any pendulum motion would be minimized. Finally he told me not to argue with the people on the net; we knew what had to be done and we could do it. Having established the ground rules, he and his new team member re-entered the TE and went to work.

Everything went just as he said it would with only one complication. The procedure was very slow because the hoist motor was being controlled by the MHT team member on the ground, and the Ogden team member was giving voice signals as he was hand feeding the cables onto the hoist. Remember he was standing on the missile inside the TE. If there was to be a hero in all this, this Ogden team member was it.

After what seemed like an eternity, the Ogden man reported that the tension had been returned to the hoist cables and he was ready to come out so we could began to raise the missile. Everyone on the net agreed that we should proceed. Our brief euphoria was disrupted by a voice coming from inside the TE. It seems that the Ogden team member was a rather stout man and with the hoist cables taut, he could not squeeze through the cables to the

access door. We had to release the tension of the hoist cables and he made his way to the access door where he was lowered to the ground by the crane boom.

Only one thing left to do. Raise the missile off of the launcher lid and hope that the pendulum oscillations were minimum. I decided that if the missile swung and hit the side of the TE, I was a dead duck anyway. So, I positioned myself so I could clearly see the missile as it cleared the launcher lid. Miracle of miracles, the missile did not swing one inch as it was hoisted into the TE. We transversed the missile onto the motor carriages and lowered the TE.

By now it was 10:00 Friday night. The MPHT net had been shutdown. I had been on the site for more than 36 hours and was ready to go home. However, there was still one last insult to common sense. It was decided that since I had been on dispatch for over 16 hours, I would not be allowed to go to the base. I was to go to Launch Control Facility Juliet-One and sleep for eight hours before going home. I was livid! I had been making on-site decision for the past 36 hours but now I was too tired to ride in a station wagon. I had a fresh driver so I was not going to drive.

I finally prevailed and was allowed to go home. On the way back to the base I was advised by Job Control that I was expected at a meeting with the wing commander and all the wing maintenance staff at 10:00 the next morning. It seems he wanted to discuss "what went wrong."

But, that is another story I will leave for another time.

December 1993, after inspecting 22-20 mobile ICBM Static Display at Kapustin Yar, Russia.

Air Force Missileers Biographies

Launch of Thor-Able.

DENNIS O. ABBEY, COL, received his AF commission in 1967 and served as a career missile officer for over 28 years. His tours of duty supported the Mintueman I, Minuteman II and Minuteman III weapon systems.

He was assigned to the 341st (SMW)—Malmstrom, 351 SMW—Whiteman, 1st Strategic Aerospace Division—Vandenberg, 91 SMW—Minot, HQ, Langley. He served in the positions supporting missile maintenance, operations, flight testing, personnel, staff and combat support. He was a maintenance squadron commander, operations group commander, base commander and wing commander. He retired in 1995 from Langley AFB.

His most memorable experience was serving as a SAC wing commander with the Minot "Roughriders" of the 91 SMW at the close of the Cold War. He was selected as the Air Force "Outstanding Missile Maintenance Field Grade Manager of the Year" for 1983 and the SAC "Outstanding Wing Commander" for 1991.

Denny and Dee, married June 10, 1967, live in Yorktown, VA. He is the corporate operations officer for Homes-Tucker International Inc. Their daughter, Gretchen, lives in Charlotte, NC; their son, Greg, in New York, NY.

CHARLES J. ADAMS, BGEN, was attending the University of Utah, when he entered the Army Air Corps, 1942. Completed flight training at Craig Field, Selma, AL, June 1943. Instructed at Craig Field until August 1994. Assigned to ATC. Ferried fighters within the US and P-39 and P-63s to Fairbanks, AK for the Russians. Flew the "hump" in C-46s, 1945-1946.

From 1946 to 1948 Gen. Adams was a pilot member of the 23 person "Ronne Antarctic Research Expedition". Flew 400 hours on skis and helped map the last unknown coastline in the world. Received the Distinguished Flying Cross helping rescue a British flight crew that crash-landed on the ice.

1st Lt. Adams joined SAC at Andrews AFB in May 1948. In May 1952, Maj. Adams graduated from AOB School. Assigned to the 91st Strategic Reconnaissance Wing, Columbus AFB, OH with RB-45s. Assigned TDY to Yokota AFB as Det. Comdr. for six months, where he flew 25 missions in RB-45s over North Korea. May 1955, assigned to "Current Operations", SAC HQ.

Transferred May 1958 to Cook AFB (later Vandenberg AFB) CA. Was site commander for the Atlas "D" missile-training during last year at VAFB. Assigned Naval War College, Newport, RI, 1961. 1962 assigned to HQ USAF as chief of the Strategic Missile Branch, R&D. Promoted to colonel 1963 and reassigned to director, Studies and Analysis, as chief, Strategic Offensive and Defensive Forces Analysis. 1967 returned to SAC HQ for two years and then to Ellsworth AFB, SD as commander of the 44th SMW. In February 1970 he was made commander, 821st Strategic Aerospace Div. at Ellsworth AFB and promoted to brigadier general. July 1971 Gen. Adams returned to the Pentagon where he retired early February 1973.

General Adams and his wife, Virginia Lee Polk-Adams were married in 1944. After retirement they

moved to Santa Maria, CA. Worked for GTE as their VAFB manager. Retired from GTE and still reside in Santa Maria, CA. General Adams is a past president of the VAFB AFA Goddard Chapter (life member), Santa Maria Breakfast Rotary Club and of the Allan Hancock Community College Foundation Board of Directors. He is a life member of the Retired Officers Association, the Strategic Air Command Association and the VAFB Museum. He is a member of the Association of Air Force Missileers.

He is a command pilot with over 7,000 hours, a charter Command Missileer. Ratings of navigator, bombardier, and observer. Displays 17 medals and ribbons including Distinguished Flying Cross, Air Medal, Air Force Commandation Medal and Antarctic "over winter" Medal.

CHRISTOPHER S. ADAMS JR., MGEN, entered the AF through AFROTC in August 1952 following graduation from East Texas State University. After pilot training, he was assigned to the 95th BW, Biggs AFB flying B-36s. After six years of flying the mighty bomber, he attended B-52 CCTS and moved to Ramey AFB, Puerto Rico with the 72nd BW. "Several" Chromedome missions later and after the Cuban Crisis, he attended missile combat crew training at Chanute with a follow-on assignment to the 44th SMW, Ellsworth AFB, as a MM I Missile combat crew commander and wing senior instructor. The Fall of 1966 and Vietnam brought a "back to the cockpit" assignment initially to C-141s Dover AFB, flying sorties to and from SEA, and eventually a tour of duty at Korat AB, Thailand, as the 388th CSG Director of Operations. Returning to the US in November 1967, he was assigned to Joint Task Force Eight (Defense Nuclear Agency [DNA]), Sandia Base, NM, as the air operations officer for atmospheric nuclear readiness-to-test planning. In 1970, he was reassigned to DNA HQ in Washington, DC, as executive to the director and later director, J-5 DNA. In August 1973, he was appointed DCO, 351st SMW at Whiteman AFB and three months later he moved on to F.E. Warren AFB as vice commander of the 90th SMW. In June 1974, he became commander of the 90th and "nurtured" the wing through the comprehensive upgrade from MM I to MM III. Selected for promotion to brigadier general in November 1975, he was reassigned as commander, 12th Air Div., Davis-Monthan AFB. In July 1976, he was directed to move the 12th Air Div. HQ to Dyess AFB, TX and to begin preparations to accept the B-1 bomber at Dyess. In July 1979, he was reassigned as deputy director of operations, HQ SAC, Omaha, followed by DCS, Operations Plans & Deputy Director, JSTPS. In June, 1982 he was appointed chief of staff, SAC.

Adams retired in February 1983 to become associate director, Los Alamos National Laboratory, a position he held until March 1986 when he joined Andrew Corporation, Chicago and Callas, as vice president for Business Development. From 1991 through 1995, he worked for Andrew extensively in Russian and the former Soviet States directing the recovery of their near-collapsed commercial communications systems. Presently, "more or less" retired, he and his wife, Alene, live in DeCordova Bend Estates near Granbury, TX and he serves a number of "pro bono" appointments, including the USSTRATCOM Strategic Advisory Group (SAG), Andrew Corporation Strategic Advisory Board, the Texas A&M University Chancellor's Council of Advisors, president, DeCordova Country Club board and vice chairman, Mayo of Texas Foundation (Texas A&M). Otherwise they enjoy some "questionable golf and skiing at their place in Ruidoso, NM."

CLARK N. (SPARKY) ADAMS JR., MSGT (RET), born in Lewisburg, PA, Jan. 10, 1950, enlisted Aug. 19, 1969. Following basic training at Lackland AFB, TX, began 22 weeks of training as a Titan II Missile facilities specialist/technician at Sheppard AFB, TX, graduating March 31, 1970. In April 1970, he reported to the 308th SMW, Little Rock AFB, AR, where he would "homestead" for the next 16 years. Following completion of Operational Readiness Training at Vandenberg AFB, CA and unit upgrade training at the 308th, he was certified combat-ready on Sept. 10, 1970.

Adams served in numerous capacities on squadron "line" crew (14 months), squadron and wing instructor crew (nine years), and wing standardization/evaluation crew (six years). He completed the 8th AF NCO Leadership School and the SAC NCO Academy in residence at Barksdale AFB, LA, in 1976 and 1981, respectively. Instrumental in training many of the 308th crews on the Coded Switch System and Universal Space Guidance System modifications, and was a member of the crew which brought the first USGS-modified Titan II to on-alert status. In August 1986, he reported to the 6595th Aerospace Test Group (Titan Division), Western Space and Missile Center, Air Force Systems Command, Vandenberg AFB, CA. Assigned to the Facilities Engineering and Operations Branch, he was heavily involved in the cleanup and reconstruction activities following the April 1986 explosion of a Titan 34D rocket directly above Space Launch Complex 4E. His branch was directly responsible for the modifications required to configure SLC-4E to support the Titan IV rocket. Adams participated in the last launch of a Titan 34B rocket; the last launch of a Titan 34D; the first launch of a Titan II Space Launch Vehicle; and the first west-coast launch of the Titan IV rocket. Having retired on Oct. 31, 1991, he lives in Mifflinburg, PA, and works as a maintenance technician at an area vocational-technical school. He and his wife, Shirley, have two children, Paul and Richard, both of whom live in the Houston, TX area.

RICHARD E. (DICK), ADAMS, LT COL (RET), born at Beaumont, TX, graduated from Sherman, TX High School and attended Oklahoma State University on a baseball and ROTC scholarship. He later transferred to Wichita State University (Kansas) where he graduated in 1954 and then served in the AF until September 1977. Duties included being the personnel officer at Forbes AFB, KS, Ramey AFB, Puerto Rico and Lowry AFB, CO. He then entered the missile program at Davis-Monthan AFB, AZ in 1962, where he was the chief of the Missile Training Branch. He later became a Titan II MCCC, an instructor and evaluator. He was awarded a Crewmember Excellence Award for numerous highly qualified evaluations and his missile crew was selected Crew of the Month. In August 1967, he attended Air Command and Staff College and was then assigned to SAC HQ, Offutt AFB, NE where he performed duties in the computer area. He developed computer programs to evaluate the status of the SIOP and the implementation of the SAC Worldwide Military Command and Control System and was the chief of the Computer Div. In 1974 he was assigned to Ellsworth AFB, SD where he became the operations officer and then commander of the 67th SMS (Minuteman II). Task force commander for a Minuteman II launch at Vandenberg AFB, CA. He became the commander of the 44th Transportation Sqdn., which was his last assignment prior to retiring in 1977.

Adams and his wife, Carolee, have resided in Sherman, TX since 1977. He has held positions in a local hospital and in the insurance business. Has served as the Campaign Chair and chairman of the Grayson County United Way, president of the Sherman Rotary Club and chairman of the Administrative Board of the First United Methodist Church. Golf, serving on other committees and grandchildren keep him busy.

GORDON E. ALBERTI, 1LT, born March 14, 1940, Seaside, OR. Attended Lewis and Clark College, Portland, OR, earning BS degree in math in 1962. Logger in his father's business, G&A Logging Co.

2nd lieutenant, December 1962, OTS/Lackland AFB; Missile School, Sheppard AFB, Spring 1963; deputy combat crew commander (DMCC), 381st Strategic Missile Wing (TITAN II), McConnell AFB, KS. As an instructor crew (S-139) they received many "high qualified" stand-board evaluations and was nominated for wing crew of the month. Later, he was nominated for "Outstanding Junior Officer." He became a 1st lieutenant in June 1964.

His crew was mentioned in October 1966 for "Outstanding" service while performing duty at the alternate command post during the SAG IG/OR I of the 381st. He received the Commendation Medal in December 1966.

Employed as computer programmer, 1967; insurance agent, 1973; and received chartered life underwriter designation (CLU) in 1980.

Married he has a daughter and grandson and lives in Redmond, WA. His daughter is a graduate of University of Washington.

WILLIAM A. ALBRO, COL (RET), graduated from Hobart College as a Distinguished AFROTC graduate and was an AF officer from September 1961 through August 1985, first as a graduate student at Massachusetts Institute of Technology followed by weather officer duties at Hanscom Field. In August 1965, after duty in Vietnam, he was an AFIT student at Texas A&M University, receiving an MS in meteorology in May 1967. Assigned to the Environmental Technical Applications Center in Washington, DC, from June 1967 to June 1970, he served as operations plans officer. His next assignment was to Hawaii in several staff positions at HQ, 1st Weather Wing and one year at the Pearl Harbor Navy Fleet Weather Central. Prior to leaving Hawaii in June 1974, he served one year at HQ, Pacific AFs.

Next, he completed Air Command and Staff College as a Distinguished Graduate in June 1975 and also earned a master public administration degree from Auburn University. He was then assigned to HQ, Air Weather Service, Scott AFB, where he was deputy director for Operational Evaluation.

He completed missile combat crew training at Vandenberg AFB in June 1977 and was assigned to the 91st SMW. After a year on line crew in the 741st SMS, he became operations officer for the 742nd SMS in July 1978 and wing assistant deputy commander for Resources in June 1979. In April 1981, he was appointed commander, 740th SMS where he served until becoming wing assistant deputy commander for Operations in November 1981. He next served as wing deputy commander for Operations from June 1982 to

June 1983. Following a year as director of safety, 1st Strategic Aerospace Div., Vandenberg AFB, he was reassigned to Florennes AB, Belgium, as deputy commander for Operations, 485th TMW and then wing vice commander from April 1985 until retirement.

His decorations include the Legion of Merit, Meritorious Service Medal w/3 OLCs, and AF Commendation Medal w/OLC. He and his wife, Judy, reside in Washington's Puget Sound area.

JOSEPH D. (JOE) ANDREW JR. CMSGT (RET), followed his father, an AF chaplain, into the AF in June, 1959. After attending control systems mechanic school at Lowry AFB, CO, and combat crew training at Orlando AFB, FL, he served a 13 month tour in Osan, Korea, as a TM-61C Matador launch team member. In February 1962, following his overseas tour, he was assigned to SAC with the newly activated 341st SMW at Malmstrom AFB, MT. There, as a missile systems analyst, he participated in bringing the first Minuteman missile in the USAF to alert status during the Cuban missile crisis. Andrew remained at Malmstrom until November 1969, working as a missile maintenance team (MMT) member, an electromechanical team (EMT) chief, and finally as a QC evaluator for EMT. He was then picked to serve as a traveling inspector with the 3901st Strategic Missile Evaluation Sqdn. operating out of Vandenberg AFB, CA. After four years with the them, he returned to Malmstrom where he served in a variety of missile management positions before being selected for a HQ SAC tour in 1979. For the next three years he was the functional manager for ICBM maintenance personnel. His notable accomplishments in that job included developing the initial personnel plan for the first MX missile wing and developing the combined ICBM career field structure that is still in use today.

He retired from the AF in 1982 following 23 years of service. Since that time he has become involved in the field of finance, working as a chief financial officer for a software development firm, and as a business appraiser. He teaches finance as well, dividing his time between Webster University's Washington, DC campus and its international campus in Hamilton, Bermuda. His latest academic accomplishments include authoring a college finance text book for Prentice-Hall and being nominated for Webster's 1997 Outstanding Teacher Award. Andrew currently lives in Woodbridge, VA, with his wife Emily, who is an AF ballistic missile defense engineer stationed at the Pentagon. He has two daughters, Denise Andrew, in Bozeman, MT and Connie Cybulski in Moline, IL.

JOHN K. ARNOLD, III, COL (RET), born in San Antonio, TX, March 30, 1938, entered the AF Feb. 22, 1960 after graduating from Florida State University in January 1960. Duties included personnel officer, 9th BW and Titan-I Missile maintenance officer in the 569th SMS, Mountain Home AFB, ID; Minuteman I and Minuteman II Missile combat crew commander 351st SMW, Whiteman AFB, MO; Operational  Readiness training instructor and assistant chief, Development Branch, 4315th Combat Crew Training Sqdn., Vandenberg AFB, CA; Air staff strategic planner, Directorate of Plans, HQ, USAF, The Pentagon; Commander, 509th SMS and deputy base commander, 351st SMW, Whiteman AFB, MO; AF Research Associate, the Mershon Center, The Ohio State University;

Chief Nuclear Planning Branch, International Military Staff, NATO HQ, Brussels, Belgium; Chief Space Branch and chairman, Department of National Security Affairs, Air War College, Maxwell AFB, AL. His professional military education included Squadron Officer School in 1965; Air Command and Staff College in 1971 where he graduated as a distinguished graduate and earned a masters in business administration; The National War College in 1977.

After retirement from the AF in August 1986, he taught junior high social studies for several years. He and his wife, Louise, have resided in Montgomery, AL since 1983. He has served and is serving on several social service boards in Montgomery. He spends his time now bragging about his four children: Rebecca, Carol, Jack and Jim and his 11 grandchildren; boating on Lake Martin and attended Auburn University football, basketball and baseball games.

ARTHUR W. (ART) AVANT JR., CMSGT (RET), born Nov. 19, 1948 in Charleston, SC started his career as a civil engineer power production specialist. Assignments: 1968-1969 Lackland AFB and Sheppard AFB, TX; training. Moody AFB, GA and Wheelus AFB, Libya, as a prime power plant operator. 1970-1973 RAF Alconbury, England, responsible for Aircraft Arresting System. December 1973 assigned to 44th SMMS Ellsworth AFB, SD. Involved with Minuteman I. 1975 remote Sparrevohn AFS, AK, returning to Ellsworth AFB in 1976 and retrained into the newly formed ICBM facility maintenance career field.

1982 was a member of the Blanchard Trophy winning team for Ellsworth AFB. July 1982 was assigned to HQ/SAC 3901st SMES, as inspector/evaluator. Involved with Peacekeeper, Minuteman I, II, III, Rail Garrison. September 1987 assigned to 351st SMW, Whiteman AFB, MO, involved with Minuteman III. 1990 as Olympic Arena Project NCOIC, Whiteman again won the coveted Blanchard Trophy. 1991 assigned to Vandenberg AFB, CA, as superintendent of maintenance for the newly activated 20th AF. Involved with SALT II Treaty deactivation and conversion of missiles. October 1992 assigned to Moody AFB, GA due to a grave illness in family.

Avant was awarded the MSM w/2 devices, Commendation w/2 Devices, AF Achievement, Outstanding Unit w/3 Devices, Longevity w/7 Devices, Good Conduct w/9 Devices, PME w/Device, AF Training, Expert Marksman w/Device, Short Tour w/Device, Long tour, Volunteer Service.

He retired as 347th Support Group, Sr. Enlisted superintendent, Moody AFB, Valdosta, GA in 1998 after 30 years with the AF. He and his wife Pam have two daughters Sylvia and Sara and continue to make their home in Hahira, GA.

DICK AYARS, LTC (RET), commissioned at Penn State in January 1959. Initial duty was hospital administrator, Homestead AFB. Served as dispensary squadron commander in Morroco, 1961-1963, and operations officer, 1st Aeromed Evac Gp., Pope AFB, and as a medivac aircrew member on C-130E. Served a combat duty tour, Dominican Republic, 1965.

Transferred to Line of the Air Force, with duty as line instructor, standboard, ACP MCCC, 509th SMS, Whiteman AFB, 1967-1970. Reassigned to Tactics Div., Operations Plans, HQ SAC, until 1973.

Ayars earned a masters in logistics at AFIT, 1974. Assigned as commander 90th FMS, Warren AFB, 1974-1976. Reassigned as assistant deputy program manager for logistics at SAMSO, San Bernardino, CA, responsible for reliability, maintainability, supportability, and life cycle cost of Peacekeeper, with emphasis on AIRS guidance, 1976-1979. He and Marty reside in Oakmont, PA. A fireman for 39 years, Ayars is now captain of Oakmont Fire Department. He is a founding member of AAFM (L-0063).

BRYCE H. BAKER, MAJ (RET), served in several areas of missile operations and safety. He worked with Titan II, Thor, target drones (BQM-34 A/F, Bomarc, and PQM-102), AIM-4/7/9 series missiles, and ATR-2A rockets. Following graduation from the University of Utah in June 1962, entered the AF through the OTS program. Between 1963 and 1967, he served as deputy missile combat crew commander and as missile combat crew commander with the 571 SMS, Davis-Monthan AFB, AZ. From 1967 to 1970, he continued his assignment as Davis-Monthan, serving as OIC Combat Crew Training Records Branch, 390 SMW. Between 1970 and 1974, he served as operations plans officer for the 10ADS, Vandenberg AFB, CA. During this period, he participated in four operational exercises at the overseas operating location, Johnston Island. Between 1974 and 1980, he served as chief, missile and Nuclear Safety Div., USAF ADWC, Tyndall AFB, FL. From 1980 to 1983, he served as weapons safety staff officer, HQ ADTAC, Peterson AFB, CO and Langley AFB, VA.

(Note: the headquarters moved from Colorado to Virginia in June 1981).

Awards include two Commendations and two Meritorious Service Medals. Baker and his wife, Sally have resided in Tucson, AZ since retirement in June 1983. They have three daughters and also one grandchild on the way. He works for the Villa Del Rio Association while Sally works for Haemonetics Corp. Sally is an international distributor for NuSkin/IDN products. Baker can be found on most weekends at the golf course.

GLENN R. BALDWIN, CMSGT (RET), born Oct. 13, 1941 at Symco, a small farming community in central Wisconsin. He entered the USAF on April 9, 1963. After completing initial military training at Lackland AFB, TX, he was assigned to the 3348th Technical Training Sqdn., Chanute AFB, IL, for electronic training as a missile systems analyst technician in the AGM-28B (Hound dog) career field.

Assignments in the AGM-28B career field included the 410th SBW (Heavy), K.I. Sawyer AFB, MI from February 1964 to May 1966 and from April 1971 to December 1974, the 6th SBW, Walker AFB, NM from June 1966 to April 1967 and the 456th SBW, Beale AFB, CA from April 1967 to March 1971.

He was then cross trained into the Minuteman II electronic technician career field, being assigned to

the 44th SMW, Ellsworth AFB, SD from January 1975 to October 1983. While assigned to Ellsworth AFB he held many positions within the 44th SMW, including missile maintenance team chief, job control shift controller, senior job controller, job branch superintendent. Facilities maintenance branch superintendent, and electrical-mechanical branch superintendent.

In October 1984 he was reassigned to the 485th TMW (USAFE), Florennes Belgium, where he played an instrumental part in the bed down of the first Ground Launched Cruise Missile Wing on the European Continent. He held positions as wing missile support superintendent and squadron missile maintenance superintendent.

In October 1985 he was reassigned to Ellsworth ADB until his retirement as of Jan. 1, 1991.

Honors include his selection as the 1985 Outstanding USAF Missile Maintenance manager of the year, while assigned to the 485th TMW Florennes. The USAF Meritorious Service Medal w/OLC, USAF Commendation Medal w/2 OLC, and numerous USAF Recognition Medals.

Since his retirement, he and his wife Sandra have resided in Box Elder, SD. He owns and operates his own roofing and asphalt pavement sealing business. He is involved in local politics and served as a city council member for the city of Box Elder, from May 1991 to May 1992 and mayor of the community from May 1992 to May 1996.

Other present and former community involvement positions include, member of the Ellsworth AFB Heritage Foundation Board of Directors, Board of Directors of New Business Attraction for Box Elder and Rapid City, SD, Joint Land Use and Transportation Coordinating Committee, Military Affairs Committee, Rapid City Chamber of Commerce Century Club, and Ellsworth AFB Task Force.

In his spare time, he enjoys playing golf, fishing and big game hunting. He and his wife have two grown children and one grandchild who reside in Colorado.

JAMES W. BARNARD, CAPT, born July 24, 1942, Chicago, IL, raised in Chicago and Winnetka, IL. Graduated Bradley University (ROTC), commissioned second lieutenant, USAFR, June 6, 1965. First assigned Vance AFB, July 1965. Reassigned to 341st Missile Maintenance Sqdn. (MIMS), 341st SMW (Minuteman), Malmstrom AFB, MT, November 1965, as a Minuteman Missile maintenance officer. TDY to Missile Maintenance Officer Course, Chanute AFB, January-April 1966. Team chief, Target & Alignment Team "Tango-Nine" April 1966-June 1967. Maintenance field supervisor (call-sign "DM-Three") June 1967 with additional duties at squadron coast reduction officer. Promoted first lieutenant Jan. 26, 1967; captain July 26, 1968. Completed tour of active duty Jan. 26, 1969.

Married Miriam Strick on April 9, 1967. Currently resides in suburban Denver, CO. Two grown children, one grandchild.

Following additional engineering course work, aero-space systems and design engineer (missiles and launch vehicles) with several major corporations until 1989. Currently owner and president of Trailrider Products, manufacturer of holsters and accessories for hunters and Old West shooters.

Most memorable experience was the dedication and professionalism of mobile maintenance team personnel. For over 30 years, sometimes working under the most arduous weather conditions, on a "flight line" up to 200 miles long, with little recognition of the importance of their job, these dedicated troops assure that the

land-based missile leg of America's strategic deterrence Triad remains at the highest state of readiness. He is proud to have been a member of that team.

Awarded the Missile Badge, earned by missile maintenance personnel after 18 months in assignment. (Launch officers were awarded their "pocket rocket" immediately upon being declared "combat ready.")

Minuteman Missile Maintenance personnel were first assigned to school at Chanute AFB, IL, 13 weeks for officers and 26 weeks for enlisted personnel. After return to their assigned bases, personnel generally received on-the-job training with a mobile maintenance team running "in the field," to the missile sites.

After qualification in the performance of various tasks, which usually took about another month, personnel were assigned to their own team. Most mobile maintenance team chiefs were in grades E-4 to E-6 (late 1960s). JCS rules required Target & Alignment team chiefs to be officers, as they had the last "physical access" to the nuclear warhead atop the missile. The officer team chief and the two enlisted team members were cross-trained in all tasks, however.

In the first several years of Minuteman, many of these officers were transferred from Atlas and Titan I, and included senior captains, and even som field grade officers. By 1968, most of these more senior officers had been reassigned to supervisory positions, as sufficient lieutenants became available to fill team chief slots.

WILLIAM E. BARNES, LT COL (RET), was a pioneer in SACs Ballistic Missile program. After commissioning through ROTC he entered active duty in 1950 and spent the next eight years in the B-36 program at Carswell and Ramey AFBs. He was sent to Vandenberg (then Cooke AFB) in 1958, assigned to 1st Missile Div. The next 15 years saw two tours each at VAFB and SAC HQ plus a stint at the Armed Forces Staff College (1964). During his second tour at VAFB he was chief, Analysis Div., 3901 SMES. On his second SAC HQ tour he was chief, Evaluation Div., BME, responsible for developing ballistic missile SIOP planning factors.

His final assignment was in 4th AD at F.E. Warren AFB before retiring in 1975. He later worked as a logistics engineers on the Peacekeeper Missile Program at Martin Marietta in Denver for eight years. He and Rosalind are now enjoying retirement in Tucson, AZ.

JORGE BARRIERE, C/LT COL, CAP, born March 10, 1978, San Francisco, enlisted in the USAF, March 1991. His military locations: Aux., CAP, Cadet, CCSF Cadet Sqdn. 86, SF, CA.

He is a member of 561st AF Band CA ANG. Chair, CA WG Cadet Advisory Council. He participated in many McClellan AFB CA Cadet Encampments; Cadet commander, Cadet Leadership Schools, Cadet commander Watsonville Antique Aircraft Fly-in 1997.

His memorable experiences include looking down the barrels of an A-10 Thunderbolt II Gatling Gun.

He was awarded the Mitchell Award, Earhart Award and numerous other CAP Activity Awards. Barriere is single and resides in Daly City, CA. He is a junior at the University of the Pacific, CA.

KENNETH L. BEATON, MAJ, (RET), born Sept. 18, 1933 in Wenatchee, WA, enlisted March 11, 1953 and sent to Parks AFB, CA assigned Flight 117, 3288 BMTS for basic training. Proceeded to Lackland AFB and Harlingen AFB, TX for aviation cadet training as part of Class 54-06C. Commissioned second lieutenant on June 2, 1954 and received wings as navigator on Aug. 4, 1954. Assigned to McClellan AFB, CA; Campion AFS, AK; Mather AFB, CA; March AFB, CA; Castle AFB, CA and Fairchild AFB, WA flying RC-121, C-130, B-47 and B-52 aircraft. In October 1965 was assigned to 446 SMS, 321 SMW at Grand Forks AFB, ND with Minuteman II training at Chanute AFB, IL and Vandenburg AFB, CA. From June to December 1965 was TDY at Miami, FL International Airport supporting HQ USAF project for Vietnam. Returned to 446 SMS as MCCC of Crew R-029 in January 1966. In December 1969 was transferred back to Mather AFB, CA flying as navigator in B-52s. Flew 100 combat missions in Southeast Asia including four over North Vietnam. Retired as major on July 1, 1973 at Mather AFB, CA and reside in Sacramento, CA. Interests include reading, computers and genealogy. He has three grown sons.

GEORGE F. BENNETT, COL (RET), born in Wheeling, VA June 3, 1924 and graduated from the University of Nebraska, Omaha. Entered the US Army as a private Dec. 4, 1942 and became a radio operator gunner on B-17 bombers. He was shot down over northern France in February 1944 and with the help of the French Underground made his way 500 miles south through occupied territory and crossed the Pyrenees mountains in Spain. He entered pilot training after the war and graduation in 1949 was sent to the 19th BG on Guam. When the Korean War started the 19th was deployed to Okinawa immediately. He flew 51 missions in B-29s before returning to the States in 1951.

Assignments followed as a B-29 aircraft commander. Then came a year of schooling to become a navigator/bombardier so he could be qualified to be a B-47 aircraft commander. After serving as a B-47 crew commander for an extended period, he spent six months in Squadron Officers School. Upon return from school he was assigned as 310th BW Director of Safety. Then came assignment to missiles.

Colonel Bennett was one of the original operational unit intercontinental ballistic missile safety officers. He started with the Thor System as it was going operational in England, working on both the safety deficiencies of the weapon system and as a training officer for the RAF. Upon returning to the States, he was the wing director of safety for the Titan Missile Wing at Lowery AFB, CO. He was next assigned as chief, Missile Safety at 15th AF HQ, March AFB, CA. His responsibilities included Atlas, Titan I, Titan II, and Minuteman. Air War College followed. After graduation, he proceeded to Vietnam in June 1968 as chief of Flying Safety for the 7th AF. During this period he flew 17 combat missions as a pilot in aircraft ranging from the O-2 to the B-52. SAC HQ as deputy director of safety was the next stop. He was promoted to colonel shortly after arriving at SAC. After two years he moved to the 90th SMW as vice commander. Four months later he was assigned as commander 44th SMW at Ellsworth AFB.

After retirement in 1973, he served two years on the governors cabinet as secretary of public safety for the state of South Dakota. He then joined the staff at the University of Texas, Arlington serving as director of safety and assistant vice president for business affairs. Full retirement came in 1985. He and his wife, Iris, reside in Horseshoe Bay, TX. Both are avid golfers and George enjoys feeding from 10 to 30 deer twice each day. He serves on the Board of Trustees of The Church at Horseshoe Bay.

TERRY JOHN BERNTH, COL, born April 17, 1941 in Omaha, NE, enlisted in the USAF Sept. 16, 1963, officer, SAC.

His military stations and locations: Whiteman AFB, MO, 1963-1967, 351 SMW/508 SMS; Vandenberg AFB, CA, 1967-1971, 4315 CCTS; Offutt AFB, NE, 1971-1976, HQ SAC/Int - Malmstrom AFB, MT, 1976-1981, 351 SMW/12 SMS/DOC/10SMS; Offutt AFB, NE, 1981-1983, HQ SAC/INT; Minot AFB, ND, 1983-1987, 91 SMW/DO 57 AD.

His memorable experiences include being chief of only non-flying command post to ever win the Ivan L. Bishop Award for Best SAC Command Post; squadron commander 10 SMS, 1980-1981.

Bernth was discharged as colonel July 1, 1987 and was awarded four Meritorious Service Medals, Combat Crew and three Air Force Commendation Medals.

Married to Kathleen, they have two daughters and a son. He is employed with Papillion Lavista School District as seventh grade Social Studies teacher at Papillion Lavista High School. He is also a soccer coach.

STANLEY I. BIELESKI JR., CMSGT (RET), joined the USAF in 1961, attended Missile Facilities Specialist Course, Sheppard AFB, and in 1962 transferred to the 548th SMS, Forbes AFB. Selected as Airman of the Month for June 1964. Moved to the 321st SMW, Grand Forks AFB in 1965 as a facility manager; graduated number one academically from the 2AF NCO Leadership School, Little Rock, AFB and in February 1966 was the Wing Airman of the Month. After promotion to TSgt., transferred to the 351st SMW, Whiteman AFB in 1969; became wing instructor facility manager. Graduated from SAC NCO Academy, Barksdale AFB and was promoted to MSgt. in 1971. Transferred to Lowry AFB in 1972 as a military training instructor. Became the NCOIC, Standardization/Evaluation Team; and developed a self-inspection program that became the model for all of ATC. Became the NCOIC Missile Crew Chief Section, 381st SMW, McConnell AFB in 1975; as NCOIC established the Support Division as a viable part of the Maintenance Deputate; promoted to SMSgt.; earned his BS in Industrial Management from Kansas Newman College. Transferred to 321st SMW in 1978 as NCOIC, Support Div., then became the NCOIC of the Training Div. In 1979 he became the 321st FMMS maintenance superintendent, was promoted to CMSgt. then moved to the 321st OMMS as maintenance superintendent in 1981. Received his masters in Public Administration from UND in 1982, became the NCOIC, Maintenance Control Div. in 1983 and retired from active duty in 1984.

He and wife, Barbara, live in Clinton, TN, where he is the senior logistics engineer and program manager for REMOTEC, Inc., the world leader in hazardous duty robots used by US Military EOD personnel as well as police departments and foreign countries around the world. He enjoys travel, fishing, golf and their seven grandchildren.

RICHARD J. BIGGS, COL (RET), born in Nogales, AZ on May 4, 1931, enlisted in the USAF July 18, 1950. He served as a career guidance specialist, at-

taining the permanent rank of staff sergeant, prior to entering Aviation Cadet Training in July 1953. Commissioned in 1954, he received wings as a navigator in 1954 and as an electronic warfare officer in 1955. He served as crewmember, instructor and Standboard Evaluator in the 509th and 301st BW aboard EB-47E Bombers until 1962 when he was selected to attend the University of Arizona under the auspices of the Air Force Institute of Technology (AFIT). He was graduated, BSEE (cum laude), in 1965 and assigned to the 44th SMW as chief, Technical Engineering Div. Colonel Biggs flew 100 combat missions in Southeast Asia as an electronic warfare officer aboard EC-121 aircraft in 1969-1970. He was then assigned to AFIT where he earned a masters of science degree in Systems Management.

Subsequent assignments included 91st MIMS maintenance supervisor, 47th Air Div. missile maintenance officer, and 15th AF director of Missile Maintenance before his final assignment as vice commander, 91st SMW where he retired in July 1979. He was promoted to the temporary rank of colonel effective July 31, 1974 and permanent colonel (REGAF) effective Oct. 1, 1977.

His military decorations incude the Legion of Merit, Distinguished Flying Cross, Meritorious Service Medal, Air Medal w/2 OLC and the Air Force Commendation Medal w/4 OLCs.

Biggs was subsequently employed by Lockheed Missiles and Space Company as the company product safety officer, retiring in 1992. He and his wife, Delores, presently reside in Spokane, WA.

ARTHUR W. BIKKER, LT COL (RET), born Aug. 25, 1939 in Denver, CO, was commissioned Aug. 25, 1962 into the USAF. His locations and stations include: Richards - Gebaur, 1962-1963; Minot, 1963-1967; Brooks AFB, 1967-1969; RAF Mindenhall, 1969-1973; Randolph, 1973-1976; AFMPC, 1976-1979; NORAD, 1979-1982; Space Command, 1982-1984; Travis, 1984-1989.

He participated at Minot AFB OPS and Security checkout of first two MCC's (Alpha and Bravo).

His memorable experiences: on 1st Stan eval was going through escape/egress check list and stepped off trap door platform in dark, but barely caught himself before falling between the outer capsule and inner enclosure. On another stan/eval. accidentally cut off status msg. processing group when he was simulating shutdown. He was on training crew two years; command post controller, six months. First director of manpower for Air Force Space Command.

Bikker retired Feb. 1, 1989 as lieutenant colonel. He is married to Kathleen and they have five children, one of which is a captain in the USAF. They also have six grandchildren. Employed with State Farm Claims for nine years as insurance claims specialist and active in LDS church.

ROBERT E. BINA, LT COL, entered active duty in 1973 as DMCCC, 12 SMS, 341 SMW, Malmstrom AFB; subsequently serving as an evaluator and MCCC in the 10 SMS. In 1977 he went to Kelly AFB, TX as program manager, C-9/T-43 as maintenance programs.

He returned to Malmstrom AFB in 1980, performing duty as MCCC; flight commander, evaluator, code controller, chief, Codes Training; assistant operations officer and operations officer of the 10 SMS. In 1986, he became director of Command Control, 1STRAD, Vandenberg AFB, followed by a tour in Korea as chief, Contingency Plans, HQ Air Component Command. He returned to missile duties in 1989 as chief, Codes Div., 351 SMW, Whiteman AFB MO; then assistant deputy commander, Resources, and deputy commander, 351 Logistics Gp.

In 1993, Bina moved to Vandenberg AFB as the chief of Logistics Plans, HQ 14 AF. He and his wife, Anne have two children, Jennifer and Kristen.

RON BISHOP, COL, graduated from the Naval Academy, entered the AF and married Barb in 1957. After training at Lowry AFB he spent six years as maintenance officer in the 4751st and 46th ADMS (BOMARC). He then became an instructor crew commander at the 351st SMW. After ACSC he was assigned to HQ USAF in Missile R&D followed by a tour in AFIT at SMU. He then became a maintenance squadron commander and assistant DCM at the 91st SMW. Following a year at the Naval War College he was DO of the 341st SMW followed by a tour at HQ AFLC before returning to SAC as vice commander, 381st SMW. He subsequently became first STRAD chief of staff and 4392nd support group commander at Vandenberg before assuming command of the 308th SMW.

Upon retirement in 1982 he worked for Martin Marietta. He and Barb are fully retired in Arizona and spend as much time as possible visiting four offspring and six grandchildren.

G.D. (MO) BLACKMORE, COL, is chief, Strategic Plans Division, HQ US Strategic Command, Offutt AFB, NE. He leads the joint service agency that develops the National Target Base, the National Desired Ground Zero List and the Single Integrated Operational Plan, the nation's nuclear war plan.

The colonel was commissioned in 1971 through the Reserve Officer Training Corps. His career has spanned all organizational levels, including assignments as a Minuteman III wing standardization crew commander, aerospace division test and evaluation launch countdown team member, MAJCOM missile flight test program manager, Air Staff program element monitor, missile operations squadron commander, Joint Staff operations officer, wing vice commander and wing commander.

Major awards and decorations include the Legion of Merit, Defense Meritorious Service Medal, Meritorious Service Medal w/3 OLCs, AF Commendation Medal, Joint Meritorious Unit Award, Combat Readiness Medal and National Defense Service Medal w/BSS. Other achievements include published research study: Strategic Conflict in Transition: Directed Energy Warfare, 1984 and authored research study, The Demise of Deterrence, 1990.

Education: BA in psychology, Miami University, Oxford, OH; MBA, University of North Dakota; Squadron Officer School, Maxwell AFB; Air Command and Staff College, Maxwell AFB; Industrial College of the Armed Forces, Ft. McNair.

Col. Blackmore and his wife Sandra have two daughters, Amy and Megan.

JOHN A. BODOVINAC, CMSGT, born in Virginia, MN entered the USAF at age 17.

His tenure was over 32 years and included Korean and Vietnam combat service. He entered the ICBM program in 1959 as a technician and combat crew member. He was on the 15AF HQ Staff. He was maintenance superintendent of the 100 SRW, 390 SMW and 321 SMW. He was senior enlisted advisor of the 390 SMW.

He retired from the AF on July 1, 1984 and was hired as test engineer by Martin Marietta Aerospace. He worked Peacekeeper, Midgetman and the Titan III Missile.

Bodovinac married the former Norma Hanela from Virginia, MN in 1953. They have three daughters, Paige, Toni and Dara, who have provided them five grandchildren.

The Bodovinac's reside in Denver, CO during the winter and at Lake Vermillion, MN in the summer.

DONALD L. BOELLING, SSGT, born May 17, 1956 in China Lake, CA. After graduation from Evergreen High School in Vancouver, WA, enlisted in the USAF June 6, 1974 for a six year enlistment with the intention of becoming one of Americas elite missile combat crew members. After almost one year of training he was assigned to the 570th SMS at Davis-Monthan AFB, AZ as a BMAT (ballistic missile analyst technician) on a Titan II ICBM operational launch crew.

His dedication and outstanding performance soon earned him a position on a squadron instructor crew and then a wing instructor position responsible for Alternate Command Post training and audiovisual training productions. He was chosen to attend the 15th AF NCO Leadership School at March AFB in 1976 as a wing instructor he was awarded outstanding performance awards from the SAC Inspector General Team and the 3901 Missile Evaluation Team.

His awards include the AF Commendation Medal, AF Combat Readiness Medal, AF Good Conduct Medal, SAC Inspector General Outstanding Performance Award, 3901 Missile Evaluation Outstanding Performance Award.

Boelling was discharged as E5, staff sergeant, Feb. 6, 1980. After leaving the AF he has worked for the Boeing Company in Seattle, WA, where he is currently the Network Security officer. From 1995-1997 he worked as the director of security for Ideon Group Inc. and has just recently returned to the Boeing Company. He also authors and operates an Internet web site dedicated to the Titan II ICBM weapon system.

Divorced, a single parent of one son, Daniel, 18 years old and a junior in high school.

ROBERT L. BOLTON, TSGT (RET), born May 30, 1940 High Point, NC. Following his January 1961 enlistment he completed basic at Lackland AFB and reported to Lowery AFB for electronic and missile guidance system training. After combat launch crew training at Orlando, AFB, in March 1962 he was assigned to the 38th TAC Missile Wing, Sembach, Germany. He was stationed at the 887th Missile Sqdn., Gruenstadt as a launch crew mechanic on the Mace.

Following the September 1966 stand down of the Mace operations at Gruenstadt he retrained into radar maintenance at Keesler, AFB, serving at numerous AF stations throughout the CONUS, SEA and Iceland.

Bolton retired January 1981. He and his wife, Ingrid reside in Lawrenceville, GA. He has worked for the Philips Electronic Instruments Co. since January 1981 in New Jersey and Georgia. He is the production manager of the Tooling Refurbishment & Manufacturing Center.

RICHARD T. BOVERIE, MGEN (RET), upon graduating from the US Naval Academy in 1954, Boverie was commissioned in the AF. He went to Lowry AFB, CO, for training in the Matador, the AFs first operational nuclear missile. He then was assigned to the 11th Pilotless Bomber Sqdn. at Orlando AFB, FL, and participated as a launch officer in Matador operational suitability testing from Cape Canaveral. He deployed to Germany with his squadron, eventually renamed the 11th Tactical Missile Sqdn., where he served as a Matador launch operations officer for three years. After graduate work at the University of Michigan, Boverie went to the AF Space Systems Division in Los Angeles, where he was a project officer in some of the Air Forces earliest space development programs: SAINT (Satellite Interceptor), satellite inspector, and manned orbiting laboratory.

Subsequently, he served as an aerospace engineer with the European Office of Aerospace Research in Brussels. Later, Boverie held various policy, plans and analysis positions at HQ USAF, in the office of the Secretary of the Defense and with the National Security Council staff at the White House. These positions included, among other things, efforts regarding strategic missile forces. Boverie retired from the AF in 1982.

Boverie and his wife Gudrun live in West Palm Beach, FL.

BILLIE B. BOYD JR., MAJ (RET), enlisted in underage from McCool, MS in May 1942. (In 1942 in order to enlist in the Army Air Forces, one had to be 18 years of age and have parental consent. He was not yet 17 when he enlisted. He served as Tennessee State Commander of the Veterans of Underage Military Service, wherein the national roster includes enlistments at ages 12, 13, 14, 15, 16, in the various branches of service in WWII and Korean eras.)

He flew 50 combat missions as engineer-gunner with 386th BG (B-26 Marauders/England/July 1943-April 1944), followed by a healthy stint on the Berlin Airlift/Frankfurt/C-54s.

Boyd was directly commissioned from the ranks in 1950 and served as aircraft maintenance officer at Barksdale AFB, LA; Keesler AFB, MS; Evereux, France and Homestead, FL.

From 1962 to 1965 he was missile maintenance shops officer in the 556th SMS (Atlas-F) at Plattsburgh, NY, where he also was instrumental in formulating the Sector Maintenance Concept adopted throughout Strategic Air Command in 1964. He voluntarily retired in the grade of regular major, March 1, 1965.

He and his wife, Juanita, reside in East Tennessee

(Rutledge), where they enjoy the sunrise (coffee), the sunset (Happy Hour) and visitors. Y'all come!

GEORGE R. BRENDLE, COL (RET), activated

the 351st MSM, Feb. 1, 1963. Held position as deputy commander of maintenance for three years. Five years as commander, 351st SMW, having the distinction of serving longer in that position than any one other wing commander in SAC. During his term as commander, the 351 SMW won the missile competition at VAFB in 1967 and 1971.

In 1992 Col. Brendle was honored at the missile competition at VAFB, having commanded the first wing to win The Blanchard Perpetual Trophy. His decorations include the Distinguished Service Medal, Distinguished Flying Cross, Air Medal w/12 clusters, AF Commendation Medal w/2 OLCs, and ETO Ribbon w/ 5 Battle Stars.

RONALD G. BROHAMMER, COL (RET), born

Feb. 5, 1945, Hillsboro, IL, enlisted July 3, 1967, USAF, entered the USAF through OTS in 1967. Initially assigned as an aircraft maintenance officer, Scott AFB, IL in 1968. In 1971 went to the 68th MASS, Cam Ranh Bay, RVN. During his tour, the C5A began in country flights. In 1972 he was the 22nd FMS supervisor, March AFB, CA and rear echelon SqCC during Linebacker II. In 1973 he entered Missile Operations, was a DG from ICBM operations training and was assigned to Whiteman AFB, MO. Serving as a line MCCC, instructor and evaluator, he became the senior standardization crew commander and had the privilege of experiencing two SMES evaluations. He also earned an MBA from MU. Assigned to the 4315CCTS in 1978, he eventually became the chief of Minuteman Modernized Training. Graduating as a DG from ACSC in 1981, he went to HQSAC in Future Concepts and later in Force Applications. In 1983, he became commander of the 68th SMS and in 1985 returned to HQSAC as chief, Missile Operations Management. This SAC tour saw many innovations in missile crew duty, upgrades to crew life and changes in operational requirements. He returned to Whiteman AFB in 1989 serving as ADO; DO; commander, 351st Support Gp., and commander, 509th Support Gp. As the last commander of the 351st Support Gp. and first commander of the 509th Support Gp., he oversaw much infrastructure development preparing for the B-2. His last assignment, vice-commander, 44th MW, saw the removal of the 44th ICBMS and inactivation of that wing on July 4, 1994.

A most memorable moment occurred when he served as an escort when the commander of Russia's Rocket Forces visited the 44th. Another was watching the last missile arrive back at Ellsworth AFB. He had the unique opportunity and privilege to participate in the buildup for the B-2 and the drawdown of the Minuteman II. Retired as a colonel on January 1, 1995.

His awards and decorations include the Legion of Merit, Bronze Star, AF Meritorious Service Medal w/3 OLCs, Joint Service Commendation Medal, AF Commendation Medal, Combat Readiness Medal, National Defense Service Medal w/OLC, Vietnam Campaign Medal w/3 Service Stars, Vietnam Cross of Gallantry w/Pam, Vietnam Service Medal, and AF Outstanding Unit Award w/Silver OLC.

He and his wife Judy live in Marysville, MO. He is the assistant city manager/director Public Works for Maryville. Judy is the executive director, Maryville Chamber of Commerce. They have three children: Kelli, Michael, and Kristi, and one grandson, Garrett Bailey.

ELMER T. BROOKS, BGEN (RET), born Dec.

30, 1932, Washington, DC, BA degree and commissioned in USAF via ROTC from Miami University, OH; MS degree from George Washington University. Completed ICAF and the AF Advanced Management Program (University of Virginia).

His missile/space-related assignments include: Dep Missile Combat crew commander (DMCCC)/MCCC/ Instructor MCCC, 551st SMS (Atlas F), Lincoln, NE; aerospace flight control technologist, NASA, Houston, TX; Exec to the director, Space Systems, OSAF, Pentagon; vice commander, then commander (July 1979-October 1981), 381st SMW (Titan II), McConnell AFB, KS; Director, International Negotiations and Dep. Commissioner, US/USSR Standing Consultative Commission, J-5/OJCS, Pentagon; Asst. Dep Under Secretary of Defense/Research & Engineering (Strategic & Theater Nuclear Forces), OSD, Pentagon; Dep Associate Administrator (Space Communications), NASA HQ, D.C. Also served as military assistant to SecDefs Rumsfeld and Brown. Retired from USAF as a brigadier general, May 30, 1985, and from the Senior Executive Service with NASA, Feb. 6, 1995.

Memorable experiences include pulling alert during the Cuban Missile Crisis, successfully recovering from a potentially catastrophic Atlas F Propellant Loading Exercise, the first Saturn V/Apollo launch, the catastrophic Titan II accident at Rock, KS, and winning the Blanchard (1980).

His awards include the Defense Distinguished Service Medal, Defense Superior Service Medal w/OLC, Legion of Merit, Mertorious Service Medal w/OLC, Joint Service Commendation Medal, AF Commendation Medal, Combat Readiness Medal, NASA Medal for Outstanding Leadership, the George Washington University, Distinguished Alumni Achievement Award.

He married Kathryn Casselberry, June 23, 1954. They have a daughter, three sons and eight grandchildren and reside in No. Bethesda, MD.

He is now a church worker, director and mentor at an alernative school and consultant.

WILLIAM RILEY BROOKSHER, BGEN

(RET), born February 1930, Turkey, AR; enlisted USAF January 1950; commissioned from Officer Candidate School September 1953; graduate Air Command and Staff College and National War College; major active duty assignments: Sergeant Major, Department Administrative and Supply Training, Warren AFB; Commander 75th Air Installations Sqdn., ROK; Commander, Titan I and Minuteman Missile Combat Crews and Beale and Whiteman AFBs; Director of Personnel, 416th BW, Griffiss AFB, Director of Operations 17th AD, Whiteman AFB, Commander 341st SMW, Malmstrom AFB and 91st SMW, Minot AFB, Chief of Security Police SAC, Offutt AFB; Commander, AF Office of Security Police and the chief of Police USAF, Kirtland AFB.

Brooksher was awarded Distinguished Service Medal, Legion of Merit w/OLC, Meritorious Service Medal w/OLC and AF Commendation Medal w/OLC.

Retired in 1981, joined Westinghouse Hanford Com-

pany as director, Safeguards and Security, retired 1994. Currently writing books: co-author Glory at a Gallop and author Bloody Hill and War Along the Bayous.

CHARLES E. BRYANT, MSGT, USAF (RET),

born between the towns of Wetumka and Weleetka, OK on March 13, 1933. He first entered US Army in November 1947 at the age of 14.

Enlisted in the USAF Feb. 8, 1951 and served in the Korean War, Vietnam War and a host of "brush fires." The first aircraft he worked with was the F-80 and F-84 at Luke AFB, AZ. He worked on the B-29, T-33, C-123, C-119, C-130 and the KC-97 and spent three years working on a multi-purpose pipeline in Spain.

Schooled on the Minuteman One Missile, Boeing Co., Seattle, WA, Fall 1961 and reported to the 394th SMS in December of that year. He started training with the contractor in January of 1962.

Sgt. Bryant participated in the first launch of the Minuteman One Missile down the western test range, then was shipped TDY to Malmstrom AFB, MT during the Cuban crisis in October of 1962. He was later assigned there on a permanent basis in Spring 1963.

He helped to posture 150 Minuteman One Missile sites, and 50 Minuteman Two sites at Malmstrom AFB and then was sent TDY to Grand Forks AFB to help posture the Minuteman Two sites there. He returned to the 394th in 1968 and helped to test the Minuteman Three Missile.

Retired and returned to his home state of Oklahoma in December 1970 and went into the business of standing quarterhorse and thoroughbred stallions and after 25 years of that his doctor retired him to fishing.

He and his wife of 43 years, Cherie, now live on their 50 acre farm near Norman, OK, where they have five mares, an old stallion and assorted cats and dogs. They have three children: Chris, Carol and Corey.

HEIDI HELLAUER BULLOCK, MAJ, com-

missioned a second lieutenant through ROTC at Saint Michael's College on May 11, 1985 and assigned as an adjutant for the 68th Tactical Fighter Sqdn. at Moody AFB, GA. In May 1987, she reported for missile duty. A distinguished graduate from IQT, ILCS 84, she was assigned to the 510th SMS. While at Whiteman AFB, MO, she held numerous positions in the 351st SMW Standardization/Eval (Stan/Eval) Div. She completed her tour as the first female senior crew commander of Stan/Eval in the history of the 351st SMW. She was also the MCCCC of the first female Stan/Eval senior crew in the history of SAC.

After her missile tour, she was an Education With Industry (EWI) student at General Electric in Mooresown, NJ. From there, she's been a contacting officer at Tinker AFB, OK and Scott AFB, IL. Major Bullock is currently the deputy, Contracts Div., Airborne Laser SPO at Kirtland AFB, NM.

JERRY M. BULLOCK, COL (RET), born June

2, 1932, in Ralls, TX. Commissioned, second lieutenant, USAF, Aug. 28, 1953. Graduated AB Defense School, Parks AFB, CA July 1954. Served active duty air/security police units at Laredo AFB, TX; Osan, Korea; and Fairchild AFB, WA. Chief of Security Police, TUSLOG, Ankara, Turkey. Served in Vietnam as resource manager for the Chief of Security Police, 7th AF. Returned to the US to an assignment with the Air Staff in Washington, DC. Follow-on assignments:

Chief, Security Police, 15th AF (SAC); Chief, Tactical Air Command; and last assignment as Deputy Chief, USAF and vice commander of the AF Office of Security Police.

Bullock was awarded the Air Force Commendation Medal w/3 OLC, AF Meritorious Service Medal, Bronze Star and the Legion of Merit w/3 OLCs. Retired 1981 and currently serving as executive director, Air Force Security Police Association.

KENNETH W. (ARCHIE) BUNKER, CMSGT, born Oct. 23, 1943, in Longview, WA, grew up in Laverne, OK and graduated high school in 1961.

Bunker completed basic at Lackland AFB, as well as Missile Maintenance Training at Lowry AFB, CO and Orlando AFB, concluding with the launch of Mace Missiles from Cape Canaveral, FL. He maintained missile systems throughout the world including, the Mace in Germany and Okinawa, the GLCM in Sicily and the Minuteman in the United States.

Decorations include the Meritorious Service Medal w/4 clusters and Air Force Commendation Medal. He was promoted to chief master sergeant Jan. 1, 1983 and culminated his 29 1/2 year missile maintenance career at Whiteman on Dec. 5, 1990.

Chief Bunker is married to Laura L. Hinnant of Micro, NC. They have sons, Bill and Jim and reside in Sedalia, MO, where they operate a Cow-Calf Farm.

EARL W. BURRESS JR., 1LT, is a 1994 gradutae of St. Cloud State University. He completed Officer Training School, a Hoya of Sqdn. 2, in January 1995 and served with the 76th RQF for two months prior to beginning Undergraduate Space and Missile Training, Class 95-06. He graduated Minuteman III WS-133B IQT, Class Deuce 166, in October 1995.

His crew tour began November 1995 as a deputy in the 447 MS "Dragon Masters" of Grand Forks AFB. He was selected to instruct and write simulator scripts for the 321 OSS in August 1996 and participated in Grand Forks Flood relief efforts of 1997 during this period. Burress returned to the 447 MS and upgraded to crew commander in August 1997. Lt. Burress is a graduate of Lieutenant's Professional Development Program, Squadron Officer School, Mobile Instructor Skills Training, and National Search and Rescue School.

ROBERT V. BUSH SR., CWO-4 (RET), born March 26, 1925 in Charlotte, NC. CWO-4 Bush joined the USMC March 20, 1943 serving in air units in the US and Pacific areas until June 1946. In 1948 he became a charter member of the North Carolina Air National Guard and was employed full time as an air technician. His unit was called to active duty at the beginning of the Korean conflict and CWO-4 Bush elected to pursue a career in the regular Air Force. His first contact with the Air Force Missile program came while serving as a senior instructor at Sheppard AFB, TX. He was assigned to a special project to investigate and write

and report on what impact the impending introduction of multiple new missile weapons systems would have on the present training structure of the AF Training Command.

At the completion of his tour of duty at Sheppard, he was assigned to the 19th BW (SAC), Homestead AFB, FL, as the propulsion branch chief. A short time later his missile "experience" caught up with him and he was transferred to the 321st BW (SAC), McCoy AFB, FL and was assigned duty at Eglin AFB, FL as one of SAC's representative to the Category II testing of the Air-to-Ground GAM 63 Bell Aircraft "Rascal" missile. The GAM-63 was a large aircraft type liquid rocket propelled missile originally designed to be launched from the underside of B29/50 type aircraft. Strapped to the side of a B-47, the "Rascal" was a sight to see. The GAM-63 never went into production. He was immediately assigned to the final operational testing of the GAM-77 "Hound Dog" and the GAM-72 "Quail" missiles also located at Eglin AFB, FL. At the completion of these tests, he was assigned to the 4135th SBW, Eglin AFB, the first SAC unit to become equipped with the Hound Dog and Quail missiles. Leaving the air launched missile program, he was among the first contingent of personnel assigned to the "Atlas F" 556th SMS, Plattsburgh AFB, NY, during the final phase of construction and missile installation. During this period he returned to Sheppard AFB to participate in the training program, for which he helped to lay the ground work. Assignments in the 556th included Quality Control, OIC Electrical and Electronic Branch and Sector Maintenance Officer (Sector 1). When the unit was deactivated, he was one of the officers selected to supervise the removal of the missiles and prepare the silos for shutdown. CWO-4 Bush's final assignment was to HQ, 8th AF where he completed his military career as a material controller on the 8th AF Flying Command Post. Retired from active duty in September 1969, he and his wife Julia, reside in Swansboro, NC, where they are active in family, church and community affairs.

PATRICK J. CALLIGAN, LT COL (RET), graduated from Buena Vista University in 1964, attended Officer Training School, and was commissioned a second lieutenant in August, 1964. His first assignment was a personnel service officer at RAF Croughton and RAF Upper Heyford, England. In 1967 he entered Minuteman missile crew training and was certified mission ready in May, 1968 and served in the 510th SMS, Whiteman AFB, MO. While at Whiteman, he was instructor, MCCC and wing operations senior instructor. During this tour, he earned an MBA from the Univeristy of Missouri. From May, 1974 to May 1975 he was assigned as the chief, Special Services at Thula AB, Greenland.

In May, 1975, he was reassigned to Whiteman AFB, MO and held the position of missile training officer, Commander Headquarters Squadron, Chief of the EWO Training Branch and chief of the Plans and Intelligence Division. In April 1981, he was assigned to HQ Strategic Air Command Inspector General as the senior ICBM operations inspector.

After completion of his IG tour, he was assigned as the commander, 532nd SMS at McConnell AFB, KS. He advanced to assistant deputy commander, operations, and in 1985 became the last deputy commander, Operations for the 381st SMW. He directed the operational activities during the Titan II deactivation, which was completed in August 1986. He remained at McConnell AFB to complete the unit inactivation and retired from the AF on July 31, 1987.

He and his wife, Fran have resided in the Wichita, KS area since his AF retirement. They have two children. He has been employed since 1987 as an internal auditor for the Boeing Company. He is a CPA, certified internal auditor and certified professional for Webster University and owns a tax and accounting practice in Wichita.

JOHN R. CAMPBELL, born Aug. 26, 1924, Perry, MO, entered service in 1943 and retired as colonel in 1974. Rated: master navigator.

Was B-47 navigator, 303rd BW, Davis Monthan, AFB; Atlas F missile crew commander, 551st SMS, Lincoln AFB; and chief of supply, Vandenberg AFB. Also had tours in operations, 15th AF, March AFB, and with inspector general, HQ, SAC, Offutt AFB.

Graduate of Command & Staff College (1961) and Air War College (1966).

His crew won B-47 phase of SAC Navigation & Bombing Competition (1953); also, his crew was chosen SAC missile crew of the Month (July 1963); and his squadron was chosen "Best Base Supply" in SAC (1996).

After retirement he managed the North San Antonio Chamber of Commerce and served six years as director of development for AF Village. Married after WWII to Grace Miller of Vallejo, CA, now retired in Tucson, AZ. They have three daughters and four grandchildren.

RUSSELL C. CASE, 1LT, born Aug. 8, 1943, Middletown, NY, enlisted July 5, 1967, USAF, 1967 to 1979: 1967-1972: Missile Combat Crew Commander and MPT operator, Minuteman IB & F, 68 SMS, 44th SMW, Ellsworth AFB, SD.

1972-1979: Auditor with USAF Audit Agency with duty at Forbes AFB, KS; Athenai Airport, Greece; Wright-Patterson AFB, OH; and Hill AFB, UT.

1980-present: Wyoming Air National Guard, Cheyenne, WY having served as comptroller, communications officer, and currently the supervisory auditor for the USPFO. He holds the present rank as first lieutenant.

ALOYSIUS G. CASEY, LT GEN, currently a technical and management consultant, Gen. Casey is an outside director serving as chairman of the board of National Technical Systems. NTS is an independent testing company with ten locations across the country to test a variety of commercial and aerospace products.

Casey retired from the AF in 1988 after more than 34 years of distinguished service, during which he held major development and managerial positions. From October 1986 to July 1988, Gen. Casey was commander of the Space Div. of AF Systems Command. In this position he was charged with managing the design, development, delivery and operation of the AF space systems and launch vehicles. During this period the Titan 34D returned to operational status; the Titan IV contract was negotiated; and two new medium launch vehicles, the Delta II and the Atlas II, were established. Several on orbit satellite constellations were refreshed. From 1982 to 1986, Gen. Casey served in San Bernardino as commander of the ballistic missile office and Peacekeeper program director. He managed the design, development, and delivery of the Peacekeeper Weapons Systems from approval in June 1983 through initial operational capability (IOC). He also directed the development flight test program noted for excellent performance.

Casey held responsible positions as a colonel in the Minuteman Program, and at Wright Patterson AFB in the A-10 and B-1 aircraft developments. He has an extensive operational background as a B47 navigator and 130 combat missions in the night flying gunship in Southeast Asia.

General Casey was awarded the 1984 Dr. Theodore Von Karman Award by the Air Force Association for science and engineering. In 1985 he received the Air Force's highest management award, The Eugene M. Zuckert Management Award for outstanding management of the Air Force Missile Programs.

He received a BS degree from the United States Naval Academy, MS Astronautics, Air Force Institute of Technology, Air War College. Registered Professional Engineer, and long distance runner; four marathons a year.

General Casey is married to the former Patricia Casey of Washington, DC and they have three sons: Matthew, Joseph and Patrick.

DAVID H. CHAGNON, MAJ, July 30, 1939 in Worcester, MA, enlisted in the USAF February 1963. His military locations and stations: Holloman AFB, NM; Takhli AFB, Thailand, F.E. Warren AFB, WY, 1967-1971; 1971-1977, Vandenberg AFB, CA; Hancock Field, NY; O'San, AB, Korea; Beale AFB, CA.

Participated in training, Chanute AFB, and Vandenberg, AFB, 1968.

His memorable experiences: Alternate Command Post (ACP) instructor. Training others: 4315th CCTS, training others beginning their missile combat crew experience.

Chagnon was awarded the Air Force Commendation Medal w/2 clusters, Meritorious Service Medal w/ cluster, Presidential Unit Citation, AFOU Award w/7 clusters, Master Missile Badge. He was discharged Feb. 1, 1983 as major.

Chagnon is married to Sandy. Employed with emergency management agencies and in restaurant management.

JOHN E. CHAMBERS, COL (RET), received his regular commission directly from Ohio State University ROTC in 1959. In the early 1960s, he served as an Atlas F ICBM Combat Crew member (577th SMS, Altus AFB, OK), and was later assigned to Strategic Air Command HQ (DORQ) to manage the Atlas F Operational Flight Test Program. In mid-1964 he was a charter member of the missile requirements team in the newly formed DPLD (later XPQM) and became the Minuteman III project officer — from earliest weapons system incep-

tion to final planning for the first flight.

From 1968 to 1971 he was on the Air Staff (XOOSS) and supported SACs ICBM requirements; he was the first OBL (Operational Base Launch) project officer. He attended the Air Command and Staff College (ACSC), Class of 1972. From 1972 to 1974 he was Squadron Commander, 571st SMS, Davis Monthan AFB, AZ (Titan II). Reassigned to the air staff in 1974 as a colonel in Air Force Studies and analysis, he was chief to ICBM Analysis.

He was a distinguished graduate of the Industrial College of the Armed Forces (ICAF Class of 1977), and then assigned to the air staff as the first chief of the Air Force Budget Issues Team (AFBIT); he was responsible for the Congressional testimony by the Secretary of the Air Force, and the Air Force Chief of Staff. Reassigned to SAC HQ (XPF) in 1979, he managed overall strategic force planning, and directly supported CINCSAC in congressional appearances. In 1980-1983, assigned to the 390th SMW, Davis Monthan AFB, AZ (Titan II), he was deputy commander for operations, vice wing commander and then wing commander. In 1983 he was selected as the Outstanding Wing Commander in the SAC. In 1984, as the final 390th commander, he cased the colors as the first Titan II Wing was deactivated. He then served as the commander, 308th SMW, Little Rock AFB, AR (Titan II) until 1986; during this assignment the 308th captured the Blanchard Trophy.

Reassigned to the Pentagon, Office of the Secretary of Defense, he retired in 1987. Among other decorations during his career, Colonel Chambers received four awards of the Legion of Merit. He was a key founder of the 390th Memorial Museum in Tucson, and the Titan II Missile Museum in Green Valley, AZ.

He and his family (wife, Pamela, daughter, Jennifer, and son, John) reside in Arlington, VA.

ROSALIE F. (BRIGGS) CHAMBERS, CAPT, born October 4, Wright Patterson AFB, OH, enlisted in the AF.

Her military locations and stations: Sicily, Belgium, Montana, training in Arizona; 302nd (TMS) at 487th TMW, 71 TMS at California, 485 TMW, 12 SMS at 341 SMW.

She ground launched cruise missiles; Minuteman ICBM. Memorable experiences include being a distinguished graduate from Vandenberg IQT. Six deployments with A-flight in Belgium.

Chambers was awarded the Outstanding Unit Award at every squadron she was stationed at. She was discharged as captain July 17, 1992. She is married to Major James (Security Forces) and they have two sons, Justin (5) and Connor (3). Today she is molding the future and investing in her children. She also does volunteer work.

BRIAN K. CHAPPELL, CAPT, is the assistant executive officer, 91st Operations Gp., 91st Space Wing, Minot AFB, North Dakota. Born Oct. 12, 1970 on Monroe, MI, he graduated from Monroe High School in 1988. After graduation from the University of Michigan, Ann Arbor, in 1992, with a bachelor of arts degree in Political Science. He was commissioned a second lieutenant through AFROTC.

In December 1993, he completed Undergraduate Missile Training, 392nd Space and Missile Training Sqdn., Vandenberg AFB, CA. Assigned to the 742nd Missile Sqdn., 91st Space Wing he served as a deputy missile combat crew commander, instructor in 91st OSS, missile combat crew commander, assistant flight commander and flight commander. In 1996, he completed a master of science degree in administration, from Central Michigan University and Squadron Officer School in 1997.

Captain Chappell has over 200 alerts in the Command Data Buffer (CDB) and Rapid Execution and Combat Targeting (REACT) Weapons systems.

He was awarded the AF Commendation Medal, AF Outstanding Unit Award w/OLC, Combat Readiness Medal, National Defense Service Medal, 20th AF Missile Combat Crew of the Quarter Award, AF Space Command Crew Member Excellence Award, AF Space Command ICBM Crew Safety Award of Distinction, AF Space Command Inspector General Professional Performance Award.

MICHAEL J. CHRISTENSEN, LT COL, received his commission at Oklahoma State University in 1978. ILCS Class 11A. Duties included Deputy Missile Combat Crew commander, Missile Combat Crew commander, and instructor at the 351st SMW, Whiteman AFB, MO. He was among the initial cadre of 2LT crew commanders of the 510th SMS—the Emergency Rocket Communications System (ERCS) squadron. He earned the 8th AF Crew Member Excellence Award for 10 "highly qualified" evaluations and a MBA through the University of Missouri-Columbia. Served as Glory Trip 146MS operations officer.

In March 1984, he reported to the 4315th Combat Crew Training Sqdn., Vandenberg AFB, CA, serving as a Minuteman II instructor and ICBM operations senior flight chief for ILCS classes 57-86. Became Command Data Buffer qualified to train the 341st and 351st SMW SMW initial cadre. He then became a warning systems controller, Strategic Air Combat Operations Staff, HQ SAC, Offutt AFB, NE to assess real-time surveillance data from both land and space-based sensor systems. In March 1990, he moved over to the Joint Strategic Target Planning Staff, Analysis and Simulation Div. as missile operations staff officer evaluating strategic CE systems for SIOP planning purposes. When JSTPS was disestablished, he became the senior C4I Modeling and Simulation Analyst at USSTRATCOM/ J614. He then reported to HQ AF Operational Test and Evaluation Center, Kirtland AFB, NM as chief, Cheyenne Mountain Upgrade Div. and Chief, Tactical Warning/Attack Assessment Div. Presently, he is assigned to USSTRATCOM/J3621 as mission commander, Airborne Battle Staff Director, Offutt AFB, NE.

He, his wife, Soon, and his five children currently reside in Omaha, NE.

MARK W. CLARK, MAJ, was commissioned in 1983 through AFROTC at the University of Louisville. Following Titan II training, he served as a senior DMCCC with the 532 SMS, McConnell AFB, KS from 1984 to 1986. Transferred to Little Rock AFB and the

373 SMS in 1986, he completed Rivet Cap, the Titan II deactivation program.

After Minuteman III training in 1987, he arrived at the 447th SMS at Grand Forks AFB, ND. He served as a flight DMCCC and MCCC before joining the 321st SMW/DO9, Missile Control Div. He served as an officer code controller, codes instructor, and Chief of Code Handler Training, from 1989 to 1991. Leaving missiles in 1991, he served the 6592 ABG, Los Angeles AFB, CA as chief of social actions. Separating from active duty in 1992, he joined the Reserves and the Civil Air Patrol Reserve Assistant Program in 1995. Recently promoted to major, he is the Deputy Wing Reserve Coordinator for the Kentucky Wing.

He lives in Louisville, KY with his family, and is in human resources. Major Clark holds the Senior Missile Operations Badge.

DEAN L. CLEMMER, MSGT (RET), enlisted in the USAF in May 1958. He spent his first four years at Forbes ABF, Topeka, KS. He worked aircraft maintenance on the KC-97G.

In 1962 he cross trained into the Titan II Missile Engine Propulsion System. Stationed in the 395 SMS at Vandenberg AFB, CA. During his career he was also stationed in the 308th SMS at Little Rock AFB, AK.

He worked in the missile engine propulsion shop. In 1967 he was put back into the aircraft maintenance field and was stationed at Korat AFB, Thailand. He worked aircraft engine on the C-121 aircraft. In 1968 he returned to the propulsion shop at Vandenberg AFB, CA. In 1976 he served in the 3901st as the NCOIC of the tech order library. He has been an ordained minister with the Assembly of God since 1979. He is at present pastoring in Ellington, MO.

Clemmer is married to Maxine and they have reaised two sons, both live and work in Springfield, MO.

RAYMOND H. (HAL) CLEVELAND, COL (RET), graduated from Clemson University in 1956, and entered navigator training at Ellington AFB, TX. Married Alice Corbett, his high school sweetheart from his hometown of Seneca, CA, December, 1956. Then, in September 1957, moved to first operational assignment at Schilling AFB, Salina, KS, in SACs KC-97 air refueling tanker in the 40th BW, SAC, and later in the 310th BW, served as instructor and standardization navigator.

Family moved to Robins AFB, GA in 1964 to the KC-135 refueling tanker, in the 19th BW. Then, in 1966, into missiles, the 321st SMW, Grand Forks, ND, after training at Vandenberg AFB, CA, 448th and then 447th Sqdn. Crew commander, instructor, wing standardization evaluator, and chief, Standardization Div. While on the crew force, earned MS in Industrial Management from the University of North Dakota, through the Minuteman Education Program. Departed Grand Forks July 1970 for South East Asia, C-130 aircraft at Naha AB, Okinawa. Naha soon closed and family moved to Clark AB, Philippines.

While in the Philippines, spent two plus years flying into and out of Vietnam for two to three weeks at a time, operating mostly out of Saigon and Cam Ranh

Bay. In November 1973, back to Grand Forks in old job as chief, Wing Standardization Div., then as commander, 446th SMS. In early 1975, moved to Vandenberg AFB, CA as deputy commander, 4315th Combat Crew Training Sqdn. In late 1975, selected for colonel and moved across base to 3901st Strategic Missile Evaluation Sqdn. as director for operations, then to Montgomery, AL for the one-year Air War College course.

Then to the Pentagon for a two-year tour on one of the five around-the-clock teams that operate the National Military Command Center for the chairman, Joint Chiefs of Staff.

September 1979, to Minot, ND as vice and soon thereafter, commander of the 91st SMW. After almost three years at Minot, on to last assignments at HQ, SAC, Omaha, as director, ICBM requirements, XPQ.

Retired from the AF, after 27 very good years, in late 1983 and began second career with GTE as Midwest regional marketing manager with an office in Omaha, concentrating on large communications and information systems.

Retired from GTE in 1993, and he and Alice moved to Warner Robins, GA, where he ran a consulting business for about three years, then, bought a truck and really retired in December 1996. Completed the University of Georgia's "Master Gardener's course in April 1997.

He and Alice have two daughters, Amy and Anna and five grandchildren. Now, spending most of his time gardening, doing volunteer work in trial and teaching gardens for local schools and universities, visiting the grandchildren, and, as time permits, golfing.

FRANK R. CLOYD, SSGT (RET), born March 21, 1921, in Glasgow, KY, began his missile training at Lowry AFB, CO and worked on the TM-61C Course. He was then stationed at Fairchild AFB in Spokane, WA as part of the 567th SMS. While stationed at Fairchild, he completed the Missile Mechanical Technical Course 430XO, the Missile Erection System Course (SM65E) ADS54250A-7, and the Pressurization System Control Course ADS44370A-2 at Sheppard AFB in Texas. Cloyd also completed the Launch Crew (Phase I) Training, Launch Crew (Phase III) Operational Training and the Missile Maintenance Management AMF 31000-1 course at Vandenburg AFB in California. He was a missile maintenance technician on the Atlas-E launch crew from activation until it was phased out in 1965.

His crew was on duty during the Cuban Missile Crisis, and he says a million things went through his mind while waiting at his station, knowing it was not a practice drill.

He retired in 1966, and worked at the Allison Div. of General Motors as a parts inspector for helicopter motors. Cloyd and his wife, Sally, eventually moved to Arcola, IL where they still reside. He is still "trying to play golf".

SEBASTIAN F. COGLITORE, BGEN (RET), was a charter member of the 448th SMS, Grand Forks AFB, ND. Between 1966 and 1970, served as both deputy and commander on instructor and stan-eval

crews. Following an AFROTC assignment, he spent the remainder of his career in space-related assignments. He was a satellite engineer, launch integration manager, and the program manager for the first DOD payload to fly on the Space Shuttle. Seb then served in the Pentagon in Space Plans and Policy and later as military assistant for Space to the SECAF. He then served successively as director, Titan IV Program, USSPACECOM/J5V, NORAD command director, and as the first commander, 30th SW, Vandenberg AFB. His final assignment was as director of Space Programs, ASAF/Acquisition. He retired in 1995 and now works for Lockheed Martin. He and his wife, Reggi, reside in San Jose, CA.

TED L. COOK, MSGT, USAF (RET), born April 19, 1934, Pittston, PA. Enlisted July 8, 1953 in the USAF and was stationed at Goodfellow AFB, TX; Hickam AFB, Hawaii; Dyess AFB, TX; Beale AFB, CA; Little Rock AFB, AR; Davis Monthan AFB, AZ; and March AFB, CA.

His crew brought the first Titan II into the EWO at Little Rock AFB in December 1963. He participated in Titan II test launch at Vandenberg AFB in 1966 and his crew successfully launched a Titan II. Also participated in Power Box.

Won academic award at NCO Academy, Barksdale AFB and was runner-up to Honor Grad in 1968; instructor, NCO Leadership School, March AFB; final assignment as 1st sergeant 1974-75, March AFB. Discharged Aug. 2, 1975, attended Cal State University on GI Bill where he graduated with honors and obtained his MA degree in education in 1981.

He and Novie have been married 43 years and have three children: Teddy, Terry and Christy. Retired from teaching in 1990 and moved from Riverside, CA to Bullhead City, AZ where he teaches management part-time at Mohave Community College. He also has a home business and conducts seminars for casino employees, Laughlin, NV.

WILLIAM T. COOPER, LT COL, born July 3, 1944, Ft. Madison, IA, enlisted in the service April 1, 1968. He is the maintenance control officer and chairperson, Wing IV Deactivation Committee, 351st MW, Whiteman AFB, MO.

A native of Ft. Madison, IA, Lt. Col. Cooper was commissioned through the AF Officer Training School following graduation from the University of Iowa.

A career ICBM missileer with maintenance, operations, and headquarters assignments, Lt. Col. Cooper brings 24 years of experience to the deactivation effort.

His initial assignment was missile combat crew duty at the 91st MW, Minot AFB, ND, after which he entered the ICBM maintenance career field.

During wing level assignments, Lt. Col. Cooper has held a vast array of positions in the ICBM maintenance arena, to include a tour as commander of the 351st Organizational Missile Maintenance Sqdn. at Whiteman AFB, MO. The squadron earned the AF Outstanding Unit Award during his tenure of command.

In 1981, Lt. Col. Cooper was recognized as the Air Force Outstanding Senior Missile Maintenance manager (Lt. Gen. Leo Marquez Award), thereby earning the Air Force Recognition Ribbon.

During assignments to SAC HQ, Offutt AFB, NE, Lt. Col. Cooper was responsible for structuring the ICBM Maintenance Standardization and Evaluation Program, and for formulating Operational Readiness Inspection criteria for the SAC inspector general.

In 1984, Lt. Col. Cooper was assigned to SAC Detachment 2, Air Force Systems Command, Norton AFB, CA, as the chief, advanced ICBM Logistics. In this capacity, Lt. Col. Cooper represented the SAC deputy chief of staff for logistics in the development, acquisition, and deployment of the Peacekeeper weapon system, and the full scale development of the Mobile Launcher for the Small ICBM.

Lt. Col. Cooper has served as the project officer for various programs.

-Wing IV Upgrade Silo Program: a three year program to harden launch facilities and upgrade operational ground equipment.

-Simulated Electronic Launch Program: a test of the operational Minuteman weapon system to validate reliability and performance.

-Follow-On Test and Evaluation Launch: the launch of a Minuteman missile from the Western Test Range. The missile and supporting hardware were randomly selected from 351st MW assets.

Lt. Col. Cooper was discharged Oct. 1, 1993 and his decorations include the Meritorious Service Medal w/4 OLC, Air Force Commendation Medal w/OLC, Combat Readiness Medal, National Defense Service Medal w/Bronze Star, AF Recognition Ribbon, AF Outstanding Unit Award w/OLC, and the AF Organizational Excellence Award. He died Dec. 18, 1997.

WILLIAM COX, MSGT, born April 5, 1949, Shenandoah, PA, enlisted in the USAF Aug. 1, 1969.

His military locations and stations: 4th TFW, Seymour Johnson AFB, NC; 57th FIS, Keflavik (NAS) Iceland; 27 TFW Cannon AFB, NM; 388th TFW, Korat RTAFB, Thailand; 52nd TFW Spangdahlem AFB, Germany; 4th TFW, Seymour Johnson AFB, NC; DET 192 Incirlik, Turkey, 31 TTW, Homestead AFB, FL; 52nd TFW Spangdahlem AFB, Germany; 42nd BW Loring AFB, ME. He participated in 1985 52 TFW Salty Demo.

His memorable experiences: in 1986 at Spangdahlem AB, Germany. They prepared a shipment of AGM-45 Shrike Missiles to be delivered to the US 6th Fleet in the Mediterranean. Several days later a message from the 6th Fleet Admiral went to USAFE and down to the 52 TFW, asking "where missiles wings and fins were", Questions went down the line where Cox was on the floor. So he said, "fastened in the missile containers lid where they belong." There were some red faces that went all the way to the 6th Fleet Admiral.

He was awarded AF Commendation Medal w/2 OLC, AF Achievement Medal w/3 OLC, AF Outstanding Unit Award Ribbon w/OLC, AF Good Conduct Medal w/5 Awards, National Defense Service Medal, Vietnam Service Medal, AF Overseas Short Tour Ribbon w/OLC, AF Overseas Long Tour Ribbon w/OLC, AF Longevity Service Award Ribbon w/OLC, NCO PME Graduate Ribbon, Small Arms Expert, Republic of Vietnam Campaign Medal, Meritorious Service Medal.

Discharged April 31, 1990 as master sergeant. He was awarded the 17th AF Maintenance Man of the Quarter July-September 1975; Maintenance Professional of the Year 31 EMS 1981.

Cox is a munitions handler, Ft. Indiantown Gap, PA.

BRUCE E. CRANMER, SMSGT (RET), upon graduation from technical school at Chanute AFB, IL he was assigned duties as a missile maintenance technician at Minot AFB, ND where he participated in the Minuteman III upgrade program. In September 1973 he was reassigned to Malmstrom AFB, MT. He was assigned duties as a technician, instructor and team chief. In May 1981 he was selected for the AF Operational Test and evaluation team at Vandenberg AFB.

There, his responsibilities included supporting the test team's role in the development of the Peacekeeper

Weapon System. He participated in conceptual, developmental, production, deployment and 17 test launches. His next assignment was with the ballistic missile office at Norton AFB, CA. He was assigned duties as an integrated logistics manager, working operational basing for future weapon systems and logistical support for peacekeeper and the small ICBM. In September 1989 he began an overseas assignment at Wueschiem AS, Germany. There, SMSgt. Cranmer was involved with intermediate nuclear forces treaty inspections, operational deployment, and deactivation of the wing. His final assignment was Malmstrom as the superintendent of the Missile Mechanical Branch.

He and his wife Iva, live in Yucaipa, CA, where he works as the purchasing manager for a major manufacturing company. He enjoys golf and restoring classic cars.

JAMES L. (JIM) CROUCH, BGEN (RET), born June 30, 1935, Hillsboro, TX, graduated from the University of North Texas, 1955. Commissioned through ROTC. Retired 1986 as brigadier general. Served as transportation officer; avionics maintenance officer in US and SEA; missile crew member; senior instructor crew commander; ICBM maintenance officer at squadron and HQ SAC; maintenance squadron commander; deputy commander for maintenance; vice wing and wing commander; ICBM plans and requirements officer at SAC HQ; deputy chief of staff for data automation, HQ SAC; J6 (director of information systems) US Readiness Command, McDill AFB, FL (only assignment in US south of Omaha).

Following retirement, was executive director of child advocacy agency in Austin, TX; executive director of state agency in Austin; and program manager in privately owned software company, also in Austin. Volunteer in church; Habitat for Humanity work and Austin Writer's League. Married in 1958 to Barbara and they have two children and three grandchildren.

WILLIAM (FRED) CRYTZER, MSGT, began his AF career on Nov. 8, 1960, shortly after graduation from high school in Ohio. Basic training was completed in December 1960 at Lackland AFB, TX. Technical training was completed in July 1961 at Chanute AFB, IL in the Aircraft and Missile Ground Support Equipment Repairman Course.

In Block IX of technical training, overseas orders were presented to the students. Being on "C" shift (1800-2400) most of the choice assignments were gone except ten slots to Korea. Ten of them from various shifts volunteered and off they went. No one knew what kind of unit or weapon system they would be assigned to. As it turned out they were all assigned to the 58th Tactical Missile Gp., 310th TMS at Osan AB, Korea, a TM61C "Matador" missile unit.

In April of 1962 just before the inactivation of the unit, most of them were reassigned to the 498th TMG, 498th MMS, at Kadena AB, Okinawa, Japan, a TM76B "Mace" missile unit. There were only two of four missile complexes operational at the time so some of them were assigned as drivers charged with transporting missile combat crews to the two sites at White Beach and

Ona Point. Eventually, they were all assigned to the Ground Support Equipment Branch.

Finally, in November 1962 MSgt. Crytzer returned to the US and was stationed at Plattsburgh AFB, NY with the 556th SMS equipped with 13 Atlas "F" missiles. This is the base where he was married to the former Jane Welch on May 1, 1965. By this time the unit had been inactivated and those left behind assigned to a caretaker unit charged with maintaining critical power and water and waste systems until a salvage contractor went to work removing equipment that was to be retained by the government.

MSgt. Crytzer stayed with this unit, part of the 380th Civil Engineering Sqdn. until June 1966 at which time he reverted to his secondary AFSC of Aircraft Ground Support Equipment and was a part of the 380th Field Maintenance Sqdn. October 1966 found the Crytzer family headed for Davis-Monthan AFB, Tucson, AZ and duty with the 390th SMW, 570th SMS equipped with the Titan II ICBM. Technical School at Sheppard AFB, TX was finished in April of 1967 and crew training done at Vandenberg AFB, CA with a combat ready date of May 1967. October of 1968 found crew E-005 off to Vandenburg AFB again, this time to participate in Glory Trip 26T a follow-on operational test launch of the Titan II. Their missile was removed from Site-2 and had the honor of being the oldest missile in the fleet at that time.

Their crew labored along with two other crews and a maintenance contingent to put the missile back to an alert configuration and launch after the prescribed "soak-period" along about the last week in October.

Upon return to Davis-Monthan, MSgt. Crytzer was assigned to duty as a squadron instructor ballistic missile facilities technician (BMFT) with the 571st SMS and later as a wing instructor BMFT. Crew time ended in May 1971 and the family was off to the frozen north at Malmstrom AFB, MT and the 10th SMS, 341st SMW equipped with the Minuteman I and II. Duty with the 10th SMS was as a facility manager with duty at Echo Launch Control Facility.

MSgt. Crytzer performed facility manager duties until 1976 when he was assigned to the 341st Field Missile Maintenance Sqdn. in the Periodic Inspection Teams section with follow-on assignments to the Facility Maintenance Teams Section as NCOIC and eventually to the 341st SMW deputy commander for Maintenance Staff as resource advisor.

He was then caught in a leveling action in November 1980 and the family was off to "El Forko Grande" or the 321st SMW at Grand Forks AFB, ND and the 321st FMMS, first as NCOIC of the Shops Maintenance Branch, then to the Facility Maintenance Teams Section, Vehicle and Equipment Configuration Branch, Unit Career Advisor, and lastly as NCOIC of the Facility Maintenance Branch, 321st FMMS.

Retirement for MSgt. Crytzer came on Dec. 1, 1986 and moving to San Antonio, TX and life as a student at the University of Texas at San Antonio from which he graduated in May of 1990 with a degree in Political Science.

Presently he is working at the United Services Automobile Association (USAA) as a plant operator and enjoying retirement and six grandchildren and is going to reunions of missile units, including AAFM.

LARRY L. CUNNINGHAM, CAPT, born April 19, 1944, Providence Hospital, Washington, DC, enlisted in the service Nov. 6, 1969. He served with the 544th Aerospace Reconnaissance Wing, Offutt AFB, August 1969-September 1972; 570th MS, September 1972-November 1976; 2nd Comm. Sqdn., Buckley

ANG, November 1976-November 1978; Cobra Dane, Shemya, AK, November 1978-November 1978; Norad, Space Computational Center, November 1979-August 1980.

He received his training at Nichols State University, receiving a BS in 1969. Currently completing masters in Systems Management with Golden Gate University of San Francisco.

His memorable experiences: while working for TRW on the Defense Support Program (DSP) in 1986 at Cape Canaveral, he witnessed the explosion of the Space Shuttle Challenger that took the lives of all the astronauts.

He was discharged during the reduction of Force (RIF) November 1980, as captain.

Cunningham presently is a Satellite Systems engineer with Lockheed Martin. He married Judy V. Basye March 6, 1970 and they have no living children. He was employed with General Electric, August 1980-August 1984; TRW, August 1984-July 1987.

GARY L. CURTIN, MGEN, is the director, Defense Special Weapons Agency (DSWA). A Distinguished Graduate of the University of Maryland's AFROTC program in 1965, he served as a Minuteman I DMCCC-I and Airborne Launch Control System MCCC-A in the 44th SMW, Ellsworth AFB, SC from 1965-1970.

Subsequently he served as an airborne targeting officer in Thailand, flying 105 combat missions. He then became an intelligence staff officer at HQ PACAF and later a Pol-Mil officer on the Air Staff. Returning to ICBM duties, he commanded the 400th SMS and served as assistant DCO of the 90th SMW at F.E. Warren AFB, WY from 1980-1982. After graduating from National War College in 1983, he became director, ICBM Requirements and CINCSAC's special assistant for M-X at HQ SAC, Offutt AFB, NE. In 1986, he became vice commander and then wing commander of the 90th SMW. During his tenure at the 90th SMW, the Peacekeeper (M-X) missile deployment commenced in the 400th SMS. Beginning in 1988, he served as director of Command Control, HQ SAC, before becoming assistant director of plans, HQ SAC. In 1990, Gen. Curtin became the JCS representative to the START I negotiations in Geneva, Switzerland. With the conclusion of that treaty in 1991, he became deputy director of International Negotiations, J-5, Joint Staff. He served as director of intelligence, USSTRATCOM from 1993-1995, before assuming duties as director, Defense Nuclear Agency (DNA). (DNA became DSWA in June 1966). MGen. Curtin holds a BS in Astronautical Engineering from the University of Maryland and a MS in economics from South Dakota State University (MMEP).

He and his wife Karen, currently reside on Bolling AFB, DC. Their son, Scott, is a 1990 USAFA graduate and an AF helicopter pilot. Their daughter, Jennifer, is married to an AF C-141B flight engineer. Maj. Gen. Curtin holds the Defense Distinguished Service Medal, Defense Superior Service Medal, Legion of Merit, Bronze Star Medal, Meritorious Service Medal w/2 OLCs, Air Medal w/2 OLCs, Aerial Achievement Medal, AF Commendation Medal and the State Department Meritorious Honor Medal.

JOHN W. DARR, LT COL (RET), former infantryman, graduated bombardier school in 1944; B-24 training at Tonopah and Langley Field followed. Recalled with onset of Korean War, he acquired triple-rating, served in SAC on RB-29, RB-36, B-52 crews. Obtained aeronautical engineering degree through AFIT, served in Systems Command with Lockheed Corporation, then SATAF at Grand Forks AFB, building and modifying Minuteman Wing VI. SATAF duties evolved from Weapon System Engineering through Detachment Commander.

Following a tour as squadron navigator in Tactical

Electronic Warfare in Vietnam 1969-1970, he returned to Grand Forks as commander of the 446th, later the 448th, SMSs. Military career concluded in 1974 at Warren AFB as missile operations officer of 4th AD. His son, Steven (q.v.), is also a missileman.

In retirement, vice president of Unicover Corporation, a world-wide direct marketing firm. He and Angeline, former WAVE, reside in Cheyenne, where he works as a research consultant.

STEVEN W. DARR, entered AD having served as cadet colonel, University of Wyoming AFROTC in May 1976. A student in MM3 CDB Class 43, June 1977, he was assigned to DMCCC in Warren's 90th SMWs 400th SMS. Competed in Olympic Arena 80, helping earn Best 15AF Operations. DOT instructor duties followed, then MCCC-ACP. He was then a command post controller, very possibly the first-ever first lieutenant so assigned. He later became an EWO instructor in DO22T. In 1983 he transferred to the 4315CCTS/DO22T at Vandenberg AFB, teaching all Minuteman and Titan II EWO. Final active duty assignment: senior instructor in the old BMSC course, called Space and Missile Orientation.

He returned to civilian life in December 1985, working first for Honeywell, then Hughes, both in Southern California. Today he manages a family of airline products within AlliedSignal Corporation, Redmond, WA. He and his wife, Carla, plus Cameron and Janessa, reside in nearby Woodinville.

BENJAMIN J. DAVIS SR., TSGT (RET), is one of the founders of the Assocation of the Air Force Missileers. Born May 12, 1921 in Abberville, SC. Attended St. Emma Military Academy, Rock Castle, VA near the town Powhatan and the James River; High School of Commerce, Detroit, MI. Drafted while working at Congoleum-Nairn, Chester, PA. Inducted at Ft. Bragg, NC Jan. 6, 1944. Attended basic at Lincoln AFB, NE, Flt. C222. Trained with 18th Avn. Sqdn., Chanute AFB, IL. Slow action, volunteer surveying at McDill AFB, FL, 36th BU, joined the 1348th Engrs. at Drew AFB. FL. Duties as team chief embrace constructing an air field at the South China Sea and bridges up the mountain to Baguio, Philippines. The team included Japanese prisoners of war. Discharged at Camp McCoy WI 1946.

Enroll at St. Augustine College, Raleigh, NC September 1950, ordered to Keesler AFB, MS, then to George AFB, CA, join the 6th Shoran at Lawson AAFB, GA. Surveys in Georgia, Mount Mitchell, NC and Mount Chehaw, AL. With the activation of the 3rd Shoran Beacon Unit at Shaw AFB, SC. Embarked at New York on General R.E. Callan for Landstuhl AB, Germany, Oct. 7, 1953. Acting First Sergeant, attended

Management School for supervisors. Team chief of Astronomical and Geodetic Surveys. (From bunkers, locate and measure bombs accuracy at Baumholder bombing range.) Challenges were: Winterburg, Kassel, Sechenhiem, Olenburg, Munster, Pforzheim, Flensburg, Nuremburg, Paris, France and Libyan Sahara Desert in Africa. While in the Sahara PCS to Loan AB France, 38th BW. Returned to Landstuhl for the family. NCOIC Bombing Targets/Computations, assist bombardiers with Shoran. (A practice run caused concern of France and Germany when a bomb fell in the vicinity of a hospital in Germany.) AFSC as survey computer technician, met a mechanical computer when he flew to Germany. In a cave near Kaiserlautern was Monorobot VI Digital Computer. Unit won the Secretary of the Air Force Outstanding Award. PCS to 1823 AACS (STRATCOMSYS Engineer & Inst), Andrews AFB, MD, 1823d became 2877th Ground Electronic Engineer and Installation then 2878th. Compute wave length in Transmission Engineer. TDY to Hawaii; Schofield Barracks, and Diamond Head Mountain, install transmission and receiver sites. Outfit moved to Olmsted AFB, PA. Changed to 2861st. TDY to Griffis AFB, NY, Westover AFB, MA and Andrews AFB, after Hurricane Hazel for Plant-in-Place. TDY to Peshawar, Pakistan and Kyber Pass. PCS to 1381st Geodetic Survey (Missile), Orlando AFB, FL. NCOIC Astronomical-Geophysical Branch. Assist. In Charge Plans & Recon Stellar Camera Branch. Attended AF Academy (MATS), Orlando, FL. Team chief: Internal Minuteman Missile team, Moble Rapcon Turntable Center. Locations include Harlowton, Lewiston, Harve, Montana. (Little excitement while working in the van. It became the center of a deer hunt). Awarded the Missile Insignia. PCS to Osan AB Korea, 6314th Civil Engineer.

Motor trip to the DMZ, returned the route a group of North Korean broke through. Next trip via helicopter to DMZ for site survey. To Japan for survey equipments. Managed Base Education Office. Attended the University of Maryland on base. PCS to HQ Ballistic System Div., then to HQ Cen Ground Electronics Engineer, Tinker AFB, OK. Immediate TDY to Missouri; Kaysinger Bluff, project areas include Clinton, Warsaw, Sedalia, and Independence. Quality Assurance on Cable Relocation (Missile) in Conrad, and Cut Bank vicinity. (Vehicle hydroplane, flip front to rear then upright near Kalispell, MT). Site checks in Sundance, WY. Cable location in Minote, Max, Bowbells, ND; Pierre, SD. Last TDY, Albuquerque, NM.

He was awarded the AF Commendation Medal and Silver Defect. Retired July 1, 1968.

Attended North Carolina State University; project manager of Raleigh, NC Urban Renewal. Started a general contractor and real estate corporation. He has two sons, Benjamin J. and Benjamin G., two grandsons and four great-grandchildren. He moved to San Francisco, CA in 1983 when his sister became ill. Space A and Frequent Flyer.

MICHAEL R. DENINGTON, COL (RET), joined the AF in 1963. Assignments included Mace Missile combat crew commander (MCCC), Germany; Weather forecaster, in Colorado, Florida and Vietnam; MCCC and flight commander in Minuteman I and Minuteman Mod/CDB, and plans officer, 319th SMS, F.E. Warren AFB, WY; missile operations staff officer, 1st Strategic Aerospace Div.; strategic operations officer, National Emergency Airborne Command Post; commander, 510th SMS and deputy commander, 351st Combat Support Group (CSG), Whiteman AFB, MO; assistant director of Support Service Inspections, SAC inspector

general; assistant, and deputy commander for operations, 321st SMW, Grand Forks AFB, ND. His final assignment was commander, 351st CSG.

His decorations include the Legion of Merit, Bronze Star, Defense Meritorious Service Medal, AF Meritorious Service Medal, Joint Services Commendation Medal, AF Commendation Medal, Vietnam Service Medal w/4 Bronze Service Stars, Vietnam Campaign Medal and Republic of Vietnam Cross of Gallantry w/ Palm.

Denington and his wife, Marilyn, live in Bartlett, TN, where he enjoys creative writing. In 1997, he was designated a Laureate Man of Letters at the World Congress of Poets in High Wycombe, England.

JAMES E. DENMAN, MSGT, born Oct. 1, 1939 in Sturgis, MI, joined the USAF, Dec. 30, 1959. A2c Denman was the guidance man on the launch of Matador missiles #56-1948 and #56-1949, launched on June 1, 1961 at Cape Canaveral, then served a combat crew tour at Osan AB, Korea. Assigned duty as a BMAT with the 44th SMW. Ac1 Derman's targeting team participated in the posturing of the 1st Minuteman missiles at Ellsworth. In 1964 he was assigned to the 351st SMW, Whiteman AFB and served as combat targeting team member and wing maintenance scheduler.

In 1969 TSgt. Denman transferred to the Defense Meteorological Satellite Program where he served at both Loring AFB as standardization evaluator for antenna systems, and at Fairchild AFB in maintenance supervision.

Retiring in 1982, he had received the AF Meritorious Service Medal and the AF Commendation Medal and wears the Master Missileman Badge.

WAYNE E. DEREU, COL., born July 30, 1947, Pipestone, MN, enlisted in the USAF January 1970. His military locations and stations: Fairchild, 1970-71; Whiteman, 1972-76; Vandenberg, 1976-79; Offutt, 1979-82; 1987-90; Ft. Leavenworth 1982-83; Remstein, 1983-85; Wueschhiem, 1985-87; Minot, 1990-93; Malmstrom, 1993-95; F.E. Warren, 1995-97; Peterson, 1997-99. He participated in ground launch of cruise missile deployment.

His memorable experiences include the first squadron commander 89 TMS (GLCM). Won best tactical missile squadron in USAFE for 1986.

Dereu will be discharged as colonel January 1999. He is married to Sally and they have a daughter, Shilo and son, Darrin. They reside in Colorado Springs, CO.

ORVILLE E. (ED) DICKERSON, MAJ (RET), graduated from UCLA ROTC in January 1956, trained as a pilot, received his wings June 1957; flew the WB-50 in Weather Reconnaissance Yokota AB, Japan. In August 1960, trained in the Mace Missile, served as instructor, Orlando AFB, FL. He volunteered for "Minuteman Education Program" Ellsworth AFB, SD, helped bring the 44th SMW to alert status in 1962, served as crew member, instructor and evaluator. Launched a Minuteman Missile from Vandenberg AFB. In 1967 received his MBA from Ohio State University and assigned to the Joint Strategic Target Planning Staff, HQ SAC to target Titan, Minuteman and Polaris missiles. Flew Looking Glass as missile operations officer; 2nd ACCS. Retired at 20 years March 1967 from F.E. Warren AFB, Cheyenne, WY, a senior pilot with 5,000 hours flying time, and a Master Missileer.

Civilian career was spent as a restaurant owner in

Evergreen, CO; with Department of Motor Vehicles in San Diego. Dickerson and Yvonne enjoy San Diego and use it as a base for their travels.

ROBERT R. DOCKUM, LT COL, born Feb. 2, 1938, Cincinnati, OH. Enlisted Oct. 2, 1961 in the USAF and attended primary and basic pilot training, survival training, 1962; helicopter pilot training, 1963; Squadron Officer School and USAF Coin Indoctrination Course, 1964; deep sea survival training and jungle training, 1967; pilot instructor training, 1968; Air Command and Staff College, 1974; pilot advanced training, 1977; Air War College, 1981.

He was director of resource management, one year; air operations officer, pilot-officer general, 16 years and instructor-assistant professor, USAF Academy and Air Command and Staff College for eight year.

Lt. Col. Dockum retired Oct. 2, 1986. Awards include DFC, BSM, MSM w/OLC, AM w/ OLC, AF Commendation Medal w/OLC, AF Presidential Unit Citation, AF Outstanding w/2 OLCs and V, NDSM, VSM w/3 OLCs, RVN Gallantry Cross w/Palm, RVN Campaign Medal and several ribbons.

Currently employed as AFJROTC and economics instructor. He and wife Jarmila have two sons, Todd and James, and three daughters: Leah, Kathy and Stephy.

PAUL T. DOELKER, LT COL, born March 19, 1917 in Columbus, OH, served with the Army Air Force and the USAF, Air Training Command, Air Material Command and SAC. Participated in operations in WWII, AMC, Germany; training, Lowery; Mace-Sembach, Germany; Missile Maint., USAFE, Wiesbaden, Germany; Minuteman, Whiteman AFB.

Discharged in 1940 as lieutenant colonel and resides in Kissimmee, FL. He was employed as school counselor and instructor. Now contentedly retired.

ORVILLE L. DOUGHTY, LT COL (RET), became a missileer with Thor in August 1959. Served with 99th ADS in England December 1959 to December 1961. Became maintenance control officer, 579th Atlas F. Sqdn., Walker AFB, NM in January 1962, then to SAC HQ in July 1963 in DM4A and later in the "Big Missile Div." Flew C130s in Vietnam June 1967 to

August 1968, then to the Pentagon, as air staff action officer in plans until November 1971. Became 390th SMW MIMS commander in December 1971.

Retired as ADCM May 1, 1974 with 34 years service in pilot, maintenance and missile career fields. Formal missile training includes Thor IRBM. Atlas F and Titan II ICBMs and the maintenance blocks of Minuteman. Doughty lives in Tucson, AZ with his wife Myrna. Both are active volunteers, she at DMAFB and he at the Titan Missile Museum, near Green Valley, AZ.

DARRELL A. DOWNING, COL (RET), grew up on a farm in Iowa, attended Iowa State College and joined the AF in March 1959. He spent most of his 27 year career in missile positions. Starting as an Atlas crew member at Forbes and continuing as a training officer and Minuteman II crew member at Grand Forks, an evaluator with the 3901 SMES, two tours with the SAC inspector general, a tour with the 4315 CCTS, command positions at Warren Murted Turkey and McConnell AFBs and retired April 1, 1986 at Offutt AFB, NE. He completed SOS and ACSC in residence, Industrial College of the Armed Forces by Correspondence and AWC by seminar.

Colonel Downing helped formulate and execute the first missile combat competition, Curtain Raiser.

He was awarded the Legion of Merit, Meritorious Service Medal w/3 OLcs, AF Commendation Medal w/ 2 OLCs, AF Outstanding Unit Award w/2 OlCs, Combat Readiness Medal, National Defense Service Medal, AF Longevity Service Award w/5 OLCs, Small Arms Expert Marksmanship Ribbon.

He is married to the former Janice Schoening and they have one daughter, Cathy, and two grandsons, Andy and Robby. He enjoys gardening, antiqueing and spending time with his grandchildren.

ALAN O. (SAM) DUNKIN, MAJ (RET), commissioned 1967. Assigned 1970 to 390th SMW, Davis-Monthan AFB, AZ, serving as a Titan II DMCCC, MCCC, and evaluator MCCC. Reassigned to 351st SMW, Whiteman AFB, MO, 1975, serving as Minuteman II (SSAS) MCCC, flight commander, evaluator MCCC and chief, operations, Training Div. into 1980. Served as chief, Evaluation Div., 15th AF Directorate of Missiles, March AFB, CA for four years and at 57th AD, Minot AFB, ND as missile operations officer. He retired in January 1987.

He married Lt. Nancy Nerenberg in 1978. They had two children, Bradley and Jessica. Upon his retirement, they moved to Astoria, OR where he worked in the family furniture business. Nancy died in December 1996.

Now fully retired, Dunkin fences, is an assistant scoutmaster and volunteers in two school libraries.

CHARLES F. (CHUCK) DWYER, born Jan. 26, 1928 Staten Island, NY, served with 3499th MTW, Chanute AFB, IL, USAF, Electronics Fundamentals, Airborne Radar Mechanics, Instructor Training at Keesler AFB, Biloxi, MS; Airborne Shanicle Guidance System, Factory School at the Glenn L. Martin Company, Baltimore, MD.

His principal military assignment was to develop course material and conduct classes in theory of operation and maintenance of the TM-61 guidance system and ground handling equipment for combat launch crew personnel of the 1st and 69th Pilotless Bomber Sqdn. at Patrick AFB, FL. These squadrons later became part of the AF first TMW in West Germany.

Upon discharge, he was employed by the Glen L. Martin Company (now Lockheed Martin) as a field engineer/technical representative on the TM-61 Matador Missile program. His first assignment was to assist personnel at the Guided Missile School at Lowry AFB, Denver, CO to activate a mock-up missile and test equipment training line. He then transferred to the Martin Engineering Field Test Crew at Patrick AFB, FL to support the flight test development program of the Shanicle guidance system.

He was then assigned to the 11th TMS at Orlando AFB, FL and deployed to West Germany in 1956 to support operations, maintenance and training at Sembach AB, Hahn AB and Bitburg AB and launch operations at Wheelus AB, Tripoli, Libya. At completion of this contract in early 1960, he returned to the ZI and transferred to Martin Denver to work on the Site Activation programs of Titan I and Titan II. When Titan II was selected to be the launch vehicle for the Gemini space program he transferred to the Martin Canaveral Launch Operations Div. where he worked as a Flight Controls and Malfunction Detection Systems pre-installation test conductor and launch checkout quality systems engineer. At completion of the Gemini program he transferred briefly to Martin Orlando and returned to Kennedy Space Center with IBM where he worked on the Apollo and Skylab programs. He transferred to IBM Federal Systems Electronic Systems Center in Owego, NY where he worked various assignments on several AF and Navy programs until he retired in April 1990.

WARREN M. DYER, MAJ (RET), enlisted in the USAF in December 1967 and graduated from Communications Maintenance School at Keesler AFB, MS in October 1968. Between October 1968 and June 1976, he maintained Minuteman and Titan II ICBM Communication Systems and SAC long haul communications. His assignments included the 90th Communications Sqdn. (SAC), F.E. Warren AFB, WY; Det. 1, 33rd Communications Group (SAC), Hickam AFB, HI; the 1891st Communications Sqdn. (AFCS), Wheeler AFB, HI; and the 381st Communications Sqdn. (SAC), McConnell AFB, KS. From June 1976 to September 1979, he served as an AF recruiter with the 3504th USAF Recruiting Gp., Richardson, TX.

He graduated from Officer Training School in December 1979, reporting to HQ NORAD/Space Command, Cheyenne Mountain Complex, CO, as a communications-computer systems officer. After completing Minuteman II (ILCS) Combat Crew Training in June 1983, he was assigned to the 341st SMW (10th SMS), Malmstrom AFB, MT. At Malmstrom, he served as a missile combat crew commander; senior flight commander, instructor crew commander (341 SMW/DOTI); and as an SCP flight commander. After completing his tour as a Missileer in July 1987, he reported to the 2147th Communications Wing (3rd AF), RAF Mildenhall, United Kingdom. There, he served as XO; deputy director, Program Implementation; director of Operations; and director of Systems Management. He also directed 3rd AFs Communications Operation Center during Operations Desert Shield and Desert Storm.

He departed RAF Mildenhall in July 1991 for Scott AFB, IL, his final assignment. At Scott, he served as chief, Communications-Computer Systems Engineering and Implementation (Air Force Communications Command); and as a chief of Systems Integration (AF C4 Systems Agency). Dyer holds the Joint Service Commendation Medal, Meritorious Service Medal w/OLC, Air Force Commendation Medal, Combat Readiness Medal, National Defense Service Medal w/Bronze Star, Senior Missile Operations Badge, Senior Communications Maintenance Badge, and a master's degree in Systems Management.

He and his wife, Kay, have resided in Lubbock, TX since 1993. He serves as director of Telecommunication Services, Texas Tech University Medical Center.

KENNETH J. DZIEWULSKI, CAPT, born Aug. 11, 1947, Chicago, IL. Enlisted in USAF Jan. 26, 1967 with basic training at Amarillo AFB, TX. 1967 Distinguished Graduate, Sheppard AFB, TX; 1967-69, aircraft mechanic, EC-47, 361st Tactical Elec. Warfare Sq., Nha Trang, RVN; 1969-71, acft. mech., C-131, 375th CAM Sq., Scott AFB, IL; 1971-73, maintenance scheduler, 3535th Nav. Trng. Wing, Mather AFB,  CA; 1973-75, Cal. State Univ., BS degree, business personnel administration.

Commissioned Duties and Assignments: 1975, OTS, Lackland AFB, 2LT; 1975, Titan II Missile Ops. School (Distinguished Grad.), Sheppard AFB and Titan II qualification training, Vandenberg AFB; 1975-79, MCCC, DMCCC Wing instructor-evaluator, DMCCC competitor (77-78), Strategic Missile Wing, McConnell AFB and completed master's degree in human relations, Webster College, 1979.

From 1979-82 he was Titan II missile instructor, Squadron Officer School, Maxwell AFB; 1982-85, director of regional recruiting AF ROTC, Kansas-Missouri and supervisor-evaluator of five area AFJROTC units, U of Kansas and completed Air Command and Staff Course; 1985-87, asst. ops. officer, 741st SMS, Minot AFB, ND.

Participated in RVN Campaign 1967-69. Retired Feb. 1, 1987. Awards and medals include VSM w/3 BSS, RVN Gallantry Cross w/Palm Device, RVNCM, MSM w/OLC, AFCM w/2 OLCs, PUC w/OLC, AFLSAR w/3 OLC, NDSM, AF Good Conduct, Combat Readiness Medal, OSR and Small Arms (Expert) Pistol.

Currently working as a special projects manager for a small charitable foundation and private investment firm in Chicago, IL. He is a single dad with three sons: Ken and twins, Brad and Darren.

JOSEPH L. ECOPPI, BGEN (RET), born in Freeman Spur, IL, July 12, 1929 and graduated from the University of Illinois in 1954. He was commissioned in the AF, served at Marana AFB, AZ; Lowry AFB, CO; Larson AFB, WA; Laon AFB, France and George AFB, CA. In 1963 he became a Minuteman combat crew commander in the 741st SMS, Minot AFB, ND.

He interservice transferred to the Army in 1965, served in FA assignments, senior staff positions, graduated from the Army War College in 1976, and served in the 1st Cav. Div., Vietnam 1966-1967.

Promoted to brigadier general (1983), last assigned as chief of staff, 4th US Army, Ft. Sheridan, IL and retired in July 1987.

His awards included the Legion of Merit, Bronze Star, Air Medals, Meritorious Service Medals and the Purple Heart. Joe and Janet live in San Antonio, TX.

MICHAEL R. EDINGER, CAPT, has been a member of AAFM since March 1994. After graduation from Southern Illinois University in 1991, he reported to the 3315th Combat Crew Training Sqdn. at Vandenburg, CA to attended Undergraduate Missile Training. Upon graduation, he was assigned to the 12th SMS at Malmstrom AFB, MT. While at Malmstrom, he served as an instructor and evaluator as both a deputy and crew commander. He was also in charge of the MPT script development section in DOT, and was among the first REACT qualified missile crew members at Malmstrom. While on crew, he completed his master's degree in public administration from the University of Montana in 1995.

In March 1996, Edinger reported to the Space Warfare Center at Falcon AFB, CO where he served as the chief of the Operational Analysis Systems Development in the Special Projects Branch. In November 1997, he was selected to be the commander of the newly-formed space tactics flight for the 21st Space Wing where he currently serves.

ROBERT F. EICHEL, MAJ (RET), joined the AF after graduating from Kent State University in 1967. He initially served as a security policeman in Taiwan. He attended Officer's Training School in 1972, and was commissioned as a second lieutenant. From then until his retirement in 1987, he held various missile operations positions. For five years he was an instructor in the Titan II weapon system at Davis Monthan AFB, AZ and Vandenberg AFB, CA and an authority on the propellant transfer systems. He also holds various positions within the Minuteman weapon systems at both Whiteman AFB, MO and Grand Forks AFB, ND.

Eichel is a graduate of Air Command and Staff College and Squadron Officer's School. His decorations include the Meritorious Service Medal, AF Commendation Medal w/OLC, AF Achievement Medal and the Combat Readiness Medal w/OLC. He is presently an elementary school teacher.

RONALD L. EISSNER, CMSGT (RET), born Nov. 20, 1953. Upon completion of basic training at Lackland AFB and the K-System Series course at Lowry AFB, he was assigned to Smoky Hill AFB (later Schilling AFB), Salina, KS, served as a bomb-navigational technician and flight chief on B-47 aircraft, and was then assigned to the 3901st Strategic Standardization Sqdn. at Vandenberg AFB. He served as a Titan I Launch Equipment inspector/evaluator. While at Vandenberg, received his degree from Allen Hancock College. Additionally, he attended the 15th AF NCO Academy where he received the Commandant's Award and was the class adjutant. In June 1965, transferred to the 321st SMW where he served as an electro-mechanical team inspector/evaluator. In March of 1966, assigned to the 3901st SMES where he was an evaluator and NCOIC of the EMT Evaluation Section. In June 1971, transferred to the 90th SMW where he was NCOIC of the Quality Control Div. and later the DCM maintenance superintendent. In December 1971, promoted to chief master sergeant.

Eissner and his wife Georgia continue to live in Cheyenne, WY. They enjoy traveling, golf and Denver Bronco football.

LESTER E. EKLUND, COL (RET), entered the AF from Forman, ND 1951 and received his commission and pilots wings through the Aviation Cadet Program in 1952. His pilot duties were flying C-119, T-29, C-131, C-47, T-39 and B-47 aircraft in the US, Germany and Korea. In 1962 Col. Eklund was assigned to the 341st SMW at Malmstrom AFB, MT and was one of the first to be certified as a Minuteman Missile combat crew commander and earned a masters degree in Aerospace Engineering through the Minuteman Education Program. From 1966 to 1971 he served in engineering and operations positions at Offutt AFB and in Korea. In 1971 he assumed command of the 68th SMS, 44th SMW, Ellsworth AFB, SD. He later commanded the 44th MMS and served as deputy commander for maintenance in the 44th SMW and 96th BW at Dyess AFB, TX.

In 1975 he became deputy program manager for Logistics at the Ballistic Missile Office, Norton AFB, CA where he was instrumental in the development of the Peacekeeper Weapon System. Colonel Eklund retired from the AF in 1979 and pursued a second career with Aerojet General in Sacramento, CA until 1991. He and his wife Elinor have resided in Fair Oaks, CA since 1979.

SAMUEL D. ELDER, MAJ, born Oct. 13, 1935, Callensburg, PA. Enlisted in the USAF in September 1961 and was stationed at Ellsworth AFB, SD; Vandenberg AFB, CA; Minot AFB, ND and Malmstrom AFB, MT.

Missions/Operations include OIC Electronic Lab, combat targeting off., maint. control off., SATAF SAC liaison, SATAF safety off., OOALC Det. comdr.

Memorable experiences include 2nd sign in to 44 SMW; coded long life I launched from EAFB N-2; helped launch long life II at VAFB and participated in 382 launches at VAFB; while in SATAF helped upgrade Wings I, II, III, IV, V, VI; TRW Project Mgr. Peacekeeper Wing V.

He and wife Phyllis have three daughters: Rolenta, Terri and Patti; and three grandchildren. Test engineer and project engineer for TRW; currently driving United Van Lines tractor trailer.

JAMES C. ENGLEHART, A/1C, born April 4, 1937 in Cleveland, OH, entered the USAF March 1, 1960. He served as 71st TMS Guidance System Mech. Matador; Mace. Stationed at Lowry AFB, CO; Bitburg Germany w/71st TMS, 585th TMG.

His memorable experiences include the alerts, Berlin Wall and Cuban Missile.

He was awarded the Missileman Badge, Combat Readiness Medal, AF Outstanding Unit Award. He was discharged Sept. 24, 1963 as A/1c.

Englehart is married to Pat and they have sons: Brad, Greg and Jeff.

He is owner/president of Select Aire Service Corp, a HVAC company. Adjutant of American Legion Post 196, Brecksville, OH.

WALLACE EDWARD (ED) ENGLISH, LT COL, born Oct. 17, 1925 in Westwood, Lassen County, CA, enlisted in RCAF April 1942 and served until 1946; USAF July 1948 through 1968.

His military locations and stations: Canada, England, Sandia Base, Sculthorpe England, Lowry AFB, CO; Camp Darby Italy, Walker AFB, NM; Bunker Hill AFB, IN; (Grissom AFB); Vandenberg AFB.

He participated in WWII, NATO, SAC and ATC. His memorable experiences: just a bunch of "war stories"

His awards include the CVSM and leaf, Commendation Ribbon w/3 OLC and the Presidential Unit Citation.

English was discharged Aug. 1, 1968 as lieutenant colonel. He was married to Audrey (now deceased) and they had a daughter, Patty and son, Gary. He has been employed 23 years with the Engineering Div., Cuesta College, S.L.O. Retired and teaching part time. His hobbies include collecting antique radios, restoring antique cars, genealogy, photography, gardening and enjoying life.

WILLIAM D. EVA, O5 born Sept. 5, 1936 in Peterborough, NH, enlisted in the USAF, Dec. 9, 1959, SAC and EVCOM. Graduated from the University of NH in 1959.

His military locations and assignments: Malden AFB, MO, 1960, flight training; Webb AFB, TX, 1961, flight training; F.E. Warren AFB, WY, 1961-65, 549th and 566th Atlas E; Grand Forks AFB, NC, 1965-70, 321st-Minuteman trainer; Offutt AFB, NE, 1970-74, JSTPS and ADIN, INTP (SAC); EUCOM HQ, Germany, 1974-77, EUDAC; Offutt AFB, NE, 1977-80, ADW/ADOIC (SAC), innumerable TDYs.

Eva was awarded the AFLSA w/4 OLCs, AFCM, JSCM, MSM, AFSAEMA. Discharged July 31, 1980 as O-5.

He is married to Carol and has sons, Dale and Donald; daughter, Debra, with son and daughter. Owner/operator of forest products business, Longview Forest Products, including the production of maple syrup. Since retirement he has taken additional graduate courses at UNH and has served on various town committees.

JOHN EVANOFF, LT COL (RET), born in Sofia, Bulgaria, was commissioned in 1952. He served four overseas tours; spent the first half of his AF career in Minuteman maintenance staff assignments at unit, 3901st and SAC HQ levels. After completing his graduate studies work at the University of Utah, under the auspices of the AF Institute of Technology, he spent the remainder of his career in staff intelligence assignments,

primarily overseas. He retired from the AF in 1979 and was employed as a department manager in the Peacekeeper Logistics Directorate, Martin Marietta Corporation, until he retired again in 1987.

He and his wife, Wanda, have resided in Littleton, CO since 1979. He served on the board of directors of the National Tuberous Sclerosis Association for a number of years, established a quarterly TS newsletter for Colorado families and is active with the Colorado TS Support Group.

DAVID G. FAAS, COL, born Jan. 16, 1947 in LaCrosse, WI, enlisted in the USAF, Jan. 6, 1970. His military locations and stations: 91 MW Minot AFB, ND, 1970-74; HQ 15 AF, March AFB, CA, 1974-75; 91 MW Minot AFB, ND, 1976-79; HQ SAC, Offutt AFB, NE, 1979-84; 351 MW, Whiteman AFB, MO 1984-86, 487 TMW, Comiso AS, IT 1986-87; 91 MW, Minot AFB, ND, 1987-89; 4315 CCTS, 310 TRTW 30SW, Vandenberg AFB, CA, 1989-94, AFROTC, Norwick University, Northfield, VT, 1994-1997; HQ 20 AF, F.E. Warren AFB, WY, 1997-present.

His memorable experiences: launch crew for GT01GM (first operational test of MM III); Task Force commander for GT130. GM squadron commander for 742 SMS, Minot AFB, ND; squadron commander for 4315 CCTS, Vandenberg AFB, CA.

His awards include the Meritorious Service Medal, JSCM, AFCM. He is presently ranked as colonel.

Faas is married to Terri and they have children: Gretchen, Erika and Dave Jr. Resides in F.E. Warren AFB, WY, serving with 20 AF, chief of Programs and Policy.

MICHAEL FABBRI, E4, SGT, born Aug. 8, 1954, enlisted in the USAF, Aug. 15, 1972, Framingham, MA as missile systems analyst specialist. His military locations and stations: 316XDG Electro Mechanical Team Member; 3702 BMTS, Lackland AFB for basic training; Chanute AFB, Tech School; 351st SMW, Whiteman AFB, duty; Vandenberg AFB, TDY, twice.

He was team training branch instructor for approximately two years. Participated in S.E.L.M. (Simulated Electronic Launch Minuteman) 1974.

His memorable experiences: being selected in 1975 and again in 1976 to represent his unit at Olympic Arena, Vandenberg AFB, CA. Came in second place in 1976 by less than a percentage point. Fabbri has received the Master Technician Award. He was discharged June 11, 1996 as E-4 sergeant and returned to school.

He married Karen Johnson of Michigan in 1987. They had met while he was stationed at Chanute AFB in 1973. They have three children. Employed for 12 years as county prosecutor and presently serves as regional administrator Middlesex County District Attorney's Office, Framingham, MA. He is active in a number of political and non-profit community activities.

JERRY D. FAGAN, MAJ (RET), graduated from Southern Illinois University in 1964, taught music in the public schools until 1967. Joined the AF (1967-OTS), then attended pilot training at Randolph AFB, TX. Fagan taught student pilots in the T-38 at Laughlin AFB, TX (1969-72). His next assignment was the 381 SMW, McConnell AFB, KS (1972-77). His duties included DMCCC, MCCC line crew, line instructor, wing instructor and Stan/Eval. From 1977-80, Fagan taught pilot instructor training in the T-38 (560 FTS) at Randolph AFB, TX. In 1980, returning to McConnell AFB, KS, he flew the KC-135 (91 ARS) and was the senior controller of the McConnell Consolidated Com-

mand Post (381 SMW/384 AREFW) his last four years, retiring in 1987.

He resides with his wife, Catherine, south of Dayton, OH. He is a technical writer with Veridian Corporation and has a home-based technical writing business.

GERALD G. FALL JR., BG, born May 27, 1923, Boston, MA, where he attended school. Graduated from Modesto Junior College, CA and attended U of Maryland, Cambridge University in England, The George Washington University, Chapman College, Orange, CA (BA degree).

Military career began as Army private during WWII. Completed aviation cadet training and earned his commission as a 2LT in USAAC and pilot wings in 1943. Attended ACSC, Maxwell AFB, 1958; 1960, Atlas Guided Missile Maint. and ops courses and the Titan Guided Missile ops and maint. courses; 1966, Industrial College of the Armed Forces, Ft. McNair.

Served as flight instructor and B-24 pilot; ops off. and instr. pilot, 829th BS, Smoky Hill AAF; B-29 instr. pilot for three of SAC's first B-29 bomber groups; 1947-54 served as select crew aircraft cmdr. on B-29, B-50 and B-47 aircraft, instructor pilot and ops staff officer in bomber groups at Smoky Hill AAF, Anderson AFB, Guam and Castle AFB.

Transferred in 1954 to SAC HQ, Offutt AFB as Chief, Combat Crew Branch in the Directorate of Personnel. He began his career in the missile field at Vandenberg AFB in 1958 where he served as guided missile ops staff officer, 704th Strategic Missile Wing. In 1959 he became deputy commander of the 564th Strategic Msl. Sqdn., Francis E. Warren AFB; 1960, commander of 706th Msl. Maint. Sqdn.; 1961-63, served at HQ SAC, Offutt AFB in various missile staff assignments; 1965, deputy commander for maintenance, 390th Strategic Msl. Wing, Davis-Monthan AFB; 1967-70, commanded 321st Strategic Msl. Wing, Grand Forks AFB; 1970, commanded 341st Strategic Msl. wing, Malmstrom AFB; 1972, commanded 4th Strategic Msl. Div. (a later realignment redesignated the unit as the 4th Air Div. with all Minuteman and bombardment wings at Ellsworth AFB, F.E. Warren AFB, Grand Forks AFB and Whiteman AFB.

His variety of assignments in the strategic missile field made Gen. Fall one of the few Air Force officers who possessed such depth of experience in both strategic bomber and missile weapons systems. He is a command pilot with more than 10,000 flying hours, most of them in SAC bombers. He also holds the master missileman rating.

Military decorations include the Distinguished Service Medal, Legion of Merit w/OLC, AFCM w/OLC, Army Commendation Medal and AFOUAR w/OLC. He was promoted to brigadier general May 1, 1971 and retired July 31, 1974.

Completed law school and is currently practicing in that profession. Married to the former Eleanore C. Lohr and they have six sons: Robert, Dennis, Vincent, Shawn, Terrance and Brendan.

RICHARD L. FARKAS, COL (RET), born May 15, 1944 in Massillon, OH, graduated from the University of Florida in 1966 and received his commission through Officer Training School in 1967. His missile assignments included: Titan II missile operations line instructor and standardization/evaluation crew commander at 381 SMW McConnell AFB, KS from 1968-1972; HQ SAC personnel plans missile staff officer and chairman of the SAC Missile Management Working Group; Special Assistant to SAC DCS/Operations; Pentagon Operations staff officer; NORAD missile warn-

ing officer and director of Missile Warning Operations at Cheyenne Mountain, CO; Commander, 16th Surveillance SQ at Shemya AFB, AK; Chief, Space C3 and Electronic Combat Div., Pentagon; first missile officer SAC senior controller in the SAC Undergraduate Command Center; wing vice and commander, 90th SMW at FE Warren AFB, WY. Graduate of Squadron Officer School (1972); Armed Forces Staff College (1980); National War College (1985).

Farkas retired Aug. 1, 1991. He and his wife, Lynda, reside at their lakeside home south of Akron, OH. He is the vice president, Underwriting and Special Risks for Executive Insurance Agency and ValMark Securities, Inc.

BERYLE M. FARMER, CMSGT (RET), enlisted on Sept. 24, 1954 at Georgetown, SC, serving until Oct. 31, 1985. Duties included: tactical and academic instructor at Lackland AFB, TX and Amarillo AFB, TX; English instructor at Dhahran AB, Saudi Arabia, OJT Monitor and Evaluator for Air Force Reserve organizations in Northern California (stationed in San Francisco, CA), Radio Relay technician at Keesler AFB, MS; Clark AB, Philippines, Altus AFB, OK; Erhac AB, Turkey; Tinker AFB, OK and Izmir, Turkey and ballistic missile analyst technician at 390th SMW, Davis-Monthan AFB, AZ, where he served as senior BMAT in the Standardization and Evaluation Div. He was a member of the Blanchard Trophy winning team in the SAC Missile Combat Competition, Olympic Arena 1979.

In November 1981, he reported to 3901st SMES, Vandenberg AFB, CA, serving as Senior Titan II BMAT evaluator, his final assignment.

From November 1985 to June 1997, he was employed as launch operations engineer by Martin-Marietta Aerospace on Titan II, Titan III and Titan IV programs at Vandenberg AFB, CA.

He and his wife, Marian, currently reside in Lompoc, CA, where he spends most of his time fishing, golfing, or restoring old automobiles.

LELAND G. FAY, COL (RET), served 30-years as a regular career officer since graduation in June 1958 from the US Military Academy until his retirement in June 1988. He was initially assigned as missile operations and maintenance officer with the 4751st Air Defense Sqdn. (BOMARC) Eglin AFB, FL and then as Electronics Development Engineer with the 6555th Aerospace Test Wing, Cape Kennedy, FL. His duties there were missile test conductor/launch complex manager for the ATLAS/ABRES military launch team, Chief Flight Controls and Electronics for Gemini/Titan II Flight Test Operations Branch, and Chief Flight Controls and Guidance for the Titan IIIC Space Launch Vehicle. While he was a member of these units they were awarded two NASA Group Achievement Awards for 11 Gemini manned space launches and three Air Force Outstanding Unit Awards for contributions to ballistic missile development and the national space program.

In 1969, he was assigned as assistant air attaché and foreign technical intelligence officer for missile, space and electronics systems at the US Embassy, Paris, France, where he was awarded the National Order of Merit from the Republic of France. He was then assigned to the Space and Missile Systems Organization, Los Angeles, AFS, CA, where he was project manager

and chief, Systems Analysis and Penetration Systems Divisions for the Advanced Ballistic Reentry Sytems Program Office. He next served as program element monitor for the E4B National Emergency Airborne Command Post at HQ USAF and then as director of Electronic Systems for HQ Air Force Systems Command, Andrews AFB.

From 1982 to 1986, he was deputy director/program manager for the Joint Cruise Missile Program Office responsible for the development, test, production and deployment deliveries of the Tomahawk Cruise Missile which resulted in the Initial Operational Capability of the AF Ground Launched Cruise Missile and the Navy's Ship Launched Cruise Missile. In 1986, he was assigned as director of Strategic and Mobility Systems, Headquarters Air Force Systems Command where he served as acquisition manager during development, test and production deliveries that achieved the initial operational capability of the Peacekeeper Missile and the B-1 Bomber.

Colonel Fay earned MSEE and MSBA degrees from the Air Force Insitute of Technology and The George Washington University and is a graduate of the Defense Systems Management College, the Industrial College of the Armed Forces and the Defense Intelligence School. He was awarded the Master Missileman Badge, the Senior Space Badge and professional certification as a level four service acquisition manager. His decorations include the Defense Superior Service Medal, Legion of Merit w/OLC, Joint Service Commendation Medal, Air Force Meritorious Service Medal w/OLC, and the Air Force Commendation Medal w/OLC.

From 1988-1995, he worked as senior system engineer and program integration manager with the General Electric and Lockheed Martin Corporations as a member of the system engineer and integration contractor team supporting the Strategic Defense Initiative and the Ballistic Missile Defense Organization.

Colonel Fay and his wife, Marion, reside in Alexandria, VA and are enjoying retirement, their family: Jennifer, Stephanie, Mark, son-in-laws and six grandchildren, travel and volunteer work.

LEWIS E. (LEW) FEUERSTEIN, MAJ (RET), born Oct. 8, 1926 in Norfolk, VA. On Sept. 26, 1944 he enlisted in the Army Air Corp as a private and was discharged Nov. 20, 1946. He remained in the inactive Resere until Sept. 4, 1950. He was a ROTC student at East Carolina College from September 1949 to May 18, 1951 when commissioned. He retired Oct. 31, 1969.

After basic training he attended Remote Control Terret Mech. School (B-29). WWII ended while in school. He was sent to Germany as a dental technician.

His commissioned duties included assignments as squadron mess officer, squadron and group adjutant, air electronics officer at both squadron and base level.

In June 1958 he entered missile training on the TM-61C Matador missile with the 4504th MTW, Orlando AFB, FL and trained and deployed as operations officer, with the 58th TMG, Osan AB, Korea. In August 1960, assigned to the Joint Bomarc Test Staff, Eglin

AFB, FL as instrumentation officer, Operations Section. Testing and developing the IM-99A Bomarc missile and later the IM-99B. In July 1963, on completion of Bomarc tests he was assigned as range safety officer, Eglin Gulf Test Range, Eglin AFB, FL and later as Range Scheduling Officer. In May 1968 he was assigned as political warfare advisor, Air Force Advisory Team-1, Tan Son Nhut AB Vietnam. His last assignment was executive officer, 22nd Air Defense Missile Sqdn., Langley AFB, VA from May 1969 to his retirement Oct. 31, 1969.

After leaving the AF he worked as a Contract Right of Way Agent for power companies. He is now widowed, has three grown children and resides in Apex, NC.

JOHN R. FORSELL, SMSGT (RET), assigned to the 351st MMS as NCOIC Mobile Maintenance Branch, responsible for four Mobile Team Sections and for the functions of the Van Configuration Section, also for the provisions of para.5-ld, Vol xvIII of SACM 66-12. He and Lt. Col. Don Auvil were responsible for the posturing of the first Minuteman ICBM's at Whiteman AFB, MO in 1963. Prior to his duties in the 351st SMW, he attended Ballistic Missile Analyst Technician School at Chanute AFB from February to June 1963.

From September 1960 to October 1962 he was assigned to the 4038th SW, B52G A&E SQD at Dow AFB, ME. NCOIC of Gunlaying Systems and responsible for training and records of over 400 technicians including newly assigned GAM-72 and GAM-77 technicians. While at Dow AFB, he attended the NCO Academy at Westover AFB, MA and was awarded the Military Achievement Award.

From August, 1957 through September, 1960 he assigned to the 10th TAC Recon Wing at Spangdalhem AFB, Germany and Alconbury, England and was authorized the "Air Force Outstanding Unit Award". He also attained the highest score in his AFSC of "158" for his technical specialty.

He was assigned to the 509th BW in June, 1949 and remained with the 509th for eight years at Walker AFB, Roswell, NM until June 1957. Transition of aircraft in the 509th began with the old B-29 Red-Tails, then B-50s and B-47s along with many SAC rotation TDYs to the United Kingdom and Guam. He loaded so many A-bombs, he's lost count. He was an (911) armorer the first five years in the 509th.

SMSgt. Forsell was assigned to the 93rd BW in September 1947 with a subsequent tour, June 1947 to June 1949 in the 4th FW at Andrews AFB and the Langley Field.

Forsell enlisted in the Army Air Force on Aug. 12, 1947 with subsequent change to the USAF in September 1947. He has seen the AF grow throughout the years into a highly competent technical force with personnel he is extremely proud to have served with.

Prior to his USAF career, Forsell served in the USN as a combat aircrewman ordnanceman gunner. He enlisted two months after his 17th birthday on Oct. 23, 1942. He graduated in the top 20% of his class and was awarded the rank petty officer, ordnanceman second class after five months in the USN. He was subsequently assigned to Banana River, FL for Combat Crew Training. After graduation he was assigned to VPB-211, Anti-submarine Patrol Sqdn. He logged over 2,000 hours flying time-day and night-encompassing Grid and convoy patrols. He was discharged on Feb. 13, 1946.

His awards include the Air Force Outstanding Unit Award, Good Conduct Medal Go 8 w/Silver Loop, Air Force Longevity Award w/3 OLC, NCO Academy Mili-

tary Achievement Award, Pacific Theater Ribbon, American Theater Ribbon and the Victory Medal.

After retiring from the AF (SAC), Forsell went to work for the US Postal Service Data Center in Wilkes, Barre, PA in August 1970. He performed duties as a telecommunications technician. He was a team member that installed Telecommunications Data Centers at St. Louis, MO; Raleigh, NC; Oklahoma City, OK and his home base Wilkes-Barre, PA. He retired from the US Postal Data Center after 22 years, combining with USN and AF of 22 years for a total of 44 years federal service.

He is married to the former Alice Javage of Port Blanchard, PA. They were married in Merced, CA on Oct. 10, 1947 when he was assigned to the 93rd BW at Castle AFB, CA. They have four children: Virginia, born Dec. 12, 1948 at Bolling AFB; John, born Dec. 14, 1949 at St. Mary's, Roswell, NM; Richard, born Sept. 12, 1953 at Walker AFB, NM. Fourteen years after Richard and after retiring from the AF along came Kenneth on Oct. 21, 1967 at Valley Forge Army General Hospital, PA. Alice and John celebrated their 50th wedding anniversary Oct. 10, 1997. They moved from Pennsylvania to Ocala Palms, FL June 1997.

CHADWICK (CHAD) FOSSEN, LT COL (RET), first entered the "missile business" in 1958 as assistant foreign liaison officer with the Royal AF on the Thor missile program at Douglas Aircraft Co. in Tuscon, AZ. While on the Thor program, he also set up support facilities for the Atlas students training at Convair Astronautics in San Diego, CA. He left missile-related duties and held navigation instructor and public relations assignments until graduating from Command and Staff College in 1962.

He then attended the Basic Missile Staff Course at Sheppard AFB and Minuteman training at Chanute AFB, enroute to Minot, NC. When he arrived at Minot there wasn't a missile wing there yet. He was soon joined by Col. Gil Friederichs and others, however, and served as combat crew scheduling officer until being assigned to the 4315th CCTS at Vandenberg AFB, where he instructed ORT in the MPTs, and the Modernized Minuteman program. He finished his tour at Vandenberg as commandant of the Ballistic Missile Staff Course. His final tour in the USAF was at F.E. Warren AFB, WY, serving as chief of the command post of the 90th SMW.

Fossen retired in 1971, but stayed in Cheyenne, where he held positions as executive director of the Chamber of Commerce, Public Affairs officer with the local TV station, Special Programs director with the Wyoming Association of Municipalities, and executive director of the Wyoming Governor's Committee for Employment of the Handicapped.

He and his wife, Enestine, moved to Boulder City, NV, in 1986, where they have been active as volunteers with the Lake Mead National Recreation Area.

GARY J. FOX, MAJ (RET), has the distinction of working with many different kinds of missiles. Enlisted in the AF in December 1954 and served from 1955 to 1957 as a student and then instructor on the Matador (TM-61C) and Rascal (GAM-63) guidance systems at Lowry AFB. From 1957 to 1959 assigned to FTD314 at both Orlando AFB and McCoy AFB as a field training instructor on the  Rascal (GAM-63). In 1959, as a staff sergeant, transferred to Amarillo AFB as a SNARK (SM-64) instructor. In December 1959 attended Officer Candidate

School (Class 60B). Graduated in June 1960. Transferred, as a second lieutenant, to 724th SMS, 451st SMW at Lowry AFB. Became the ground electronics officer on senior instructor launch Crew S-12. In 1965, as a captain, attended Electronics Staff College at Keesler AFB and spent three years at Torrejon AB in Spain as chief of maintenance, Spanish Communications Region. In 1969, assigned to 90th Communications Sqdn. as chief of maintenance. In 1970 was transferred to 2ACCS, Offutt AFB and as a major became chief of the Airborne Communications crews flying Looking Glass. In 1972 was transferred to HQ SAC DOKOC as chief of standardization of evaluation of the Post Attack Command and Control Systems (PACCS). During his five years at Offutt AFB flew "Looking Glass" on a regular basis. Also supplemented the SAC IG when it traveled to other PACCS bases where he had an opportunity to fly other PACCS aircraft. Also during these five years performed a dual role as an airborne launch control officer (ALCS) for minuteman missiles. He retired in February 1975 and began graduate work at the University of Nebraska at Omaha. He and his sons own Foxware Computer Systems Inc. and serve the needs of over 400 businesses in the Kansas City area. Spends most of his leisure time flying. A member and past president of the Heart of America Chapter of The Retired Officers Association (TROA), member and Past President of the Kansas City Chapter of the Missouri Pilots Association and member and past commander of the Kansas City Chapter of the Military Order of World Wars (MOWW). With seven children and 11 grandchildren, he and his wife Terrie are kept pretty busy.

ROGER FRAUMANN, TSGT, born Nov. 15, 1951 and joined the USAF Jan. 29, 1970. Basic training; academic counseling course, 1972; operation and maintenance of code inserter verifier course, 1971; missile elect. equip. spec. course, 1970; technical instructor course, 1972; management for AF supervisors course, 1974; NCO Leadership School, 1974; management training program for AF supervisors, 1974; AF weapons accident prevention management course, February 1975; SAC nuclear safety course, July 1975; instructional system development course, 1977; reading improvement course, 1977; test and measurement course, 1979; OJT manager/supervisor course, 1979; training supervisor course, 1979; AF effective writing course, 1981.

Left the service July 28, 1981 as E-6. Awards include Outstanding Unit Award, AF Outstanding Unit Valor, AF GCM w/3 devices, AFCM, AF NCO Professional Military Education Ribbon, AFLSR w/device and NDSM.

Currently is a business development director for NCR Corporation. He and wife Elizabeth have a son Alexander and daughter Kimberley.

GARES GARBER JR., LT COL (RET), graduated in the first class of the AF Academy in June 1959. Following a flying assignment at Biggs AFB, entered the missile field in 1962 at Malmstrom AFB where he served as a combat crew commander, instructor, and OIC of the Missile Procedures Trainer.

In 1966 he was assigned to the DOD Staff for Manned Space Flight Support at Patrick AFB, serving in Recovery Operations during the late Gemini and early Apollo programs. Following missile staff assignments at Vandenberg AFB in 1968 and SAC HQ from 1969-1971, he served at the Air Force Academy as air officer commanding, Exec for Honor and Ethics, and alumni secretary from 1971-76.

He then returned to the missile field at Ellsworth AFB as operation officer and, later, commander of the 67th SMS. In 1979 he was reassigned to the Academy as director of professional ethics. He retired in 1982 and moved to San Antonio with his wife Joan and two sons. He worked for USAA for ten years and now tutors children with learning disabilities.

JAY P. GATLIN, LT COL (RET), after receiving his bachelor of science degree from Texas Agricultural and Mechanical University in May 1960, he served as an AF officer from July 15, 1960, until July 31, 1980. Duties included seven years in the supply career field with such "coveted" remote assignments as Thule AB, Greenland (in support of the Ballistic Missile Early Warning System and the Nike missiles which protected it) and logistics advisor to the Royal Saudi AF in Dhahran, Saudi Arabia. He received operational ICBM training at Chanute AFB, IL, and Vandenberg AFB, CA, in 1967, and served as a missile combat crew commander in the 10th, 12th, and 490th, SMS at Malmstrom AFB, MT. He obtained his master of science degree in Aerospace Operations Management and served as instructor, evaluator, and staff officer of the 341st SMW.

In 1969, he reported to Offutt AFB, NE, as a member of the SAC Inspector General's traveling team. His areas of inspection included both Minuteman and Titan operations, as well as logistics. In June 1973, he reported to 8th AF HQ, Barksdale AFB, LA. He served as chief, Missile Div., Operations Plans. In 1977, he became Director of Missiles, 8th AF. This included responsibility for the Titan missiles at McConnell AFB, KS and Little Rock AFB, AR, plus the Minuteman missiles at Whiteman AFB, MO. He retired July 31, 1980.

He and his wife, Glenda, have resided on a farm near Mexia, TX, since 1980, and have established a private business known as Gatlin Enterprises. They have two sons and three grandchildren.

GETTY J. GEORGE III, LT COL, born June 23, 1939, Pittsburgh, PA, enlisted Aug. 25, 1961, USAF Auditor General HQ USAF.

His military locations: USAF Aud. Gen. Warner Robins AFB, GA, 1961-63, Tachikawa AB, Japan, 1963-66, and F.E. Warren AFB, WY, 1967-68; 91st SMW, F.E. Warren AFB, WY, MCCC, senior instructor and senior evaluator, 1968-73; Air Force Command and Staff College, Maxwell AFB, AL, 1973-74; 15th AF HQ, March AFB, CA, Chief of Missile Test and Training Div., 1974-79; 90th SMW, Minot AFB, ND, chief, Evaluation Div., 1979-81.

His memorable experiences: responsible for distributing tickets to the 64 Olympics in Tokyo to US forces in Far East. Enjoyed teamwork involved in transitioning

from MMI to MMIII-CDB at F.E. Warren AFB, WY; visiting FEW AFB after he retired and seeing the MMI, MMIII, and PK Missiles at the Main Gate knowing he pulled alert in MMI, put the first MMIII-CDB Missile on alert with Tom Jenkins and helped TRW and the AF deploy Peacekeeper.

He was awarded the Outstanding Unit Awards for USAF Auditor General, 90th SMW and 91st SMW, Marksmanship, AFCM, AFMS.

George was discharged Oct. 1, 1981 as lieutenant colonel. He has been married to Lori Lovely Lee for 37 years. They have daughters, Beth Ann and Joani Lynn. Joani Lynn has five children; Son, Getty IV, married, all living in Riverside.

Employed at TRW/SSD, San Bernardino, CA. He and Lori are enjoying these golden years together, enjoying their family and grandchildren, serving Our Lord and working for TRW.

WILLIAM C. GILBERT, born May 17, 1945 in Cloverport, KY, enlisted in the USAF/SAC, Aug. 22, 1963. His military locations and stations: Walker AFB, NM; 579th SMS, 6 SAW. He participated in Atlas F, MFT, Combat Crew.

He was discharged Aug. 22, 1967 as sergeant. He married Joan Gilbert and they have children, Kristopher and Nickolas.

Gilbert was employed with the Joint Commission on Accreditation of Healthcare Organizations.

THOMAS A. GILKESON, LT COL (RET), graduated from Virginia Polytechnic Institute and entered the USAF in 1959. As an AF officer from 1959 to 1981, his duties included launch authentication officer, 99th MMS, RAF Feltwell, England, deputy missile crew commander, 374th SMS, Little Rock, AR and crew commander, instructor, 490th SMS, Malmstrom AFB, MT. He was assigned to HQ SAC as a SIOP planner. Attended Air Command and Staff College and earned a masters in Public Administration. Assigned to Joint Strategic Planning Staff, Chief Combat Target Team, SAC. In 1975 assumed command 742nd SMS, Minot AFb, ND, followed by Naval War College, Newport, RI., His final assignment was with the Leadership Management and Development Center, Directorate Management Consultation, Maxwell AFB, AL.

He is currently research, monitoring and evaluation director with the Department of Corrections. Tom and Judy currently reside in Montgomery, AL.

JOHN E. GILMORE, LT COL, born in 1924 on a farm near Fox Lake, MN. Entered service in April 1942 at Ft. Riley, KS; completed bombardier training in February 1943 and was commissioned. RTU in B-24s and assigned to 93rd BGH 330th Sqdn. 8th AF.

He was shot down over Friedrichaven in March 1944 and was POW for 13 months and 13 days in various camps including Stalag Luft III.

Completed pilot training and flew primarily trans-

port and bombers, also site support test and operations while with the 549th SMS Omaha. Assigned to crew commander, Chief of Standardization, Chief of Operations. While conducting a dual propellant loading at Omaha site, the missile experienced a pneumatic explosion. 30 days later, launched from Vandenberg and the missile exploded at 44,000 feet. No crew error on either incident.

Received a degree from U of Omaha in 1964. Transferred to Vandenberg AFB, 576th SMS as asst. operations B site commander and retired in 1966 after an assignment as chief of maintenance for the 4392nd Communications Sqdn. He was with General Electric in Space at Vandenberg; jet engines at Evendale, OH; computers at Phoenix, AZ for a total of 4.5 years. Retired as L/C, command pilot, navigator, bombardier and missileer.

Entered the real estate field in 1970 and is active in the field of Lots Land. Resides with his wife of 54 plus years in Scottsdale, AZ. Their six children all in that area.

RONALD D. GRAY, BGEN (RET), born in Dallas, TX in 1942. After graduation from Texas A&M College in 1964, he was assigned to the 351st SMW at Whitman AFB, MO. Served as a Minuteman I and Minuteman Modernized crew member. In November 1967, he was transferred to Minot AFB, ND as part of the initial cadre for the Airborne Launch Control System, as a deputy and missile combat crew commander.

In February 1970, transferred to the 4315th Combat Crew Training Sqdn. In June 1972, Gray was transferred to SAC HQ and spent four years in the SAC Underground working on integration of warning systems into SAC Command Control and Emergency War Operation Plans.

In June 1976, Gray was assigned to the 381st SMW as a maintenance officer. Over the next eight years Gray served as chief, quality control, maintenance supervisor, maintenance control officer, squadron commander and deputy commander for maintenance.

In 1985, assigned to Air Force Space Command HQ as director, Missile Warning Plans. In 1987 was selected as commander for the 1st Space Wing. In 1989, assigned as director of Operations for AF Space Command.

Gray's final assignment was a commander, 21st Space Wing, at Peterson AFB, CO. Gray retired in September 1993.

He and his wife, Pat, live in Scottsdale, AZ. He works as president, CEO and chairman of the board of Republic Western Insurance Company.

LEROY V. GREENE, born 1936, Philadelphia, PA. Commissioned US Military Academy, 1959. Retired colonel 1982. Instructor launch control officer, 385th SMW, Offutt AFB. Instructor Missile Combat Crew Commander, 351st SMW, Whitman AFB. MBA University of Missouri. Commander, Combat Control Team, 4410 Combat Crew Training Wing, England AFB, LA. Psychological operations advisor, Vietnam. Commander, 319th SMW, F.E.Warren AFB, National War College. Chief, Missile Branch, DCS Plans & Operations and Chief, Missile & Nuclear Programs Div., DCS R&D, HQ USAF. Base commander, Whiteman AFB. Deputy for ICBM Acquisiton Logistics, BMO, Norton AFB. Business Development Manager, TRW, 1982-1997.

Launch Control officer for first Atlas missile ("Big John") launched by combat ready, operational SAC missile combat crew, February 1962. Project officer, Olympic Arena 1973, 90 SMW, won Blanchard Trophy—"best in SAC."

He married Gail Knapp in 1961 and they have four children and five grandchildren.

WALTER EARL GREENE, CAPT, born Sept. 11, 1929 in Indiana, enlisted in the US Army Air Corp, Aug. 27, 1947 and also served in the USAF. His military locations and stations: Dew Line ACFW, Germany and Alaska AACS, Minot, ND; 455 SMW Missiles SAC.

His retirement Aug. 31, 1967 was his memorable experience. He achieved the rank of captain USAF.

He was awarded the National Defense Service Medal, AF Commendation Medal, Army of Occupation, Germany, Armed Forces Reserve Medal, AF Outstanding Unit Award, AF Longevity Service Award w/4 OLCs, AF Good Conduct w/2 Bronze Loops.

Married to Ruth and they have four children: Donna, Donald, Diane and David. Employed with the University of Texas, Pan American, teaching international business.

JAMES A. GREENLAW, MAJ, born June 26, 1933 in Boston, MA, enlisted in the service May 9, 1951, at age 17, after graduation from high school. He entered USAF as an airman basic. Promoted through the ranks to technical sergeant working in aircraft maintenance. Returned from Laon AFB in France to attend Officer Candidate School in November 1957. As 2/LT was a/c maintenance officer (Chennault AFB, LA and Selfridge AFB, Michigan). Completed bachelors degree through Operation Bootstrap at University of Nebraska, Omaha (1962). Trained at Chanute AFB, and Vandenberg AFB as a Minuteman I launch officer (1963). One of first operational crews at Minot AFB (Senior Missile Badge); 455 SMW (August 1963). Received an MS in Industrial Management AFIT/University of North Dakota (1967). Assigned as contract negotiator AF Systems Command Hanscom AFB MS (1967-1971).

Greenlaw was discharged as major Nov. 1, 1971. He is married to Carole Greenlaw and they have a son, Mark and daughter, Kim and six grandchildren, with one on the way. Fulltime management professor at Northeast Louisiana University (11,000 students) Monroe, LA. During the summer they fly the beaches of Florida, hoping to buy a condominium in the future.

CHARLES F. GREGORY, LT COL, born in Walton, NY, Dec. 6, 1954, married to the former Sandra Seeberg of Rockford, IL. Sandy, a USAF captain and "Top Hand" officer currently assigned to the 576th FTS. They have a daughter, Michelle, age 13.

Service history: present, 14th AF inspector general, Vandenberg, AFB; 1997, commander, 30th Range Sqdn., Vandenberg AFB; 1995-96, Chief, Strategic Planning, 30th Operations Gp., Vandenberg AFB; 1994-95, wing executive officer, 341st MW, Malmstrom AFB; 1994-94, deputy team chief, Military Liaison Team, Ministry of Defense, Albania; 1992-94, operations officer, 10 Missile Sqdn., Malmstrom AFB; 1991-92,

chief, Missile Control Div., Malmstrom AFM; 1990-91, chief, Wing Inspections, Malmstrom AFB; 1989-90, assistant operations officer, 490 SMS, Malmstrom AFB; 1986-89, chief, Navy Tomahawk Mission Planning, HQ EUCOM/J5N, Patch Barracks, Germany; 1983-86, GLCM Mission Planner, High Wycombe AS, England; 1982-83, GLCM Training Development Manager, 868 TMTS; Davis-Monthan AFB; 1981-82, Titan II DMCCC, MCCC, instructor MCCC; 390 SMW, Davis-Monthan AFB.

Education: PME: Air War College, Air Command and Staff College, Marine Corps Command and Staff College, Squadron Officer School.

He received a MS in public administration, Troy State University; 36 graduate hours in psychology, University of Northern Colorado; BA in history and secondary education, SUNY Cortland.

Significant career accomplishments: Meritorious Service Medal, 2nd OLC, AF Commendation Medal, Second OLC, Joint Service Commendation Medal, Meritorious Service Medal, 1st OLC, Duane W. Hollis Memorial Award for best Missile Control Div. in ACC, Defense Meritorious Service Medal, two AF Commendation Medals, 1st OLC, Meritorious Service Medal

SANDRA M. GREGORY, CAPT, born in Rockford, IL, Sept. 6, 1957, married Lt. Col. Charles Gregory of Walton, NY. Chuck is the 14th AF inspector general.

"Top Hand", Space and Missile Test Operations Officer, 576th FTS, Vandenberg AFB, August 1997-present; Quality Assurance flight commander, 576th FTS, Vandenberg AFB, March 1996-August 1997; maintenance supervisor, 30th Maintenance Sqdn., Vandenberg AFB, December 1995-March 1996; Peacekeeper maintenance flight commander, 30th Maintenance Sqdn., Vandenberg AFB, October 1995-December 1995; Logistics Management Graduate Student, AF Institute of Technology, Wright Patterson AFB, May 1994, September 1995; graduated top of her class. Distinguished graduate and the 1995 winner of the Louis F. Polk Award; Chief, Shops Maintenance Flight, 341st Field Missile Maintenance Sqdn., Malmstrom AFB, November 1993-May 1994; Chief, Team Training Flight, 341st Maintenance Support Sqdn., Malmstrom AFB, January 1993-November 1993; Chief, Maintenance Programs Flight, 341st Maintenance Support Sqdn., Malmstrom AFB, December 1991-December 1992; Instructor Crew Commander, 341st MW, Malmstrom AFB, July 1991, November 1991; Flight commander, 341st MW, Malmstrom AFB, July 1991-November 1991; Flight commander, 490th SMS, Malmstrom AFB, March 1991-June 1991; Missile combat crew commander, 490th SMS, Malmstrom AFB, February 1990-February 1991; senior evaluator deputy combat crew commander, 341st SMW, Malmstrom AFB, July 1989-January 1990; evaluator deputy combat crew commander, 341st SMW, Malmstrom AFB, October 1988-June 1989; deputy missile combat crew commander, 490th SMS, Malmstrom AFB, February 1988-September 1988.

AFIT: MS degree in Logistics Management (GPA=3.987), 1995, Graduate School of Logistics and Acquisition Management, AF Institute of Technology. PME: Squadron Officer School in residence, 1992.

Received masters in business administration (GPA=4.0), 1990, University of Montana. Bachelor of science degree in accountancy, 1979, Northern Illinois University.

1997, Squadron Company Grade Officer of the Quarter; 1995, Distinguished Graduate, Missile Maintenance Officer Course; 1997, AETC Commander's Award (Missile Maintenance Officer Course); 1996, Nominated for Lieutenant General Leo Marquez Award as the Outstanding Missile Maintenance Company Grade Manager of the Year; 1996, AF Achievement Medal; 1996, Maintenance Sqdn. Company Grade Officer of the Quarter; 1995, Louis F. Polk Award, Air

Force Institute of Technology; 1995, Distinguished graduate, Air Force Institute of Technology; 1994, AF Commendation Medal, first Oak Leaf Cluster; 1992, Maintenance Group Company Grade Officer of the Quarter; 1991, Air Force Commendation Medal; 1991, 15th AF Crew Member of the Quarter; 1991, Squadron Company Grade Officer of the Quarter; 1987, Distinguished Graduate, Missile Combat Crew Initial Qualification Training.

WILLIAM J. GROSSMILLER III, COL (RET), born 1925, enlisted 1943, commissioned 1944. During WWII, he flew 17 missions in B29s, then B50s and B47s in SAC. In 1958 he trained as a Jupiter MCCC. In 1959, he was assigned to 15 AF as chief, missile standardization. He and a plans officer, delivered the ELC checklists and codes to Malmstrom so that two crews could be the first 10 Minuteman on alert in 1961.

In 1963, he was the Chief Control Div., 451 SMW, Minot. After graduating from the USAWC in 1966, was assigned to the DCS, Programs and Resources, USAF as special assistant for JCS and Congressional Matters. In 1971 through 1972, was the 351 SMW commander, which won the Blanchard and operations trophies in successive years. After 30 years service Grossmiller retired in 1973. Retired in 1981 as vice president finance, Public Service Satellite Consortium,

JAMES S. GROVES, MAJ (RET), born in Viroqua, WI on March 29, 1942, graduated with honors from the University of Wisconsin-Madison in June 1964. He received a commission in the AF on Sept. 13, 1966 upon completion of Officer Training School in San Antonio, TX. He received his missile operations officer training at Vandenburg AFB, CA in 1973 and was assigned to the 67th SMS of the 44th SMW at Ellsworth AFB, SD. During that tour of duty he participated in a Minuteman II operational test launch at Vandenberg. He also served as an emergency actions officer in HQ US European Command in Stuttgart, Germany. He holds the Defense Meritorious Service Medal, AF Meritorious Service Medal, AF Commendation Medal and Combat Readiness Medal.

He retired from active duty on Sept. 1, 1987. Currently he is the guidance department chairman at Westview High School, Phoenix, AZ.

JOHN W. HALEY III, LT COL, born Feb. 3, 1935, Wichita Falls, TX. Enlisted as aviation cadet May 28, 1954; ADC radar observer, 1956-57, 46 FIS, DE; MATS navigator, 1959-61, Dover AFB, DE; outstanding student (#1) Titan training, Sheppard AFB, TX, 1961-62; 390 5 MW, 571 5 MS, Davis Monthan AFB, 1963-67;

MCCC S-117; S-120 C-130 Nav., RVN, 1967-69; HQ SAC, Msl. Plans, 1970-73; Commander, Minot AFB, ND, 1974-75; HQ SAC, MSL/Mun Plans, 1976-77; Chief Log. tests AFTEC, Kirtland AFB, NM, 1978-82; Curtain Raiser, 1st Missile Competition, Commander 5-120;

Missions/Operations include 1967, Commando Vault (Drop 10K Bomb to helo bases in RVN from C-130); UN Congo Airlift Relief, 1961, in C-124.

Memorable experience: VP Humphrey's Secret Service Detail; when prompted by a question from Mr. Humprey, Haley replied he wasn't a regular officer because he was too old (age 33) and a few months later he was made a regular. It pays to be blunt.

Retired June 1, 1982. His medals and awards include three AFLA, Reserve Medal, Expert Markmanship, ARVN Cross of Gallantry w/Palm, two CRM, AGCM, NDSM w/OLC, AFEM, RVN Medal w/ seven Air Campaigns, MSM w/3 OLCs, Air Medal w/5 OLCs, AFCM, AFOUA w/OLC and the Navy/Marine PUC.

From 1982-88 he was a Logistical test and eval. director and program engineer, B-2 Bomber, Northrop Corp., and he is currently doing research toward ancient civilizations use of modern technology.

Haley and his wife of 41 years, Viola, have three sons and four grandchildren. They reside near Albuquerque, NM.

FRANK N. HALM, COL (RET), born June 8, 1922 in Whittier, CA, enlisted in the Army Air Force June 1942 as aviation cadet. He served with the 8th AF, 94 BG, UK, France, Germany, Bolling AFB, JRS, 451 SMW, HQ SAC, PAS Oregon State University, B-17 pilot, 30 missions (20 leads), ETO WWII.

He participated in Cuban Missile Crisis, Titan I site commander, Dep Co 725 SMS, 451 SMW 1960-65.

His memorable experiences: memorable 30 year career, from PT-22 to Titan I, 457 SMW; 10 years in DC area; four in Germany; two UK/France; eight years in SAC. Final four plus years as commander AFROTC Oregon State University.

He was awarded the Distinguished Flying Cross, Air Medal w/4 OLC, several lesser. Retired Sept. 1, 1972 as colonel, USAF. Halm is married to Dottie and they have two adult children. He is employed in real estate investments. Founder 94 Bomb Group Memorial Association, co-editor Quaterly Nostalgic Notes (94BG) in 1975 through 1999.

ALONZO (AL) HALL III, SMSGT (RET), is a member of the Air Force Association of Missileers. After attending the Baltimore Polytechnic Institute in Baltimore, MD, he enlisted in the AF in October of 1974. Upon completion of basic trainng at Lackland AFB, TX, he attended the Missile Systems Analyst Course at Chanute AFB, IL. His first assignment was the 44th SMW, Ellsworth AFB, SD. His duties were: Site Security Maintenance Team member and chief; Noncommissioned Officer in Charge (NCOIC) of the Site Security Maintenance of Electro-Mechanical Team (EMT) Instructor Section, and technical engineering technician.

In January 1987, he was reassigned to the 485th TMW, Florennes, Belgium, as NCOIC Specialist Training Branch. He was responsible for conducting initial and proficiency nuclear operations and certification training for the first Ground-Launched Cruise Missile deployment on the European continent. While assigned he was promoted to the grade of master sergeant through the Stripes for Exceptional Performance (STEP) program.

He was reassigned in January 1988, again to the 44th SMW as assistant, NCOIC EMT and Job Control Sections; superintendent of the Job Control and Quality Assurance Flights. He was selected in 1989 to be a team member for the US On-Site Inspection Agency, performing numerous treaty verifications, inspection, and observed the elimination of various types of ballistic nuclear missiles while on location in the former Soviet Union.

Upon the deactivation of the 44th MW on July 4, 1994, he was reassigned to the 90th Maintenance Sqdn., F.E. Warren AFB, WY as superintendent, Aerospace Ground Equipment Flight. He retired June 1995.

He has an associate in science in Liberal Arts, University of the State of New York; associate in Applied Science in Electronic Systems Technology, Community College of the AF; bachelor of science in Applied Management, National College and bachelor of science in finance, University of the State of New York.

His decorations include the Meritorious Service Medal w/2 OLC, AF Commendation Medal w/OLC, AF Achievement Medal w/OLC. He is an Intercontinental Ballistic Missile Master Missileman, instructor, team chief, and technician. In addition, he was selected as the 44th Missile Wing's Senior NCO of the Year in 1990 and 1994.

He and his wife, Nancy and daughter Tracy, reside in Rapid City, SD. He is the technical support and information systems manager at Mikohn Gaming Corporation, Rapid City division.

M. KENNETH HAMILTON, MAJ, born June 17, 1961, Wheeling, WV. Upon graduation with honors from the University of Akron in 1984, and receiving a regular commission as the AFROTC Distinguished Graduate, he entered Undergraduate Missile Training at Vandenberg AFB. After UMT, he was assigned to SACs 90th SMW, F.E. Warren AFB, WY, home of the "Mighty Ninety." While holding numer-

ous missile combat crew commander positions, he received a variety of awards, including 90th SMW Junior Officer of the Year (finalist, twice), 90th SMW Top Warrior Competition (finalist), 90th SMW Best Operations Crew (runner-up), 90th SMW Best Operations Crew Commander, SAC Olympic Shield Missile Combat Operations Competition (finalist), and SAC honor roll of most professional crew commanders.

Following crew duty, he served in various acquisition and logistics positions at Griffiss and Robins AFBs. Stationed at Wright-Patterson AFB since June 1995, he has served on the AF Materiel Command headquarters staff in the following capacities: team leader, space and ICBM System Requirements: Chief, Acquisition & Logistics Officer Assignments; and his current position as Acquisition Reform Staff Officer, Team Leader for Logistics.

Awarded the Meritorious Service Medal w/OLC, AF Commendation Medal, AF Achievement Medal, Combat Readiness Medal, Humanitarain Service Medal w/ Bronze Star, National Defense Service Medal, Organizational Excellence Award w/2 OLCs, and the AF Outstanding Unit Award.

His most memorable experience while a SAC missileer was the highly successful launch of a wonderful relationship with his future wife, Lori Lovelass, whom he met in Cheyenne, WY and married on June 15, 1990, in Watseka, IL. They presently reside in Huber Heights, OH. They have two children, Sara Brooke (five) and Andrew Kenneth (two).

JOHN F. HAMPTON, COL (RET), entered the Missile Field in 1959 as an instructor in the Titan I System at Sheppard AFB, TX. Following Titan I, he instructor courses in Orbital Mechanics.

He commanded FTD 530 S (Missile and Space Systems) at Vandenberg AFB, CA, served as the commander, 446th SMS, and deputy wing commander for

the operations of the 321st SMW at Grand Forks AFB, ND. Colonel Hampton's final missile assignment was as wing commander, 381st SMW, McConnell AFB, KS.

He is a graduate of the University of Rhode Island, Air Command and Staff College, and the Naval War College. He is presently a member of the board of directors, Strategic Air Command Museum, Ashland, NE.

Hampton and his wife Barbara reside in Bellevue, NE and Port Richey, FL.

WAYNE N. HANSEN, COL, is the vice commander of the 341st Space Wing, Malmstrom AFB. He graduated from Augsburg College, MN, in 1968. After OTS he was a MMIII crew member at F.E. Warren AFB. He is served in missile staff assignments at Vandenberg AFB, HQ SAC, and the Pentagon. He served as deputy group commander at Eglin AFB, squadron commander at Suwon AB, Korea, and as group commander at Osan AB, Korea. After Air War College in 1992, he commanded the Support Group at F.E. Warren AFB. In 1994 he was deputy director of plans at HQ AFSPC, Peterson AFB. He assumed his present duties in July 1995 and is currently a mission ready commander in the Minuteman III/CDB/REACT/Deuce weapon system pulling monthly alerts . . . right back where he started.

He is married to the former Lois Batalden of Lamberton, MN. They have one son, Zachary.

BRIAN A. HANSON, MAJ (RET), born Oct. 19, 1957, at Cincinnati, OH, graduated from the University of Nebraska-Lincoln in May 1981, commissioned through AFROTC and served from Aug. 27, 1981 to Aug. 31, 1998. Completing initial training (CDB-78) at Vandenberg AFB, CA, he reported to the 321st SMS, F.E. Warren AFB, WY, serving as DMCCC, deputy flight commander and MCCC. He also

served as 90th SMW Standardization/Evaluation DMCCC and MCCC in both Minuteman III and Peacekeeper systems.

In 1984, he helped train the victorious "Olympic Arena" operations team and significantly contributed to the 90th SMW winning the SAC "triple crown" (Omaha, Riverside and Blanchard trophies). Selected by CINCSAC as one of ten original members of the Peacekeeper control center to meet initial operation capability as directed by President Reagan. As Peacekeeper Senior Standardization/Evaluation MCCC he selected, trained and evaluated crew members and evaluators and amassed a wing-record nine consecutive "highly qualified" evaluation ratings.

In August 1987, he returned to Vandenberg as an operations evaluator assigned to 3901st SMES. While at SMES, he participated in unit assessments, serving as operations evaluator and support staff during 1988-91 "Olympic Arena" competitions, provided developmental interface with Peacekeeper Rail Garrison, Small ICBM and REACT missile programs, and will sole MAJCOM technical order review/approval authority for all modifications and upgrades to ICBM systems as chief, technical services/support branch. He was one of three officers qualified to evaluate in all Minuteman configurations and Peacekeeper.

In August 1991, he transferred to the White House Military Office, Washington, DC, serving both the Bush and Clinton Administrations as a presidential emergency operations officer, director of the President's Emergency Operations Center and contingency Military Aide. In November 1994, he transferred to US Strategic Command at Offutt AFB, NE, where he served as chief ABNCP Planning Staff, operations planner/SIOP advisor, Standardization/Evaluation Team, completing 150+ missions on the EC-135C "Looking Glass" aircraft.

He completed Squadron Officer School, Marine Corps Command and Staff College and MBA coursework at the University of Wyoming. He currently resides in Omaha, NE.

BOBBY J. HARALSON, MSGT, born Sept. 15, 1933, near Adona, AR. Entered USAF in 1952, retiring in 1977 as master sergeant. Assigned 97th BW at Biggs AFB, TX, 1953-56. Separated October 1956. Recalled active duty June 1958-62 at Schilling AFB, Salina, KS.

Assigned Little Rock AFB, AR, December 1962-August 1971 with the 308th SMW. He served on a launch crew until June 1968. He was a maintenance chief until August 1971 when he was selected as a missile facilities technician evaluator with the 3901st at SMES, Vandenberg AFB, CA. He transferred back to Little Rock AFB in December 1975 with the 308th SMW.

Haralson was employed at the University of Arkansas for Medical Sciences and retired Jan. 3, 1997, as associate director of physical plant. He lives near Jacksonville, AR, and has been married almost 40 years, has two sons and one granddaughter.

BRUCE S. HARGER, COL (RET), began his missile career May 1964 as a deputy crew commander in the 68th SMS, Ellsworth AFB. In 1966 he served as chief of training for the Titan IIIB Military Launch Team, 6595 ATW (AFSC) Vandenberg AFB. In 1967 he went to the USAFA physical education instructor and coach. From 1969-71 USAFA and AFIT sponsored his doctorate in exercise physiology at The Ohio State University.

He returned to USAFA from 1971-75 as director of the Human Performance Laboratory. In 1975 he attended the Armed Forces Staff College in Norfolk, VA and then returned to missiles as the chief of EWO plans at F.E. Warren AFB. In this assignment he also was chief of EWO plans and training (DOX), Transportation Squadron commander, and assistant deputy commander for Resource Management. 1979-80 he commanded the USAFE/NATO Detachment at Murted, Turkey. During 1980-83, he was chief of basing for GLCM at HQ USAFE. Promoted to colonel in 1983 he served at Malmstrom as ADO, DO, and vice commander 341SMW. He was director, personnel programs, HQSAC, until retirement in 1988.

He and his wife, Cleo have resided in Springfield, MO since 1988. He is currently the director of Athlet-

ics and Chair, Exercise Science Department at Drury College.

RAYMOND C. HARLAN, MAJ (RET), is a communication consultant/trainer and technical writer. He entered the AF after completing an MA in English from the University of Texas in 1968.

He began his first assignment, at Malmstrom AFB as a deputy crew commander. He progressed to wing senior instructor crew deputy, then upgraded to a crew commander/instructor position. From 1972 to 1976 he taught AFROTC at Bradley University in Peoria, IL. While there he earned an MA in speech and Theatre Arts from Bradley. From 1976-1981 he taught in the English Department at the Air Force Academy, progressing from instructor to assistant professor and course director of Advanced Composition. In 1981 he returned to missiles at F.E. Warren AFB, where he was chief of the Instruction Systems Div., then chief of the Codes Div. While he was in that position, his organization was selected as the best codes division in the 15th AF. In 1985 he moved to AFIT at Wright-Patterson AFB to become the program manager of the Minuteman Education Program.

Since 1988 he and his wife, Linda, have lived in Aurora, CO, where he is president of ComSkills, a consulting and training firm. He has published three books: Telemarketing That Works (co-author, Probus, 1991; Malaysian edition, Golden Books Centre SDN.BHD, 1992); The Confident Speaker (McGuinn & McGuire, 1993), and Interactive Telemarketing (co-author, McGuinn & McGuire, 1996). His editorial, "The Need to Study War", was published by Newsweek, Sept. 1, 1980. His biography is carried in Marquis Who's Who in the Media and Communications, and Who's Who In Finance and Industry. In his spare time, he skis and does volunteer work for the Lutheran Church.

HOLMES L. HARTLEY, MAJ (RET), graduated from the University of South Carolina in August of 1954. He entered the AF on Dec. 29, 1954 in the pilot training program. After completing his training he had the following duties: bomber pilot, Walker AFB, NM; pilot, Moran AB, Seville, Spain; Weapons controller instructor and pilot, Charleston, AFB, SC; Missile combat crew commander, 556th SMS, Plattsburg, NY; computer systems analyst, HQ ADC, Colorado Springs, CO; air combat operation officer, Da Nang AFB, Republic of Vietnam; Systems Analyst HQ, NORAD, Colorado Springs, CO.

He has a BS degree from the University of South Carolina, an MA degree from Lacaze Academy of Language, Washington, DC; and associate degree from Pike's Peak Community College, CO. He completed Squadron Officers School, the Command and Staff courses. He retired Feb. 29, 1976 and resides in Colorado Springs, CO.

LAWRENCE HASBROUCK, COL, born May 20, 1931, New York City, entered the service in 1953, retiring as a colonel in 1983. Master missileman and senior officer aircrew member classifications.

He was an instructor and the senior instructor for the 44th SMW, Ellsworth AFB, SD, 1962-68. Became the senior instructor of the Airborne Launch Control System (ALCS during its activation at Ellsworth, 1967-68. Was chief of the ALCS system at SAC HQ, Offutt AFB, NE, 1968-72. Participated as launch director for

37 ALCS launches from Vandenberg AFB during this time frame. Became operations officer and commander of the 510th SMS and assistant director of operations and director of operations, 351st SMW, Whiteman AFB, MO, 1972-75. Commanded the AFROTC Detachment at Lehigh University 1975-80. Final assignment was as the AFROTC Northeast Area Commandant, 1980-83. This area expanded by seven new units during this period.

Awarded the Air Force Legion of Merit w/OLC, Meritorious Service Medal, Commendation Medal w/2 OLCs and Combat Crew Medal w/OLC.

Upon return to civilian life he remained in the academic arena serving in faculty and administrative positions at colleges and universities. Is presently the manager of a Prudential Real Estate Affiliate.

Married in 1971 to Betty Crowson and presently residing in Ocean Springs MS.

DENNIS M. HEITKAMP, COL, born Dec. 8, 1934 in New Braunfels, TX, enlisted in the USAF, August 1956. He was selected as part of the Cadre of Officers assigned to establish the Nation's second Minuteman I Wing, the 44th SMW, Ellsworth AFB, SD as deputy missile combat crew commander, crew commander, instructor and evaluator from January 1963 to December 1967. Assigned 774th Tactical Airlift Sqdn., Mactan Island and Clark AFB, Philippine Islands. Flew over 320 combat missions over South Vietnam in C-130B aircraft. In 1974 assigned as operations officer 571st SMS and in 1976 squadron commander 570th SMS (Titan II), Davis Monthan AFB, AZ. In 1977 assigned as deputy commander for operations in 1978 vice commander 341st SMW (Minuteman II and III), Malmstrom AFB, MT. Assumed command of the 341st on June 5, 1981 and held the position until retirement June 1938.

He was awarded the Legion of Merit, Distinguished Flying Cross, Meritorious Service Medal w/OLC, and Air Medal w/8 OLC.

Discharged June 1983 as colonel. He is married to the former Jacqueline Bouquet of Iowa, LA. They have two children, Mark and Denise and four grandchildren: Michael Christy, Erika Heitkamp, Michele Christy, and Mara Heitkamp.

Employed as vice president and trust officer Chase Bank, New Braunfels, TX. President, Wurstfest Assocation, first vice chairman, Greater New Braunfels Chamber of Commerce, life member and Blue Coat Greater New Braunfels Chamber of Commerce. Trustee, Comal Healthcare Foundation and Chairman Community Gifts, New Century of Caring Campaign, Trustee, Braunfels Foundation Trust. Recipient of the Bessering Award for an individual providing significant contributions to the New Braunfels Community through civic activities.

GERALD (JERRY) D. HELLINGA, CAPT (RET), began his military career in the US Army in 1972. After leaving the Army, he returned to college and received his BA from Northwest Nazarene College. After graduation, he went to Officer Training School, then to Vandenberg AFB, and then to the 321 SMW, Grand

Forks AFB, where he was a missile launch officer. In 1984, he took command of Detachment 6, Minuteman Education Program at Ellsworth AFB. From there he was assigned to HQ Civil Air Patrol, Maxwell AFB. While there he co-authored the textbooks, Aerospace: The Challenge and Aerospace: The Flight of Discovery, later becoming chief of senior programs until retirement in 1992.

In addition to his BA, he has an MA from Central Michigan University, an EdS from Troy State University, and PhD (ABD) from Trinity College and Seminary. Military education includes the Army Basic Armor Officers Course, the AF Squadron Officers School (C), and the Marine Corps Command and Staff College (C).

Since retirement, he worked for the CAP (1992-1996) and is currently the director of Military Programs for Barton County Community College, Ft. Riley, KS. He also teaches as an adjunct for Ft. Hays State University and the University of Phoenix.

He and Linda reside in Manhattan, KS. Their children: Sunny, Montgomery, AL; Amy, Rochester, NY; Brandy, Prattville, AL; and Paul, Troy State University.

SIR BENJAMIN S. HENDRICKSON III, born in Springlake, NJ, Jan. 23, 1931, served with USAF, Department of Defense, Rocket Propulsion Laboratory, Edwards AFB, CA; Ballistic Systems Div., Norton AFB, CA.

He was awarded Command Wings, Senior Missile Badge, Distinguished Flying Cross, Air Medal, AF Commendation Medal, AF Good Conduct, Army Good Conduct Medal w/3 knots, Army of Occupation Medal w/Japan Bar, National Defense Medal with Bronze Star, Foreign Service Bar, Korean Service Medal w/Arrowhead and two Battle Stars, Armed Forces Expeditionary Medal, United Nations Medal, AF Outstanding Unit Ribbon, AF Longevity Ribbon, AF Small Arms Expert Marksman Ribbon, AF Overseas Ribbon, AF Presidential Unit Citation Ribbon w/ 3 OLCs, Republic of Korea Presidential Unit Citation Ribbon.

AF Inspector, assigned to Douglas Thor IRBM 75 test facility and followed program from inception to deployment in Europe.

Reassigned to North American Rocketdyne for Atlas IBM Sustainer Development and Retrofit. Subsequently assigned to Convair General Dynamics for Atlas SM 65 "B" "C" & "D" Programs. Assigned position of project manager for A.F. Technical Orders Publication(s), Aero Jet General Titan & Titan II SM 68 missiles.

Upon completion returned to Edwards AFB as a tenant from Norton AFB, Ballistic Systems Div., with position of director, Liquid Rocket Propulsion Inspection and Instrumentation Category I Calibration, Certification Laboratory, with Test Facilities for Atlas Propulsion upgrade & overhaul. Commanded an Elite Liquid Missile Propulsion Trouble Shooting and Crises Intervention Team.

Cancellation of Atlas programs brought reassignment to NAS where he served on the propulsion project for the 'lifting body' ie M-2, MF-2 and X-24 and preparatory research for the Space Shuttle.

GLEN E. HENNESS, CMSGT (RET), born in Guthrie, OK, April 4, 1930, enlisted in the AF, Feb. 9, 1949 and completed basic training at Sheppard AFB,

TX. He stayed at Sheppard to complete the aircraft and mechanic school and May 15, 1950 he was assigned to the 42nd BS at Carswell AFB, TX.

There he served as a mechanic and crew chief on the B-36 aircraft until November 1957 at that time the 11th BW phased out the B-36 and the entire wing transferred to Altus AFB, OK. Here he became a crew chief and flight chief on the B52. The Atlas Missile System was installed at Altus and in June 1962 he began his missile career.

He was assigned to the 577th SMS and served as senior controller and NCOIC of job control for the next three years. When the Atlas Missile was phased out he was transferred to the 321st SMW at Grand Forks AFB, ND.

At Grand Forks he was the NCOIC of the Inspection and Evaluation Section of Quality Control. In June 1969 he was transferred to Richard Gerbour AFB, MO, where he was slated to become the NCOIC of the 10th AF inspector general team.

However, the team had been transferred to upper New York, so he convinced those in charge he had been in Arctic weather long enough and was subsequently assigned to the 381st SMW at McConnell AFB, KS, there he served as the Titan II NCOIC of quality control until December 1971. At that time he was sent to the 10th Air Defense Commands Detachment at Johnston Island where he was the missile maintenance superintendent on the Thor Missile System.

In December 1972 he was assigned to the 351st SMW at Whiteman AFB, MO. There he served as NCOIC of the Field Maintenance Sqdn. until retiring on Sept. 30, 1975.

He and his wife Della, their two daughters and five granddaughters reside in the Warrensburg area. After retirement he worked on the Harrier fighter training Spanish maintenance personnel.

When this was completed he began college in March 1978 and graduated in March 1980. He then worked in Kansas City as a loss control engineer for travelers and CNA Insurance Companies until January 1993. Since his final retirement he has been busy with his church, Masonic lodge and lots of golf.

OSCAR H. HERMANN, LT COL (RET), born March 1, 1922 in Berkeley, CA, enlisted in the USAF 1941. Trained rec. wings Foster Field, TX; Luke AFB, AZ; McDill, FL; Mtn. Home AFB, 13 years; Torrejon, Spain, Guam, Temphoff, Germany.

SAC Missions: Bruneau, ID, Titan Missile Site commander (60s) (He bought his same missile site after retirement.) Berlin airlift 1948.

Memorable experiences: a civilian was angry at the AF for slow payments to him for maintaining a communications phone line to missile site. Threatened to cut lines. This would (could) have caused havoc, perhaps even causing missiles to be launched. The shocking part was the ease that appeared to exist to nullify operations; security training pilots at Luke AFB, AZ, 1944; flew his favorite B-29 as command pilot B-47s.

Hermann retired October 1968 as lieutenant colonel. He was married and had a son and daughter. Deceased April 1994.

TOMMY H. HESTER, A2C-E3, born Nov. 16, 1942 in Rome, GA, enlisted in the USAF December 1960, serving with missile combat crew, BMAT, Shilling AFB, KS, 1962-64, 550 SMS; Sheppard AFB, TX, 1961, Tech. School.

He was awarded the Missile Badge, Senior Missile Crew S-50 and S56.

Discharged from service December 1964 as A2C - E3. He has been married 35 years to Shirley and they have a daughter, Jennifer; son, Jeffrey and grandson, Matthew. Hester retired from General Electric Oct. 1, 1997 after 32 years. He resides in Rome, GA.

LELAND (LEE) HIGLEY, MSGT, born Sept. 2, 1929, enlisted in the USAF Oct. 2, 1950. His military locations and stations: Lackland, Lowry, Kelly, Nouasseur, Smoky Hill, Hickam, Malmstrom, Truax Field, Hahn, Homestead, Bitburg, Portland International Airport, K-9 Korea, Taipei Taiwan, Spandahlem, Myrtle Beach.

Participated in British Christmas Island nuclear tests; Cuban Crisis, Homestead AFB, FL.

His memorable experiences: interface experiment, USAF F4E aircraft and Navy Walleye Missile, Bitburg, AB. Various tests with A1M9 Sidewinder Missiles and F104 aircraft directed by Warner Robins Depot.

Higley was discharged as master sergeant, Oct. 31, 1976. He is married to Amy and they have three children. Employed as supply inspector, General Electric Company, 1948-1950. Presently a realtor with Choice One Realty, BH&G, residing in Tacoma, WA.

JACK G. HILDEN, COL (RET), spent his AF career in missile development, command and management positions. In 1954, after receiving his bachelor of science in chemical engineering from the University of Wisconsin and his masters of science in chemical engineering from the University of Michigan he worked on development of liquid rockets for Atlas, Thor and Navaho missiles at Wright-Patterson AFB Power Plant Laboratory.

He was Minuteman assembly and propulsion officer at Cape Canaveral from clearing palmetto in the launch area in July 1959 through the first 30 development launches. He later was project officer for Minuteman III first, third and fourth stages at the Ballistic Missile Organization and chief of the Air Staff Ballistic Missile Requirements Office. In 1972 he became deputy commander for Maintenance of the 341st SMW and later vice commander of the 44th SMW. Before retiring in 1978 he was Ballistic Missile System manager and director of Material Management at Hill AFB.

As vice president Advanced Programs at Thiokol Corporation he then managed Peacekeeper first stage and Castor 120 solid rocket development and production programs. Since retiring from Thiokol in 1994, he and his wife Jeri have enjoyed summers in Ogden, UT and winters in Yuma, AZ.

JOHN R. HILLIARD, COL (RET), is retired and currently resides in Fort Washington, MD. He is currently the vice president for programs for the National Military Intelligence Association.

After graduation from the Virginia Military Institute in June 1960 with a BS/EE, he spent four years at F.E. Warren AFB in Cheyenne, WY. He was assigned as a launch control officer with the Atlas D and later the Atlas E ICBM. He was assigned to the 389th SMW as a standardization crew member for the Atlas E evaluating crews of the 549th and 566th SMS.

In 1964, he was transferred to the 6595th Aerospace Test Wing as a launch controller. He was involved in 31 launches from Space Launch Complex-1 and 3 at Vandenberg AFB with the Thor-Agena, Atlas-Agena and Thor-Delta. During the period, he received his MS from University of Southern California.

In June 1968, he was assigned to SAF/SS with duty in Rochester, NY. In 1971, he reported to Andrews AFB assigned to Intelligence and later the AFSC Command Secretariat. In 1975, he was transferred to HQ AF/IN as chief, missile and Space Section. He worked for the Director Central Intelligence as chairman of the Weapon and Space Systems Intelligence Committee, Space Subcommittee.

In 1979, he transferred to the Defense Intelligence Agency as Chief Ballistic Missile Systems Branch. He was also assigned as the OSD advisor to the US-USSR Standing Consultative Commission in Geneva, Switzerland. His last assignment was the director, Threat and Technology at Andrews AFB until his retirement in 1983. His awards include Defense Meritorious Service Medal, Meritorious Service Medal w/3 OLCs.

From 1983 to 1995, he worked as a manager and vice president for Aerospace Systems for ANSER in Arlington, VA. He retired in 1995.

The most memorable experiences included: first dual-propellant-loading of an Atlas missile; first launch of a satellite aboard a Thor-Agena; opening up a classified payload capsule from space; setting across the table from the Soviet Union negotiating arms control; briefing/discussing the possible compromise of a classified document with Senator Sam Nunn.

He and his lovely wife, Linda live in Maryland. They plan to relocate to Satellite Beach, FL in the near future. They have four children and seven grandchildren.

HUBERT T. (HUGH) HINDS, COL (RET), born March 5, 1938 in Springfield, IL. After graduation from DePauw University in June 1960, he began active duty on Aug. 4, 1960. He served in a variety of AF assignments, mostly in nuclear weapons operations, until his retirement in September 1983. He served first in NORAD/Air Defense Command and in December 1963, he began his career in missiles. He was a DMCCC and an instructor MCCC for Minuteman II at the 510th SMS at Whiteman AFB, MO where he also earned an MBA degree from the University of Missouri. He was also the 351st SMWs Chief EWO instructor and deputy chief, Plans Branch. He survived the infamous "Bloody August" in 1968 when the highly complicated Controlled Time Launch EWO procedures almost caused the wing to fall below minimums of qualified launch crews. After Air Command & Staff College in June 1971, he served as a missile operations planner at HQ SAC and the Joint Strategic Target Planning Staff (JSTPS). In that position, he developed tactics for ICBM and SLBM employment in the Single Integrated Operational Plan (SIOP). In 1975, was assigned to the Air Staff as a missile operations officer, developing operational requirements for ICBM systems R&D - especially the MX system. In 1979, he was selected to be a visiting scholar at MITs Center for International Studies in the Air Force Research Associate Program. After MIT, served again on the Air Staff as an operations planner in Long Range Plans and in early 1981, was selected as the deputy director, Strategic Forces Policy, within the Office of the Under Secretary for Police, Office of the Secretary of Defense. In that

final assignment, he developed employment policy and acquisition policy for US nuclear weapons systems and their associated C3I systems.

After retirement in 1983, he has worked as a defense analyst in the Washington, DC area as a systems analyst and project manager. He is also a registered commodity futures broker, specializing in defined risk trades and stock index futures.

Hinds and his wife, Ann, have lived in Great Falls, VA since 1984, and have four grown sons and three grandsons. He is active in his church choir as a tenor and serves on the board of a directors of a small company that manufactures solar water heaters and a Ski Condominium Association in Snowshoe, WV. He remains active in stock and mutual fund investments and in commodity futures trading for his own account.

BENJAMIN W. HINES, born May 14, 1933 in Alexander County, NC, enlisted in the USAF Nov. 20, 1952. He served as aircraft mechanic, 1953-55; B-47 crew chief and flight chief, 1955-62; Missile (Minuteman) team chief (1962-66); missile maintenance supervisor (1966-78); Wing Senior Enlisted Advisor (1978-83).

He received his associates degree in business administration from Wilkes Community College in 1986 and graduated from all available professional military education schools.

His memorable experiences: attended and took part in SAC Bomb Comp 57-60 (B47 crew chief) and SAC Missile Comp 1969-78 (trainer-evaluator NCOIC) a total of 14 years in competition, likely a record.

His awards consisted of: Airman of the six month; 1st Master Crew Chief in 2nd AF and Strategic Air Command; B-47 Crew Chief of the Year 8th AF; AF Commendation Medal three times; Meritorious Service Medal; Outstanding Unit Award five times.

He was married to the late Gwen Hope Kerley Hines for 40 years, nine months and seven days. They have five children, all have college degrees and 14 grandchildren.

Community involvement: member and chairperson of American Cancer Society, 1984-94; member and deacon of First Baptist Church, Taylorsville, NC 1984-present; member and president, Ellendale Ruritan Club, Taylorsville, NC, 1984-present; member and officer of Taylorsville Lions Club, 1986-present; member and committee leader, Alexander Sesquicentennial, 1996-97; work with Boy & Girl Scouts of America—Alexander troops, 1986-present; recipient of Girl Scout Community Commitment Award, March 1997; member and leader in the Bicentennial of U.S. Constitution, 1987-91; conducts presentations in local schools on life and times as Register Deeds and a military career with USAF 1986, present; conducts a history of our Flag Ceremony, 1986-present.

North Carolina Association of Registers of Deeds: was instrumental in acquiring a NCAROD lapel pin; the prime mover to obtaining a NCAROD Banner and Flag; obtained and distributed the NCAROD flag desk sets to North Carolina Registers of Deeds; served as District II chairperson for (2), two-year terms; on the Executive Committee has served as historian, secre-

tary, treasurer and currently serving as second vice-president.

RONALD W. HIRTLE, CAPT, is a contract specialist, 50th Contracting Sqdn. 50th Space Wing, Falcon AFB, CO. Currently works on the Wing Communication and Operation Maintenance Contract. This effort is the largest outsourcing effort in the AF Space Command. His unit received an excellent rating during an IG ORI, and an Outstanding Performer Team Award.

Captain Hirtle was commissioned through the Reserve Officer Training Corps at the University of New Hampshire in 1988. His career has included an assignment in missile operations, as a combat crew commander/instructor. In this capacity he received recognition as an outstanding instructor from the inspector general, and his unit attained an outstanding rating during an IG ORI. Hirtle was then selected for the AFs Education With Industry program. He performed his duties as a contracts manager on a Department of Justice solicitation and other AF contracts and was then assigned to a Systems Program Office as a contract manager/negotiator. He worked on AF-wide Air Combat Command and Air Mobility Command contracts, including the Global Command and Communication System, the Theater Battle Management Core System, and the Command and Control Information Processing System Contract. Captain Hirtle was also a member of the Patriot Honor Guard. The only all officer honor guard flight in the Air Force. He assumed his current position in July 1997.

Received his bachelor of arts degree in Political Science and psychology, University of New Hampshire, New Hampshire, 1988; master of science degree in administration, health care, Central Michigan University, Michigan, 1992; Squadron Officer School, Maxwell AFB, AL, 1996; master of business administration, Acquisition and Contracting, Western New England College, Massachusetts.

Hirtle was awarded NCMA-Certified Professional Contract Manager (CPCM); NCMA - Simplified Acquisition Specialist (SAS) designation; Air Force Commendation Medal; Combat Readinesss Medal. He has one child, Alexander.

JOHNNY HONEYCUTT JR., MSGT (RET), born Dec. 24, 1926, Lynch, KY, enlisted in the USAF, 1950, pay clerk, Kirtland AFB, NM. 1952, trained as a stock records clerk, supply, Kelly AFB, TX, 1952, 75th Air Depot Wing, Korea, 1953, FEAMCOM, Tachikawa, Japan, 1955, Kirtland AFB, NM, 1957, personnel clerk, SAC-Mike, AF Ballistic Missile Div., Inglewood, CA, 1959, Lowry AFBG, CO, Tech School,
TM-61 Matador Missile, Guidance Systems Tech., 1959, Asst. Crew Chief, 71st Tactical Missile Sqdn., Bitburg AB, Germany, 1961, 351st SMW, Whiteman AFB, MO, TDY, Tech, school, Chanute AFB, IL, LGM-30 Minuteman Missile Course: Missile Systems Analyst, Senior Controller, Job Control (HOTEL TANGO), 1968, Operation Bootstrap, BGS. University of Nebraska at Omaha, 1968, Senior controller, Job Control 394th SMS, Vandenburg AFB, CA, 1971, MBA, Master of Business Administration (Management), Golden Gate University, San Francisco, CA, 1972, Asst. NCOIC EMT, 44th Missile Sqdn., Elsworth AFB, SD, 1973. Retired and received the AF Commendation Medal w/OLC, SAC Outstanding Education Achievement Award, Senior Missileman Badge. Civilian occupation, senior superintendent of physical plant, University of California, Riverside.

Retired 1991. He and his wife Agnes reside in Riverside, CA. He is a volunteer worker, March Field Museum, Riverside, CA.

GARY A. HOSELTON, born May 2, 1938 in Salem, OR, entered USAF Oct. 27, 1958, discharged March 27, 1964 as airman first class. Served with 9th

BW (Medium) at Mountain Home AFB, ID, repairing radios on B-47 and KC-97 aircraft, and with 569th SMS, again at MHAFB, as a guidance system mechanic maintaining the ground guidance system for the Titan I ICBM (SM-68). Subsequently earned degrees in applied math and electrical engineering, designed semiconductor test equipment and medical products. Presently own and manage commerical income property in Portland, OR.

WILLIAM B. HUEY, LT COL (RET), graduated from Georgia Tech in 1962. Entered USAF in 1963, retiring in 1986. His initial assignment was in procurement at Robins AFB. In 1964, he was selected as the procurement officer best qualified for transfer to Minuteman. In Minuteman at Whiteman AFB, he specialized in performing no-notice alert standby crew duty when SAC Missile Competition crews required on-base train-

ing. In 1968 he was assigned to HQ SAC as a computer programmer. While there he produced printouts that were used to test the SAC document destruction facility. In 1972, he was assigned to the Airborne Launch Control System. He served as the senior Ellsworth ALCS crew member on satellite alert at Minot AFB. Those extended periods of satellite alert were largely responsible for his retiring in the grade of lieutenant colonel. In 1975, he was assigned to Vandenberg AFB. As chief of training aids fabrication, he produced hundreds of plaques for distinguished visitors. In 1978, he joined Air Command and Staff College. As wing chief, he created officer effectiveness reports for faculty members who conducted social and athletic programs. His final assignment was at Air War College where he frequently traveled outside the CONUS to deliver quarter-hour briefings to seminar program students. He and his wife reside in Montgomery, AL. He is the CEO of an organization that specializes in the preservation of buildings and grounds in rural Montgomery County.

WELLS E. HUNT JR., COL (RET), graduated from the Pennsylvania State University in June 1960 and entered the AF at Little Rock AFB as administrative officer for B-47 aircraft maintenance squadron. In 1961 he volunteered for missile duty and was assigned to the 551 SMS (Atlas F) at Lincoln AFB, NE, where he was an Alternate Command Post Instructor DMCCC. In 1965, he was assigned to Torrejon AB, Spain as a squadron administrative officer and earned an MS in Systems Management from USC. While in Spain, he commanded a search team looking for one of the "lost" nuclear weapons from the B-52/KC-135 crash over Palomares. In 1968, he volunteered for duty in Vietnam and was XO for the DCS/Intelligence at 7th AF in Saigon. Returning to missile duty in 1969, he was assigned to the 533 SMS at McConnell AFB as a Titan II Sector Commander. The next year, he reported to SAC HQ in the Future Space, Missile and Reconnaissance Systems Div. (XPFR) where he helped write the requirements document for the Peacekeeper. Subsequent assignments included director of Missile Test

at 1STRAD, Vandenberg AFB; commander, 570SMS (Titan II), Davis Monthan AFB; Air War College; Chief of Nuclear Policy, OJCS; director of Nuclear Policy, SHAPE, Belgium; director of Space Operations Planning, US Space Command, Peterson AFB, CO.

He and his wife, Wendy, live in Evergreen, CO. He works for Lockheed Martin as the program planner for Atlas and Titan guidance subcontracts with Honeywell.

BURTON J. HURSH, CMSGT (RET), born Aug. 25, 1923, San Franciso, CA. Inducted into the USAAC, Jan. 14, 1942, and served as flight engineer gunner (B-24). Training at Tyndall Field, Panama City, FL; Gowen Field, Boise, ID; shipped overseas to 14th AF, Flying Tigers, China, Burma, India Theater. In 1946, re-enlisted and served in Air Sea Rescue Units worldwide until 1958. Entered ICBM/IRBM Missile Program to include Thor and Atlas Models D, E and F. Was member of 3901st SMES as NCOIC of Quality Control and Evaluation Inspection Team traveling throughout the USA for SAC HQ, Omaha, NE. In 1964, transferred to 8th AF HQ to perform same duties. Retired in 1973, Westover AFB, MA.

Married Julie Ann (Jepson) May 7, 1952, and has three children: Carol, Carl and Marlene, living in Chicopee, MA. Fully retired. Working as officer manager RAO (Retired Activities Office) volunteer. Westover AFB, Chicopee, MA. (Other aircraft C-47, C-54, C-97, B-36). Worked for Friendly Ice Cream Corp, North Wilbraham, MA after retiring from USAF. Retired from there in 1988.

BRAD J. HYNDMAN, MAJ, born Dec. 22, 1956 in Jefferson, IA, enlisted December 1975 as an Airborne Electronic Intelligence Specialist aboard RC-135s, with the 6949th Electronic Security Sqdn., Offutt AFB, NE. After graduation from the University of Nebraska at Omaha in May 1984, he attended Officer Training School, graduating in April 1985. He served as a missile combat crew commander and flight commander at the 320th SMS, and command post controller at the 90th SMW, F.E. Warren AFB, WY. In August 1990, he reported to the 21st TFW, Elmendorf AFB, AK, serving as air operations officer and training officer. After his release from active duty in December 1992, he began his current assignment as a Reserve Individual Mobilization Augmentee Logistics Plans Office, with the 611th Air Support Sqdn., Elmendorf AFB, AK.

He and his wife, Kimberly, have resided in Chugiak, AK, since 1996. He is the training records manager and an instructor at Peninsula Airways in Anchorage.

WILLIAM R. INTIHAR JR., MAJ (RET), born July 3, 1950 in Jamestown, NY, enlisted in the USAF in October 1970 and was sent to Lackland AFB for basic training. Upon completion, he was assigned various security and law enforcement duties over an eight year period. In January 1980, he entered OTS and graduated in April 1980 as a second lieutenant. After graduation from Minuteman III (CDB) UMT in August 1980, he was assigned as DMCCC to the 742nd SMS at Minot AFB. He became a combat crew commander as a second lieutenant and then flight commander of "Kilo"

Flight as a first lieutenant and finally wing training combat crew commander. In graduation he was assigned to the 89th TMS at Wueschheim AS, Germany as "Bravo" flight's assistant flight commander, then as flight commander. From Germany, Capt. Intihar was reassigned as instructor, senior instructor and then chief, Minuteman III (CDB) and Peacekeeper training, 4315th CCTS, Vandenberg AFG. He was promoted to the rank of major in April 1992. In March 1993, Major Intihar was reassigned to the On-Site Inspection Agency, Washington, DC as mission commander in support to the INF and START Treaties. Over a period of five years he participated or led teams in eight US escort missions, 12 former Soviet Union inspection missions, was posted to the American Embassies in Belarus and Kazakhstan as Arms Control Chief and led State Department humanitarian aid teams to the Republics of Georgia and Armenia. Major Intihar retired Feb. 28, 1998. He and his wife, Dee, now reside and work in Fort Myers, FL.

ARLEN (DIRK) JAMESON, GEN, born Nov. 11, 1940, Vernon, TX, gained an ROTC commission into USAF out of the University of Puget Sound, June 3, 1962. He entered active duty in July at Homestead AFB, FL. Three months later the Cuban Missile Crisis unfolded.

In March 1963 2nd Lt. Jameson was ordered to Ellsworth AFB, SD to begin a career in missiles interrupted only by a tour as an aide-de-camp in the Azores, Portugal, a year as an advisor to the South Vietnamese AF in Vietnam, an assignment to the Pacific AF at Hickman AFB, HI and a Pentagon tour where he helped bring on board the Peacekeeper ICBM.

Colonel James went on to "the field" serving as the missile wing deputy commander for operations and vice wing commander at Malmstrom AFB, MT and wing commander at F.E. Warren AFB, WY.

BG Jameson was director of command and control at Offutt AFB, NE, commander of the 4th AD at F.E. Warren AFB, and commander of the 1st SMW at Vandenberg AFB, CA, where he earned his second star in time to command the renamed Strategic Missile Center. Next came a tour at Offutt AFB, NE as the chief of staff for SAC.

Gaining a third star, LTG Jameson assumed command of all the ICB forces under 20th AF and F.E. Warren. Returning one last time to Offutt AFB, NE, he served as deputy commander-in-chief, US Strategic Command, retiring on Feb. 29, 1996.

General Jameson's most memorable experiences included his wing, the "Mighty 90," and its accomplishment of a "Triple Crown" which was the 15th AF Riverside Trophy, the missile competition Blanchard Trophy, and the Omaha Trophy for the best wing in SAC all in the same year, 1984. Two other memorable events were hosting Gen. Sergeyev, head of the Russian Strategic Rocket Forces, touring US missile bases, and a return visit to Russia where he briefed the Russian Strategic Rocket Forces on the Nuclear Posture Review.

He married Betty Strobel of Tacoma, WA on Aug. 13, 1960. They have two sons and two grandsons. The Jamesons reside in Austin, TX where Dirk does consulting work.

ASHLEY D. JAMESON, COL (RET), born on April 4, 1916, Moran, TX, joined the USAAC Sept. 13, 1943. He graduated from McMurry University in 1939 with a BS degree and later received BD degree from Southern Methodist University. He was ordained May 1942.

Chaplain Jameson began his service career in 1943 attending the Chaplain School at Harvard University. During the years of WWII he served as a group chaplain of the 488th BG which flew B-17s. After the war he returned to the civilian ministry and served parishes in North Texas.

Recalled during the Korean War he served at a number of bases throughout the AF. From 1958-60, he was staff chaplain, 11th AD and ministered to personnel stationed in northern Alaska and the Arctic Ocean. He served as chief of the Professional Div., HQ TAC from 1960-62. He assisted the chief of AF chaplains, 1962-65, as personnel procurement officer. From 1965-67 he served as staff chaplain, HQ 12th AF, Waco, TX. He was command chaplain, Pacific Air Forces 1967-71. He was the command chaplain, SAC, 1971-74. As a sky pilot, he was qualified in three types of aircraft. He was an ardent supporter of Aero Clubs.

Memorable experiences: when he swore his son, Arlen Dirk, in to the USAF in the spring of 1962; a rocket attack on Pleiku AFB, Vietnam; ranking as one to remember a below minimum; out of fuel landing an announced at Fairchild AFB followed by a spread eagle face down in the snow search by security.

Retired May 1, 1976, as colonel. He received all the usual awards plus Legion of Merit and OLC.

Married Mary Hickox of Jonesboro, AR, Dec. 19, 1952. A son, Lt. Gen. Arlen Dirk Jameson recently retired from the Air Force. The three daughters are: Jana G. Nave, San Antonio; Lynnis E. Hammel, Kansas City, MO and Kadi Willis, San Antonio.

RICHARD P. JAQUES, COL (RET), is a career missileer, a life member of the Association of Air Force Missileers and the Society of Strategic Air Command. He was born and raised in Indiana, graduated from Purdue University in 1959 and entered active duty as a second lieutenant in October 1959. His retirement was effective Aug. 1, 1987.

His first duty station was F.E. Warren AFB where he served as training officer for the 389th Combat Defense Sqdn. He then "transferred to missiles" and was assigned to Det. 4, 99th MMS, RAF Hemswell, UK as an authentification officer with the Thor IRBM. Follow-on assignments included Schilling AFB, KS as an Atlas "F" DMCCC and MCCC; Whiteman AFB, MO, as MCCC and EWO instructor/command post controller; and he earned an MA is Business from Central Missouri State College (University). From Missouri, it was on to SAC HQ, Offutt AFB, NE where he was an emergency actions officer (SACM 55-18), and executive officer for the SAC Dir. of Command Control. He was then assigned to the Pentagon as an officer controller in the National Military Command Center and an action officer, Emergency Actions Div., Organization of the Joint Chiefs of Staff. He attended the Naval War College 1976-77, served as 447th SMS commander and

Assistant DCM for the 321st SMW at Grand Forks, ND between 1977 and 1979. The colonel was then transferred to Minot AFB, ND, where he was the 91st SMW DCM for a year, vice wing commander for about two years and wing commander in 1982 and 1983. He then commanded the 3901st Strategic Missile Evaluation Sqdn. from 1983-86, and served his final assignment at SAC HQ as the command's Director of Missile Operations.

Awards and decorations include the Legion of Merit w/OLC, Meritorious Service Medal w/2 OLC, the AF Commendation Medal, Small Arms Expert Marksmanship Ribbon, Combat Readiness Medal and the Humanitarian Service Medal. He is the recipient of the Joint Chiefs of Staff Badge and is a holder of the Master Missileman Badge.

Following retirement, Col. Jaques joined the BDM Corporation as a company representative on contract to assist the AF in developing an operations concept for the Peacekeeper Rail Garrison Program. He later joined the Science Applications International Corporation as program manager for a civilian team contracted to develop procedures for employing the Peacekeeper Rail Garrison weapons system. He was subsequently hired by Martin Marietta as manager of the Corporation's Omaha marketing office. He remained with the company and moved to Hampton, VA when the AF reorganized SAC and TAC into Air Combat Command headquartered at Langley AFB, VA. He retired in 1996 from the corporation that had been reorganized into Lockheed Martin.

He and his wife Judy returned to Indiana in March 1997. He is a partner in a small construction company and sells cars at a dealership in nearby Crawfordsville, IN. The couple is embarked on a long-term project to restore their large transverse frame barn that was built in 1900.

RONALD W. JAYNE, COL (RET), graduated from Ohio State University in June 1969. His first assignment was to Ellsworth AFB, SD where he served with the 44 SWM as a combat crew deputy and commander and as both a combat crew instructor and evaluator. While there he completed a masters degree in economics through the Minuteman Education Program.

He then began a series of logistics and acquisition assignments. Duty stations included Homestead AFB, FL; Kelflavic NS Iceland; Tinker AFB, OK and Wright Patterson AFB. While at Tinker AFB, he was the system program manager for the ground launched cruise missile and oversaw the return and destruction of the weapon in compliance with the INF Treaty. He was awarded the MGen Frederic J. Dau System Program Manager of the Year Award in 1988, and the USAF Outstanding Senior Logistics Planner Award in 1989.

Upon return to Wright Patterson AFB for his third tour there, he was assigned as the system program director for the Depot Maintenance Management Information System. His last assignment was as the chief, inspections and assessments, AFMC Inspector General. He retired in 1995, and is presently the director and manager, Logistics Systems Div., CACI International, Inc. He and his wife Sandy live in Dayton, OH.

ARTHUR E. JOHNSON, BGEN (RET), is a former commander of the 44th SMW at Ellsworth AFB, SD. He attended law school after retiring in 1992 as the deputy director for Strategic and Space Systems in the Office of the Secretary of Defense. He is an attorney in Indianapolis, IN, where he and his wife Penny reside.

Following his first assignment in 1964 as an administrative officer with the 564th SMS (Atlas E) at Fairchild AFB, WA, he was trained among the first cadre of missile crew members for the Modernized Minuteman II at Whiteman AFB, MO. He subsequently served as a missile fight test analyst in Ballistic Missile Evaluation at SAC HQ, as a missile maintenance officer with the 90th SMW at F.E. Warren AFB, WY, as the assistant deputy commander for Operations and the Combat Support Group Commander with the 91st SMW at Minot AFB, ND, and as the vice commander of the 351st SMW at Whiteman AFB, MO.

He served three assignments at the Pentagon, and was principally involved in arms control and strategic operations and planning. He had two tours with the Joint Staff, first as a military advisor to the SALT II Delegation and later as the JCS representative to the Defense and Space Talks with the former Soviet Union. He also served as an air staff action officer, joint planner, and the executive officer to the vice chief of staff.

ERIC A. JOHNSON, 1LT, entered the AF in September 1994 after earning undergraduate degrees in history and geography, and his commission from South Dakota State University. At that time he reported to Vandenberg AFB, CA for Undergraduate Space and Missile Training followed by Initial Qualification Training in the Minuteman III CDB-B weapon system, better known as "Deuce". Lt. Johnson began his initial crew tour

by being assigned to the 448th Missile Sqdn., 321st Missile Gp., Grand Forks AFB, ND in June 1995 and played an active part in the realignment of sorties to Malmstrom AFB, MT. While at the unit he was awarded the AF Outstanding Unit Award, the Combat Readiness Medal, and the Humanitarian Service Medal for his work during the Red River Valley Flood of 1997.

He also served the base community as curator and OIC of the base museum, specializing in missile history. In January 1998, Lt. Johnson returned to Vandenberg AFB, CA for IQT in the new REACT system and soon after began his second crew tour in the 321st Missile Sqdn., 90th Missile Wing, F.E. Warren AFB, WY. He and his wife, Lt. Lisa Johnson, currently reside in Cheyenne, WY and have no children.

RICHARD M. JOHNSON, LT COL, born March 3, 1939, Omaha, NE, enlisted in the USAF on May 3, 1962. Military locations, stations were: Davis Monthan AFB, AZ (1962-69); SOS Maxwell AFB, AL; Sheppard AFB, TX (1970-71); F.E. Warren AFB, WY (1971-77); Davis Monthan AFB, AZ (1977-83). Discharged in November 1983 with the rank of lieutenant colonel.

Memorable experiences: (1) project officer for Minuteman III modification which resulted in 33% reduction of Minuteman III crew force. (2) Instructor in both Titan II and Minuteman III weapon systems. (3) Chief of Operations Training at DMAFB when initial woman crew officer came on board. Awarded the Meritorious Service Medal w/OLC and Commendation Medal.

Johnson married Roxana June 6, 1964, and they have two children, Colleen and Katy. Presently president of Tele-Plus Services, Inc. (a telemanagement consulting firm in Tucson, AZ).

JAY W. KELLEY, LT GEN, is commander, Air University, Maxwell AFB, AL and director of education, Air Education and Training Command. As such, he ensures that AF needs in the areas of enlisted and officer professional military education, professional continuing education and graduate education are met, as well as officer commissioning through Officer Training School and the Air Force Reserve Officer Training Corps.

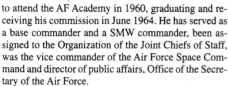

He enlisted in the Air Force Reserve in 1959, and was selected to attend the AF Academy in 1960, graduating and receiving his commission in June 1964. He has served as a base commander and a SMW commander, been assigned to the Organization of the Joint Chiefs of Staff, was the vice commander of the Air Force Space Command and director of public affairs, Office of the Secretary of the Air Force.

He served with the Strategic Offensive Branch, Operations Directorate, Organization of the Joint Chiefs of Staff; assistant deputy commander for Operations, 351st SMW; commander, 351st CSG, Whiteman AFB, MO; commander, 381st SMW, McConnell AFB, KS; assistant chief of staff, HQ SAC; senior military advisor to the director, Arms Control and Disarmament Agency; deputy chief of staff for Operations, HQ AF Space Command; deputy chief of staff for Requirements, HQ AF Space Command; vice commander, AF Space Command, Peterson AFB; director of Public Affairs, Office of the Secretary of the AF; commander, Air University.

His awards include the AF Distinguished Service Medal, Defense Superior Service Medal, Legion of Merit, Defense Meritorious Service Medal, Meritorious Service Medal w/OLC, Joint Service Commendation Medal and the Combat Readiness Medal.

He developed and orchestrated "futurist" business with major corporations and government organizations using the Third Wave principals and concepts of Al and Heidi Toffler. General Kelley and his wife, Marty, both natives of Frankfort, IN, are the parents of two daughters.

MICHAEL J. KELLY, CAPT, born Nov. 6, 1933, enlisted in the USAF, June 5, 1955. His military locations and stations: Lackland AFB; Hondo AB, Shephard AFB, Hahn AB and McGuire AFB.

He participated in the Suez and Hungarian Crisis. His memorable experiences: being involved in the development of the 69th TMS into the 586th TMG and the 701st TMW. To this day they still reunion with members of that unit (1955-58) from Hahn, Germany.

Kelly was discharged Oct. 25, 1972 as captain. He is married to the former Trudy Short (1959). They had five children. He was employed as area manager, Ashland Oil, Inc., 1969-89. He is presently president of Ark Petroleum, Inc. Former director, president; currently treasurer of the Chicago Oil Men's Club, vice president of the Classical Arts Society, director, Sauganash Community Association, director of Bel Canto Foundation, owner/director Sauganash Sounds.

DONALD E. KELTNER, MAJ (RET), entered active duty through ROTC at Baylor University in 1974. He served in the 91st SMW, as a Minuteman crew member and instructor until 1978. After assignments in Public Affairs at Minot AFB, ND and Moody AFB, GA, he returned to ICBM operations in 1980 in the 341st SMW as flight commander, Emergency War Order instructor as chief, Instructional Systems until 1985. He was assigned to the 90th SMW as the assistant chief, Codes Div. where he was involved in Peacekeeper deployment. In 1989, he reported to 16th AF as the chief, International Negotiations monitoring compliance of agree-

ments in the Mediterranean area, including compliance of the Conventional Forces in Europe treaty. He finished his career as chief, command and control at Laughlin AFB, TX in 1994.

He and his wife, Linda, live in Hewitt, TX and recently became grandparents. He is with the Waco Police Department and Linda is the director of Music and Fine Arts, Methodist Children's Home.

GEORGE B. KENNEDY, COL (RET), graduated from the University of Southern California in 1962. He entered the AF and, after training, served as a Mace A missile crew commander with the 38th TMW, Sembach AB, Germany. In 1967, he qualified as a Titan II missile crew commander with the 381st SMW, McConnell AFB, KS and was later assigned as the chief, Plans and Intelligence Div. In 1973, he reported to the 3901st Strategic Missile Evaluation Sqdn., Vandenberg AFB, CA serving as chief, EWO Evaluation, until 1977. The next assignment was with the 390th SMW at Davis Monthan AFB, AZ where he served as a sector commander; Chief, Plans and Intelligence Div.; operations officer; Chief, Training Div.; and commander, 571st SMS. In 1982, he was transferred to the 44th SMW, Ellsworth AFB, SD, and served as the deputy commander for Operations. His next assignment was in 1985 as the deputy commander for operations with the 485th TMW, Florennes, Belgium. Following that assignment in 1986, he was transferred to Vandenberg AFB, CA where he served as the SAC director of Resource Management/Logistics and director of Environmental Restoration until 1992, when he completed his 30 year career.

Since AF retirement, he has been employed by Jacobs Engineering in Pasadena, CA as a project manager for the Environmental Restoration of Federal Facilities.

He and his wife, Wilma, have resided near Lompoc, CA in their permanent home since 1988. They have two sons, John, a criminologist, and Matthew, a Navy lieutenant and nuclear engineer serving aboard the Trident Submarine, the USS Nevada.

WILLIAM A. KIDD, SSGT, born Feb. 15, 1926 in Detroit, MI, enlisted in the USNR May 1944; USCG, 1947-52; USAF, 1953-67. Stations include: Tachikawa Japan; Selfridge AFB, MI; Lowry AFB, CO; Dow AFB, MA (SAC). He was Mace Missile instructor 1957-1962, Lowry AFB, CO. Retired Feb. 1, 1967 from Dow AFB, MI as staff sergeant.

Kidd is married and lives in Detroit, MI. He retired as letter carrier from the US Post Office, Jan. 2, 1991 and resides in Detroit, MI. He enjoys being retired.

WILLIAM KIME, LT COL (RET), enlisted in 1956 and was commissioned from OCS Class 60B. His first missile duty was as a missile safety officer, 351st SMW, Whitman AFB, MO. He transferred as a missile maintenance officer, 341st SMW, Malmstrom AFB, MT. He competed in Curtain Riser, the 1st Missile Competition, as a member of a Target and Alignment Team. His next assignments were to the Directorate of Missile Maintenance, 15AF at March AFB, CA, the 3901st SMES at Vandenberg AFB, CA and ended his career as commander, 91st FMMS, Minot AFB, ND. He worked on all series of the Minuteman Missile.

After retirement he worked for Martin Marietta on the Peacekeeper missile system. He now works for Intel as tool install project manager.

He and his wife, Sarah, live in Rio Rancho, NM, where they are enjoying life with their daughter and four grandchildren.

CARL L. KING, MAJ (RET), born in Texas, graduated from North Texas University. First missile duty, 1964, DMCCC, Minuteman I, 66th SMS, 44th SMW, Ellsworth AFB, SD. 1966, Minuteman II instructor, 4315th CCTS, Vandenberg AFB, CA. MPT operator for the first Missile Competition, Curtain Raiser. He and Darrel Downing, as a crew, set all base times for Minuteman II exercises. 1971, Chief, MPT Branch, 321st SMW, Grand Forks AFB, ND. 1973, wings plans officer there. 1974, operations controller/MCCC-airborne on the SAC Airborne Command Post, 2nd ACCS, Offutt AFB, NE. While there he launched a Minuteman III at Vandenberg with ALCS. 1978, Chief, Minuteman II CDB training, 4315th CCTS. He retired in 1983, while overseas as Base Disaster Preparedness Officer, 401st TFW, Torrejon AB, Spain.

He and his wife, Anne, live near Lompoc, CA; she is a public accountant, he plays tennis and teaches microcomputer operations.

RAY W. KING JR., MSGT (RET), entered the USN in August 1941. From 1941 until 1956 he was an aircraft mechanic, hydraulic technician, structural mechanic and flight engineer on Navy seaplanes also a flight examiner for MATS. He served on the aircraft carriers USS Hollandia and USS Attu in the Pacific earning two Battle Stars during the battles of Iwo Jima and Okinawa.

He entered the AF in 1956 as an aircraft hydraulic technician and was stationed at F.E. Warren AFB in Cheyenne, WY. He reported to Vandenberg AFB, CA, September 1958 and was assigned to the 5776th SMS as a missile pneudraulic technician privileged to be a member of the crew on the first Atlas D Missile Launch Crew. September 9, 1959 his crew launched this nations first Intercontinental Ballistic Missile from Vandenberg AFB. After months of alert duty on the Atlas D missile he was assigned to the Atlas E complex and then to the 576th Quality Control. In July of 1961 he was transferred to the 3901st SMES as part of the first missile evaluation team. This team covered the Atlas F missile. This assignment lasted until his retirement in October 1963.

After retirement he was an instructor in Metals Technology at Hartnell College in Salinas, CA from 1965-70 and Cuesta College, San Luis Obispo, CA from 1971-89.

He and his wife Norma have been married 52 years and now reside in Templeton, CA and enjoy fishing, gardening and life in the country close to their son, David and his family.

SIDNEY SHEPPARD KLEVENS, born Dec. 17, 1925 in Boston, MA, raised in Chelsea, MA. Drafted in his senior year of high school in March, 1944. After infantry basic training was assigned to HQ Co., 1st Bat., 303rd Inf. Regt., 97th Inf. Div. for amphibious training on the west coast of California. The division was sent to Germany, after the Battle of the Bulge. Saw combat in the Ruhr Valley and ended the war in Pelsen, Czechoslovakia. Returned to the US for R&R and the division was shipped overseas in preparation for the invasion of Japan. Due to the Japanese surrender they became occupation forces. He was discharged from the Army, May 1946.

After attending Aircraft Mechanics School in California and getting married, he worked for the airlines, aircraft manufacturers and Aerojet-General Corp. He tested the liquid engines on the Bomarc "A", fired a Nike liquid engine into a four feet diameter curved tube with pickup points hooked onto mercury manometers for reading pressures. They were testing one phase of the missile silos that were under development.

In 1957, he transferred from the AF Reserve to active duty and cross-trained into missiles. After completed TM-76A, Mace training at Lowry AFB, CO he was assigned to the 4504th MTW, Orlando, FL. Became an instructor on the 76A single launch concept. When the multiple launch concept was developed, there was a need for TM61-C, Matador, maintenance instructors and he was assigned as a primary maintenance instructor. When the 61C program closed down he was reassigned to the Standardization & Training Branch, under the director of maintenance.

In May 1962 was transferred to the Student Squadron and went through TM76-A maintenance training and after graduation was assigned to the 586th Missile Sqdn., Hahn AB, Germany. In June 1963 was transferred to Det. 1 38th TAC Missile Wing, Hahn AFB as a senior controller in Job Control.

December, 1963 received a humanitarian transfer to Elmendorf AFB, AK and assigned back into aircraft maintenance on Fairchild C-123s. Shortly after the 1964 earthquake in Anchorage he was promoted to technical sergeant. Later transferred to Standardization and Training, under the director of maintenance.

June 1967, transferred to Dover AFB, DE. Served one year in transit maintenance and in August, 1968 received orders to Nakon Phnom, Thailand. Served the first 90 days with the 1st Special Operations Squadron and the remainder of his tour with the 21st Special Operations Sqdn. They flew old Douglas Sky Raiders, A1-Es and Hs. Search and rescue was a major function. They also covered the Ho-Chi-Min Trail through Laos. Took part in the end of TET One and some of the TET Two offense.

Returned to the US in August 1969, assigned to the 42nd SW, Loring AFB, ME. Performed aircraft maintenance on KC-135s and B-52s. Was promoted to master sergeant in July 1970. In December, 1971 was reassigned to the 6th SW, Eielson AFB, AK on KC-135 tankers. When it was discovered that he was losing his hearing in his left ear he was assigned as NCOIC Training Management (November 1972), a position he held until he retired in May 1975.

June 1972 having been a FAA licensed aircraft mechanic since August, 1948, he took the Inspection Authorization Test and worked civilian general aviation after duty hours. Upon retirement from the AF he held positions as director of maintenance or chief inspector for a small Alaskan commuter airlines or air taxi service until he finally retired in August 1997.

WILLIAM A. KNAPP, COL (RET), born Sept. 29, 1926, Jersey City, NJ. Infantryman with 29th and 78th Divs. in ETO, 1944-46. As sergeant, completed AAF OCS and commissioned, 1946. Administrative officer, counterintelligence officer, 10th AF, Brooks Field, 1946-47. Completed airborne training, Ft. Benning, qualified parachutist/gliderman, 1947. Administrative positions in Germany, HQ OMGUS, Berlin, 7200th Air Depot Wing, Erding Air Depot, 60th TCW, Fassberg RAFB (Berlin Airlift), and as wing adjutant, 7050th Air Intelligence Wing, Wiesbaden, 1947-51. Adjutant, assistant operations officer, 3201st Air Police Sqdn., Eglin AFB, 1951-53. Aircraft maintenance officer, 306th BW, MacDill AFB, 1953-57. During this period acquired BS degree from Florida Southern College through "Operation Bootstrap". Senior flightline maintenance officer, senior field maintenance officer, maintenance control officer, 1503rd Air Transport Wing, Tachikawa AB, Japan, 1957-61. Attended Command and Staff College, 1961-62. Maintenance control officer, 3635th Flying Training Wing (advanced-helicopters); Completed sur-

vival training course, Stead AFB, 1962-63. MCCC, instructor, OBO, 68th SMS, 44th SMW, Ellsworth AFB, 1963-67. While at Ellsworth earned MBA in management from Ohio State University, 1967, participated in a missile launch at Vandenberg AFB and test launches during project "Long Life".

Maintenance control officer/chief of maintenance, 355th TAC Fighter Wing, Takhli RTAFB, Thailand, 1967-68. Director of Missile Training, assistant DCO, DCM, 91st SMW, Minot AFB, 1969-72. Chief, Department of Missile Training, Deputy Commander, USAF School of Applied Aerospace Sciences, commander, 3345th Resources Management Group, Chanute Technical Training Center, 1972-75. Presented Legion of Merit upon retirement, Sept. 1, 1975.

After retirement spent 16 years in managing various staff activities for two insurance companies in Colorado.

With wife, Joyce, presently residing in Denver, where six of their seven children live.

RICKELL (RICK) D. KNOLL, COL (RET), born in Webster City, IA, Aug. 19, 1943, attended the USAF Academy, graduating in 1965. He spent the first nine years of his career in the computer career fields with assignments at Cannon AFB, NM, Purdue University, U Tapao RTAFB, Thailand and the Pentagon. In 1974, entering the missile career field, he attended missile training at Vandenberg AFB enroute to F.E. Warren AFB as a MMIII CDB MCCC in the 320 SMS. While at Warren, he served as Papa Flight commander in the 400 SMS and chief of EWO training. Leaving Warren in 1977, he became the chief of SIOP timing JSTPS at Offutt AFB. He was selected commander of the 742 SMS at Minot AFB in 1979 where he served until 1980, moving to Vandenberg AFB as 1STRAD director of Test and Evaluation. Selected for National War College in 1983, andmoved back to Washington DC. After graduation became assistant to the Deputy Undersecretary of Defense for Policy and Assistant Deputy Director for Resources on the Air Staff. He returned to Vandenberg AFB in 1985 as the base commander (4392 Group). Moving in 1986, he was vice wing commander, 90 SMW at F.E. Warren AFB until his selection as wing commander 321 SMW, Grand Forks AFB.

He retired in 1989, taking a job as director New Product Development for the Sunglass Div. of Bausch and Lomb. He and his wife, Kathe, have four adult children and live in Bloomington, MN where he is currently the Director of Research, Development and Engineering for Jostens.

WALTER KUNDIS, CMSGT (E-9) (RET), born Jan. 26, 1925, Diamond Town, PA. Graduated high school 1942. Went to work for Bethlehem Steel Co. prior to enlisting in the USN. As a gunnersmate aboard PT 524 Sqdn. 36. During the battle of Leyte Gulf Philippines in the Surigao Straits. Personally launched a torpedo towards the leading Japanese battleship the Yamashiro. Results were obscured by their smoke screen, it did absorb most the enemy gunfire directed at them. After his discharge from the USN went back to

Bethlehem Steel Co. and after trying to cope with civilian life, reenlisted into the USAF and continued to make war his profession. Actively participated in the Berlin Blockade, Korean and Vietnam War, and in between wars, was a combat crew member of a Titan II (ICBM) Intercontinental Ballistic Missile for four years, 308th MW, 374th MS at Little Rock AFB, AR and launched the most accurate shot recorded. He was awarded the Air Force Commendation Medal. While he was on a crew pointing a Titan II at the Soviet Union, his close cousins in the Ukraine were pointing a SS-19 (ICBM) at the USA. In August of 1995, they found this out, when he and his family visited his fathers birthplace. They drank a lot of vodka to celebrate the fact that they didn't have to launch at each other.

During the Vietnam War, served as NCOIC of the Missile Maintenance Branch, 412th MMS, 12th TFW. Cam Rahn Bay, Vietnam. Awarded the Bronze Star Medal and Armed Forces Honor Medal Vietnam second class. His next assignment he served as PACAFs missile chief, prior to his retirement from the AF as a CMSGT (E-9) missile electronic maintenance superintendent with over 30 years active duty, March 1, 1975. He was awarded the Meritorious Service Medal. To a place called Mililani Town, HI with wife Josephine, they have two daughters, Laural Ann (Mrs. Michael Naven), Geraldine and son, Warren. They enjoy their two grandchildren, Nicholas and Alexander.

ROBERT W. KUTULIS, CAPT (RET), served in the US military as both an enlisted member and AF officer from July 23, 1956 until June 30, 1980. He served in the USMC Reserves in Scranton, PA, his home town, before enlisting in the AF on May 22, 1958. Duties included Missile Maintenance at Lowry AFB, CO on the Matador TM-61C, MACE TM-76A, and MACE TM-76B missiles; Missile operations as a missile combat crew chief on the MACE TM-76A at the 38th TMW, 887th TMS, Sembach, Germany; and Missile Maintenance at the 455th SMW, later designated as the 91 SMW, Minot AFB, ND as an electro-mechanical team (EMT) crew chief and missile job controller. He attained the rank of master sergeant.

He attended the University of Nebraska, Omaha under the AF Bootstrap Commissioning Program and was commissioned as an AF officer in June 1970. Duties included missile maintenance officer at 15th AF HQ, March AFB, CA, during which time he earned an MA at Chapman University; AF Data Design Center, Gunter AFS, AL; Pentagon, LGY staff; AR Logistics Management Center, Gunter AFS, AL; and as a transportation officer at the 341st SMW, Malmstrom AFB, MT.

He and his wife, Carol, the former Carol M. Deininger from Scranton, PA, have five children and ten grandchildren. They reside in Yardley, PA. He is currently director of Retirement Plan Services for BT Alex. Brown, a subsidiary of Bankers Trust Company. He served as a founding member of the Board of Directors of the National Defined Contribution Council sharing the Legislative Committee. Prior to his banking associations, he was an executive manager with Sun Company, Inc.

ROY E. LADD, COL (RET), enlisted National Guard, 1937. Air Corp's 1941, Armament, and Bombsight Schools 1942, Pilot 1943, then B-24s delivered B-24s to bases all over US and one to England.

CBI Theater 1944. Flew B-25H to India. Tezpur, A Hump pilot C-87s, B-24s, C-109s. 608 Combat hours a Distinguished Flying Cross and Air Medal later, Love Field B-29s. 47 to 53 Air Weather Reconnaissance B-29s Snoptic Weather, Typhoon's, atomic testing. 53 to 58 B-36s Carswell AFB. 704th SMW, Cooke AFB Chief Missile Site Selection, and Training Aids Section, 59 to 64 assistant operation officer, then operation officer 13th Missile Div., Warren AFB, 64 to 69 priority officer AF. The Pentagon 1969-70 vice commander 321st MW. 1970-71 commander 24 Air Defense Sqdn. the Anti-Satellite Unit. 1971 to 1973 deputy then D.M. 14th Aerospace Force. Retired 1973.

PHILLIP A. LAYMAN, CAPT, is commander of 8th Flight, 742d Missile Sqdn., 91st Space Wing, Minot AFB, ND. He graduated with a bachelors degree in Aerospace Engineering from the University of Cincinnati, was commissioned through AFROTC and married his wife Mary in June 1991. After earning Honor Graduate at basic communications officer training at Keesler AFB, MS, he served as communications program manager with the 38th Engineering Installation Group at Tinker AFB, OK from July 1992 to January 1995. While at Tinker AFB, Capt. Layman earned a masters of business administration degree from Oklahoma City University.

He earned honor graduate at Undergraduate Space and Missile Training and Distinguished graduate at CDB-187 Initial Qualification Training. Captain Layman reported to the 742 MS in August 1995 where he served as a missile combat crew commander, Mike Flight Commander and acting operations officer before his current position.

JOHN C. LEBER JR., LT COL, commander of the 90th Services Sqdn., F.E. Warren AFB, WY, since November 1985, today relinquishes command of the squadron to Maj. James P. Bloom.

He was born Sept. 22, 1937, in Goochland, VA and graduated from Goochland High School in 1955 and attended college and Virginia Polytechnic Business Administration - accounting. He was commissioned a second lieutenant in the AF through Reserve Officers Training Corps (ROTC) in June 1959. He earned a master of science degree in Industrial Management from the University of North Dakota through the Minuteman ATIT program.

His first assignment was pilot training at Moore AB, Mission, TX. In 1960, he was assigned to the 99th Mu-

nitions Maintenance Sqdn. (Thor missile system), as a launch officer with the Royal AF, at RAF Feltwell, England.

From July 1963, he was assigned to the 556th SMS (ATLAS missile system) at Plattsburg AFB, NY. He served as missile combat crew member and instructor.

In May 1965, he was assigned to the 321st SMW at Grand Forks AFB, ND, as missile combat crew commander instructor and standardization crew commander.

In June 1969, he was transferred to Offutt AFB, NE working as a computer programmer/system analyst for HQ SAC Data Processing Division. He attended computer programming at HQ SAC and Honeywell, Inc., Phoenix, AZ. He worked in the development of computer software for the SAC missile and bomber target assignment, flight plans, timing on target and missile data tapes.

From July 1974 until June 1978, he was assigned to the Joint Strategic Target Planning Staff (JSTPS) at Offutt AFB, NE as chief, Missile Timing and Single Intergrated Operations Plan Development (SIOP) section, producing all assigned NATO and US Forces missile target and timing documents.

In June 1978, he was assigned to HQ US European Command at Stuttgart, Germany as Iranian desk officer for military and security assistance to Iran. He traveled on occasion with deputy commander, USEUCOM, Gen. Huyser, for conferences and procurement discussions of military sales and assistance to the Shah of Iran and the Iranian Armed Forces Staff. In 1979, he was assigned as chief, Human Resources Management Branch, HQ USEUCOM. This duty included monitoring and coordination of all personnel equal opportunity, drug and alcohol abuse programs for all AF, Army and Navy personnel in the European theatre.

In July 1981, he was assigned as the chief, Launch Control Management Div., 90 SMW Operation Deputate, F.E. Warren AFB, WY.

Lieutenant Colonel John C. Leber assumed command of the 90th Services Sqdn., F.E. Warren AFB, WY, Nov. 22, 1985.

Lieutenant Colonel is a master missileman with nine years of combat crew time in three weapon systems. His military decorations include the Defense Meritorious Service Medal, Joint Service Commendation Medal, AF Commendation Medal w/OLC, Outstanding Unit Award and Combat Crew Readiness Medal w/OLC.

He is married to the former Diana Sydney Froste of Cambridge, England. They have three children: Richard, Rebecca and Edward.

BARBARA L. LEAVITT, born in San Francisco, June 16, 1939. A life member of AAFM, founding executive vice-president of the Association of Civil Air Patrol Rocketeers. Retired teacher, San Francisco Unified School District, Middle Schools (1998). A member of the USAF Auxiliary, Civil Air Patrol since 1991. A computer specialist and an aerospace educator, she has held progressively more responsible positions as SF Cadet Sqdn. 86, AEO, Northern California Area coordinator for Aerospace Education, HQ CA Wing and California Wing assistant director of Aerospace Education for external programs.

Her most memorable experience includes observing the aerial refueling of an F-117 Stealth Fighter over the Nevada Desert from the "Boomers" pod in a KC-135, while enroute to the National Congress of Aviation and Space Education and networking with teachers from all over the world at these yearly Congress'. She is a national runner-up for the: A. Scott Crossfield AE Teacher of the Year Award, AE Crown Circle, and National Brewer AE Award. In addition, she is the 1997,

Pacific Region winner of the Brewer Award for Excellence in Aerospace Education. During her retirement, she plans to engage in Aerospace Education Curriculum Development. Her current project is the revision of CAP Manual 50-20, The CAP Model Rocketry Program, and incorporating it into a kit called "Goddard Force" which combines Model Rocketry with the history of rockets, missiles and space. When completed, the kit will be user friendly to the AAFM, the CAP, and local schools.

ROBERT M. (BOB) LEE, MAJ (RET), born Oct. 6, 1941, Oakdale, CA, entered service Feb. 14, 1964, Officer Training School, Lackland AFB, TX.

Military locations and stations: May-December 1964, Chanute AFB, IL, Aircraft Maintenance Officer Course; Aircraft Maintenance officer, England AFB, LA, 401 TFW, Jan-December 1965; 3rd TFW, Bien Hoa AB, RVN, December 1965-November 1966; Williams AFB, AZ, 4315th Combat Crew Training Sqdn., January 1967-August 1968. Air intelligence officer; Bitburg AB, Germany, 36 TFW, May 1969-December 1971; Spangdahlam AB, Germany, 52TFW, January 1972-June 1973; Nakon Phanom AB, Thailand, USSAG, June 1973-May 1974; Defense Intelligence School, Anacostia NAS, Washington, DC, August 1974-May 1975; Defense Intelligence Agency, Pentagon, June 1975-June 1978; Vandenberg AFB, CA MMII, ILCS Course, August-November 1978; Minuteman II Missile Operations, AFB, CA MMII, ILCS Course, August-November 1978; Minuteman II Missile Operations, Malmstrom AFB, MT, 341 SMW, 12 SMS, November 1978-August 1982, combat crew commander, flight commander India and Fox Flights, OIC Combat Crew Training Branch, standardization evaluation crew commander. Castle AFB, CA, 93 BW, chief, Intelligence Div., August 1982-May 1984.

Missions/Operations participated in: Operations officer and primary launch crew commander, Glory Trip 142M, Minuteman II Operational Test Launch, July 19, 1982.

Memorable experience: proposed, planned and implemented "Line Swine" crew competition between the 12 SMS, 10 SMS, 490 SMS and 546 SMS at Malmstrom.

Awards/Medals: Meritorious Service Medal w/OLC, Joint Service Commendation Medal w/OLC, Air Force Commendation Medal.

He was retired as major May 31, 1984. Married for 30 years to Judi and they have a son Bill, 28 and daughter, Melissa, 24 and three grandchildren. Employed as manager, Human Resources with UTMC Microelectronic Systems in Colorado Springs, CO, 1987-present.

His education includes: BA history, Pasadena College, Pasadena, CA, 1963; MA, Executive Development in Public Services, Ball State University 1973; MS, Human Resources Management and Development, Chapman University, 1987.

HARRY A. LEHMAN, LT COL (RET), served in the AF from June 26, 1963 to July 1, 1983. After graduation from the Pennsylvania State University in March of 1963 and a short stint teaching high school English

at Tyrone, PA, he attended AF Officer Training School at Lackland AFB, TX. Receiving his commission in September 1963, he reported to Columbus AFB, MS where he served as the manager of the Officers' Club. He then served as the food services officer at Misawa AB, Japan and Da Nang AB RVN. Upon his return to the States in July 1968, he was assigned to the 90th SMW at F.E. Warren AFB, WY where he served as a missile combat crew commander; EWO/Plans and Intelligence Instructor; and command post controller. In March 1974, after attending Logistics Management School at Wright Patterson AFB, OH, he was assigned as the director, Plans and Intelligence, 22nd AF, HQ, Military Airlift Command, Travis AFB, CA. While at Travis AFB, he earned his MPA from Golden Gate University, San Francisco, CA. His next assignment was to MAC HQ, Scott AFB, IL where he was the chief of Logistics Plan Inspection, MAC IG. Upon completion of this tour, he volunteered to return to missile duty and was assigned to the 44th SMW, Ellsworth AFB, SD. While at Ellsworth, he saw duty at operations officer, 66th SMS, Chief Missile Training, 44th SMW, commander, 44th Transportation Sqdn., and deputy commander Resource Management, 44th SMW. In June 1982, he was selected to become the director, Resource Management Inspection, SAC Inspector General, Offutt AFB, NE. In this final assignment he was deeply involved in developing SAC's mobility mission. He, his wife Shirley, and their three boys retired Cody, WY where he is vice president, Shoshone First Bank and enjoys skiing and playing golf.

STEVEN T. LIDDY, CAPT, born in St. Louis, MO, attended Saint Louis University High School. He graduated from Harris-Stowe State College in St. Louis with a bachelor of science degree. He received his masters degree from the University of Missouri, Columbia. He was commissioned through the Reserve Officer Training Corps in 1984.

After completing Initial Qualification Training, he was assigned to Whiteman AFB, MO where he was certified in the Emergency Rocket Communication System. After holding flight commander and instructor duties, Capt. Liddy was assigned to Vandenberg AFB, CA as a command instructor in 1989. He served in both operations and academics flight commander positions for the Improved Launch Control System. He is a graduate of Squadron Officer School and Academic Instructor School.

Decorations include the Meritorious Service Medal w/OLC, AF Commendation Medal, AF Achievement Medal, AF Outstanding Unit Award w/3 OLCs, Combat Readiness Medal w/OLC, the National Defense Service Medal.

He is married to the former JoAnn C. Brinkmann of St. Louis, MO and they have two children. He is currently serving as the commander, Weapons and Tactics Flight in the 321st Missile Gp., at Grand Forks AFB, ND.

DUFFY LIENBENGUTH, MAJ (RET), graduated from St. Louis University and was commissioned as an AF officer on June 6, 1971. After first serving as an administrative officer and squadron section commander at Altus AFB, OK, he volunteered for duty with the Titan II Weapon System. Upon completion of missile launch officer training at Sheppard AFB, TX and Vandenberg AFB, CA he was assigned to the 533rd SMS for the 381st SMW at McConnell AFB, KS in July 1975. While serving as the senior alternate Command Post (ACP) crew commander he earned a rating of "Outstanding" from the 3901st inspection team. In January 1979 he returned to Sheppard AFB where he served for three years as an instructor in the Titan II Missile Launch Officer Course. In December 1982 he was selected as squadron commander for the 3763rd Student Training Sqdn. (aircraft maintenance).

From November 1983 until June 1988 he was a range safety officer/destruct officer at Eglin AFB, FL where he was involved in the testing and evaluation of such new weapon systems as AMRAAM, Tomahawk,

Hellfire, Firebolt, IR and laser guided Maverick, GBU-15 and AGM-130. In June 1988 he was assigned as the chief of safety at Comiso Air Station, Sicily. His final assignment was at the AF Inspection and Safety Center at Norton AFB, CA where he served as chief of Traffic Safety and Off Duty Safety Programs for the USAF.

Major Liebenguth, his wife, Pat and their four children have resided in Pensacola, FL since 1992 where they enjoy swimming and snorkeling in the crystal clear waters of the Gulf of Mexico.

His most memorable experience as a Titan II Missile launch crew commander at McConnell AFB occurred one morning while his crew was on alert at one of their northeastern launch complexes. They were scheduled for an RV Yo Yo that day; a maintenance operation where the nuclear weapon onboard the missile is removed and exchanged for another one which is brought out from the base under armed escort. By 1,000 hours that morning the RV team on site had already removed our reentry vehicle and they were all waiting for the arrival of the convoy transporting the new weapons.

At approximately 1100 hours the phone rang and he found himself talking to a colonel from the SAC Command Post. The colonel was very upset because he had not called to inform him of the arrival of the convoy at his site. He listened patiently while he got chewed out and reminded him how important communications were during missions of this nature. He then explained that the reason he had not made the call to headquarters was because the RV convoy had not yet arrived at his site. Upon hearing that news the colonel became a bit flustered and hung up. About 20 minutes later a very angry brigadier general called demanding to know where "his" nuclear weapon was. He informed him that he did not know where "his" nuclear weapon was and then told him that "my" nuclear weapon, the only one for which he currently had any responsibility, was still located at his lauch complex and he kew exactly where it was. He suggested that if his was lost he should call the wing command post back at McConnell and ask them what had happened to it. His response was unintelligible but he did hear the receiver on his end of the line slam down.

About an hour later the convoy finally arrived and he found out what happened. While trying to make a tight turn going through a small town the clutch had burned out on one of the transporters. The maintenance team went to a local Western Auto store to get parts and fixed the clutch. Then the battery died while getting under way again. During the two hours it took to correct these problems no one, either the security team, or the maintenance team had thought to call the wing command post. Meanwhile a few folks at the SAC Command Post were going nuts because as far as they knew a hydrogen bomb was missing somewhere in Kansas and no one seemed to know where it went. He thought the whole situation was hilarious. A feeling obviously not shared by SAC HQ.

ROBERT (BOB) L. LIVINGSTON, MSGT, after high school graduation, York, PA in 1969, he enlisted into the USAF at age 18. In March of 1970, was sent to Lackland AFB for basic training. Upon completion of basic training, he was ordered to Sheppard AFB, TX to attend a 36 week training course on Titan II missile maintenance. Following tech school, he was permanently assigned to the 381st SMW, McConnell AFB, KS, arriving on station in September 1970. Initially performing duties as missile mechanic and pad (site) chief, he also retrained in missile engines as career fields combined. In late 1971, he was assigned to the staff

functions of the missile squadron. After serving as a top maintenance scheduler in Plans and Scheduling, was later assigned to Job Control, monitoring complex status and maintenance crews daily as well as coordinating with the Battle Staff on critical issues. In 1975, he voluntarily attended another technical school for Maintenance Management and Information Control System (MMICS) at Chanute AFB, IL. Upon graduation, he was assigned to the Documentation Branch and became OPR for all squadrons using this system at McConnell AFB. Maintained this position throughout remainder of career. He was honored as 381st SMW Maintenance Man of the Month. December 1977, was credited with the lowest percentage of working backlog within the missile wing as a maintenance scheduler, was singularly responsible for an Outstanding rating for his documentation section during an annual CAFI inspection, spearheaded development and introduction of the Maintenance Dispatch Data System (MDD), was awarded monetary compensation for approval of suggestion to improve computer system reporting in MMICS, was instrumental in contributing significantly to section being virtually error-free during a SAC ORI.

Honorably discharged from active service in December 1980 at the rank of staff sergeant. Currently resides with wife in Grand Prarie, TX, working with computer support consulting and has been a member of the USAFR since 1986, performing duties of first sergeant of the 931st Civil Engineering Sqdn. holding rank of master sergeant.

JOHN R. LONDON III, LT COL, born June 8, 1953, Rock Hill, SC, enlisted Aug. 8, 1976 in the USAF. Military locations and stations: Tinker AFB, OK, 1976-1980; Cape Canaveral AFS, FL, 6555th ASTG, 1980-1984; White Sands Missile Range, NM, 1984-1989; Los Angeles AFB, CA, 1990-1992; Pentagon, Washington, DC, 1993-1997, Space Shuttle operations at Kennedy Space Center, FL; Secretary of the Air Force Special Projects Office, Space Systems Div., Ballistic Missile Defense Organization.

His memorable experiences: participating in the launch of the first space shuttle on April 12, 1981, building a grand station for a critical classified satellite system, being space based laser program manager.

He was awarded the Defense Meritorious Service Medal and Meritorious Service Medal w/2 OLC.

London was discharged Dec. 1, 1997 as lieutenant colonel. He is married to Joyce Louellen Low of Oklahoma City and has children, Elizabeth and Joshua. Employed as NASA, Marshall Space Flight Centers, AL. Deputy program manager, Advanced Space Transportation Program.

PHILIP G. MACK, MAJ (RET), born Oct. 6, 1923, Knoxville, TN, enlisted in the US Army Air Corps/USAF, July 17, 1942. Served active duty 1942-1945/USAF Reserve 1946-1964.

B-17 pilot, 8th AF, England, 1943-44. Test pilot, operations officer, Special Weapons Field Test Unit, Air Materiel Command, 1944-1945, Dayton, OH, Tonopah, NV, Wendover, UT.

Completed combat tour of 26 missions March 1944.

Conducted tests of early guided missiles, vertical bombs, glide bombs, JB-2 and JB-3, 1944-45.

Memorable experiences: getting shot up by German fighters and flak. Flying chase on JB-2 in P-51 Mustang.

He was awarded the Distinguished Flying Cross and Air Medal w/3 OLCs. Retired from Reserve, 1964 as major. He is married and has four sons and seven grandchildren. Mack is the president of Mack Aviation Company, Inc., large jet aircraft sales, selling airplanes.

ELBRIDGE ALFRED (JAY) MALONE,
SSGT, born Oct. 6, 1952, Portsmouth, OH. Attended Ohio University and the University of Arizona. He enlisted Feb. 6, 1972 in the USAF, 390th SMW, 571st, 390th MIMS, 390th HQ, Davis Monthan AFB.

Malone has specialized in all facets of missile handling as well as propulusion. He is the only dual-qualified missile maintenance tech assigned to maintenance. He served on a Titan operations crew as a ballistic missile analyst tech and was an instructor in the maintenance training branch.

SSgt. Malone was team chief of the 390th Strategic Missile Wing Propulsion Team for Olympic Arena 1980 where they won as best missile maintenance team.

Currently lives in Tucson, AZ where he owns Arizona Rangers, Desert/Mountains Tours.

MICHAEL WILLIAM MANNING, SGT, born
April 13, 1956, Allegan, MI, enlisted in the USAF D.E.P. October 1974, MIANG 1989.

His military/classification/division: USAF, sergeant, 570/571 SMS, BMFT, DMAFB instructor, alt command post crew, 1975 to 1978. MIANG, SSGT, SP, assistant flight chief, Standboard, 1989 to 1995.

Military locations and stations: Lackland AFB, 1975; Sheppard AFB, 1975; Vandenberg AFB 1975; DM AFB, 1975 to 1978; Battle Creek ANGB 1989 to 1995.

Missions: Launch crew readiness DM AFB. Operation Rivet Hawk, DM AFB Operation Desert Storm.

Titan II Memorable experiences: As a youth the awesome responsibility placed on him and meeting the challenge. Career guidance/advice by Commander Thomas W. Fisher. Dating/relationship advice, from Lt. Carlos Odel Cherry. Victory Cigars from Capt. Rick Cradock. Becoming one of Col. Joe Cernys "Tigers". Mo Hollow.

His awards and medals include Honor Grad Basic Training, Combat Crew/Missile Badge, Outstanding Unit w/Device, AF Good Conduct Medal w/1 Device, National Defense Service Medal w/Device, AF Longevity w/Device, Small Arms Expert w/Device, Life Saving Medal for heroism. Honor graduate Security Police, Security Police Badge, four Titan HQ Pins. He was discharged from the USAF 1978 as sergeant MI ANG 1995 sergeant.

Manning lives in Scotts, MI with his wife Susan, and daughters, Lacey and Heather and son, Andy. He received a BA degree and is a shift commander, Dowagiac MI Police.

GARY E. MARSH, COL (RET), graduated from
the University of Denver, June 1954. He served in the

AF from November 1954 until July 1980. Major assignments included Titan II crew commander and instructor crew commander, 395th SMS, Vandenberg AFB, CA, where his crew launched a Titan II, B-1 "Gentle Annie," August 1964; AF Institute of Technology, training with industry, Martin Company, Denver, CO; chief, Minuteman Production Div., Boeing Plant 77, Hill AFB, UT; sector commander, operations officer and squadron commander, 374th SMS, Little Rock AFB, AR; a graduate of the National War College, Class of 1975; deputy commander of Operations, 351st SMW, Whiteman AFB, MO, where he received a masters degree from Central Missouri State University; his final assignment was wing commander, 308th SMW, Little Rock AFB, AR. He and his wife, Jane, reside in Parrish, FL, playing golf and traveling.

SCOTT D. MATTSON, born in McAllen, TX, Oct.
31, 1959, received his BA in history from Louisiana State University and his MA in military story from Kansas State University. He served as missile combat crew commander and November flight commander with the 68th Missile Sqdn. He served as code controller/instructor in the 321st MW and assistant operations officer with the 446th Missile Sqdn. He was also deputy chief, START Treaty Office, 319th BW. His final assignment was deputy chief, Missile Control Flight, at F.E. Warren AFB, and he separated from the AF in April 1995.

Matson and his family currently reside in Cheyenne, WY. He is completing his teaching credential in Social Studies. He is a member of Phi Alpha Theta, the International Honor Society for history, and frequently contributes to military history projects. He is archivist for the AAFM, an accomplished plastic modeler, and collects missile models, replicas and memorabilia.

Most memorable experience: He was privileged to serve as crew commander at the primary keyturn capsule for Giant Pace 88-1, a simulated electronic launch-minuteman (SELM), while serving at Ellsworth AFB. They were in the final keyturn sequence, and the computer in their missile was taking its own sweet time completing a final preparation sequence. As it approached the end of the allotted time to complete the sequence, they feared they would have to stop that portion of the test and start again from the beginning. Thirty seconds from "the moment of no return" the computer reported ready and they were able to get the checklist complete and turn keys on time. Given their training, it was really eerie to actually turn keys in an operational launch control center and watch the indicator panel show "missile away." All this in spite of their knowing the missile was configured to reflect only the electronic sequences, and never actually left the silo.

ALBERT A. MATZAT, born Nov. 8, 1934, Laona,
WI, enlisted in the AF, Feb. 9, 1955, awarded the Master Missile Badge.

He attended basic training at Lackland AFB, TX. F.E. Warren AFB, WY, for training on Aircraft Support Equipment. He was pulled out of his class after two weeks and placed in an instructor position with the course.

Presque Isle AFB, ME, 702nd SMS Snark Missile Program. He worked with the Aerospace Ground Support Equipment until the missile was phased out. Plattsburg AFB, NY, 556th SMS Atlas "F" missile combat crew. Missile phased out.

Vandenberg AFB, CA, Titan II missile. NCOIC of Maintenance Training Control, scheduling and teaching classes.

Kadena AFB, Okinawa, Mace "B" missile crew member. After a back injury he cross trained into recreational services. Andrews AFB, MD, lounge manager.

Requested assignment: AF Recruiting Service, Ft. Wayne, IN. He retired March 1, 1975 as E-6 after 20 years and 22 days.

Memorable experiences: during a practice Atlas "F" Missile tanking exercise a LOX valve hung and the liquid oxygen from the missile could not be down loaded.

Because he worked with GDA contractors during the entire building of the missile silo including wiring and installation of the equipment, he knew all the equipment and circuits. He donned an air pack and ran into the silo unaccompanied and down to level 7 with a short jumper wire. He jumpered the valve open, saving the missile, silo, and all personnel on the complex.

On the return trip up the spiral staircase his air pack went empty at about level 3. When he stepped through the blast doors, his chest felt like he was kicked by a horse. He and his crew commander were both called on the carpet for him violating the two man concept. But the missile was saved. Until now, he has not told anyone what he did in the silo because it would have been a violation of the "need to know" policy.

VANCE G. MATZKE, MAJ (RET), graduated from
the University of Idaho College of Forestry in June, 1965, and spent a short time with Potlatch Forests, Inc., in Lewiston, ID, and with the Aeronautical Chart and Information Center in St. Louis, MO. He served in the AF from Oct. 21, 1966 to Oct. 30, 1986. Duties include the missile maintenance officer in the 341st MIMS at Malmstrom AFB, MT and at ISTRAD and the 394 SMS at Vandenberg AFB. He also served on a missile combat crew in the 564 SMS at Malmstrom AFB, as a missile maintenance officer in the 341st Civil Engineering Sqdn. at Malmstrom AFB, and as chief of the Civil Engineering Section and OIC of the Training Management Section in the 3901 SMES at Vandenberg AFB. He was on the 3901 SMES Scoring Team for five SAC Missile Competitions. He further served as maintenance supervisor in the 91 OMMS, and 91 FMMS, and chief, QA Div., 91 SMW at Minot AFB. He earned an MS in Systems Management from the University of Southern California in 1976. Following retirement, he worked for Rockwell International at Vandenberg AFB and as a district executive for the Boy Scouts of America in Santa Maria, CA. He and his wife Jerri have lived in Santa Maria, CA. Since retirement in 1986, he is presently enjoying a rewarding mini career as a naturalist at an outdoor school. They have three sons: Scott in Shakopee, MN; Brett in Nanuet, NY and Mark at the USAFA in Colorado Springs. He enjoys Boy Scouting, golf, hiking, and biking. Aced hole #17, Minot Country Club, June 3, 1985.

WOODROW W. MCCLELLAND, MSGT, born
Aug. 18, 1930, Abbeville, LA, enlisted in the USAF Jan. 28, 1947.

His military locations and stations: FEAF Tokyo Japan, 1947-52; Barksdale AFB, 1952-54; Yokota AFB, Japan, 1957-60; VAFB 1960-66, L.R.A.F.B 1966-69.

Arrived at Vandenberg AFB the day after 395A launcher 1 was launched. Participated in all receipt to

...unch in the CAT 2 and 3 programs. During Cuban Missile Crisis, they put three Titan missiles on alert in only seven days. (empty holes to EWO).

When the missile program became all military management, they organized unique all blue suit launcher refurb team for Titan I. They reduced 280 civilians to 20 blue suits, 180 days to 5 days to do same job. After Titan I program, became refurb manager for Titan 2. After transfer to L.R.A.F.B took over as head Titan 2 maintenance team at 308 M.M.S. until retired in Abilene, TX near Dyess AFB.

He and his wife Hamako have eight children. They own and operate FAA repair station for airborne A.P.U. In spare time they travel and visit with all their children and grandchildren.

SAMUEL (SAM) M. MCGOWAN, after commissioning was assigned as an Atlas E crew member with the 548 SMS, Forbes AFB, KS. Following Atlas phase out in 1965 was transferred to the 321 SMW, Grand Forks AFB, ND, serving as a Minuteman II line, instructor and standardization crew member and trainer operator. Transferred to the 4315 CCTS, Vandenberg AFB, in 1969 as an instructor. Assigned to 91st SMW, Minot AFB, ND, in 1974 as a maintenance officer. Assigned to the 5 BW, Minot AFB, as the logistics officer in 1976. Liaison officer with Royal Saudi AF 1978-79. Logistics staff officer, HQ AFLC, from 1979-1985. Engineer with Rockwell Autonetics 1985-93 on the Peacekeeper, Small Missile, Peacekeeper Rail Garrison and Ground Based Interceptor programs; installed the ICBM Code Processing Systems at HQ SAC and each MM Wing. Consultant preparing proposals 1993-95. System engineer with TRW Credit 1995-1997, currently with TRW working the National Missile Defense program.

KATHRYN M. MCGUIRE, MAJ, came on active duty in January 1980 and received her commission as a graduate of Officer Training School, Lackland AFB, TX on April 23, 1980. She attended both the Titan II Missile Launch Officers Course at Sheppard AFB, TX and the Titan II, SAC Missile Combat Crew Initial Qualification Training at Vandenberg AFB, CA. In September 1980, she was assigned to the 571st SMS at Davis-Monthan AFB, AZ as a deputy missile combat crew commander. She became an alternate command post deputy combat crew commander. She became an alternate command post deputy combat crew commander and an instructor deputy missile combat crew commander in June 1981. In December 1982, she upgraded to missile combat crew commander.

Major McGuire was then assigned to the 44th SMW, Ellsworth AFB, SD as the second female officer code controller, Minuteman II in March 1994.

In July 1985, Maj. McGuire was specifically selected for initial cadre of (10) women selected for Minuteman Missile Crew duty. She attended the Minuteman Modernized Improvement Launch Control System Initial Qualification Training at Vandenberg AFB, CA in December 1985. Upon graduation, she was assigned to the 509th SMS, Whiteman AFB, MO. Her vast potential and in-depth knowledge led her to immediate upgrade as crew commander, Minuteman II, ICBM.

In September 1987, Maj. McGuire became the chief, plans and programs for the 351st Transportation Sqdn., Whiteman AFB, MO. While in this position, she served as the interim squadron commander.

In September 1988, Maj. McGuire served a remote tour at Osan AB, South Korea as the traffic management officer for the 51st TFW.

In September 1989, Major McGuire became the officer in charge, air freight at the 436th Aerial Port Sqdn., Dover AFB, DE. She was deployed from August 1990 through March 1991 to Dhahran and Riyadh, Saudi Arabia with the 1681st and 1680th Airlift Control Sqdns. Provisional in support of Desert Shield/Storm. April through July 1992, she was deployed to Dhahran, Saudia Arabia as the 4404th Composite Wing Transportation Sqdn. Commander.

In October 1992, Maj. McGuire was assigned to HQ, Allied Forces Southern Europe (NATO), HQ Command, Naples, Italy as the chief, Transportation Branch. She managed and supervised US and Italian AF, Army and Navy personnel, as well as, Italian civilian labors and Italian NATO employees.

In October 1955, Major became the Cargo Movement Operations System program manager, Transportation Systems Div., Logistics Support Systems Directorate, Headquarters Standard Systems Group, Maxwell AFB, Gunter Annex, AL. April through July 1997, she was deployed to HQ US European Command, Joint Movement Div., Stuttgart, Germany in support of Operations JOINT GUARD. Major McGuire is scheduled to report to her next assignment on Nov. 26, 1997 as the commander, Aerial Port Flight, 640 Air Mobility Support Sqdn. (AMC), Howard AFB, Panama.

Major McGuire's military decorations include the Defense Meritorious Service Medal, Joint Service Commendation Medal, AF Commendation Medal w/2 OLCs, Joint Service Achievement Medal, AF Achievement Medal, Joint Meritorious Unit Award, Combat Readiness Medal, Southwest Asia Service Medal w/3 Bronze Stars, Kuwait Liberation Medal (Saudi Arabia), and Kuwait Liberation Medal (Kuwait).

Major McGuire is married to Mr. James I. McGuire. They have a daughter, Kelly.

JAMES J. MCLAUGHLIN, LT COL, born Nov. 29, 1921, in Weekhawken, NJ, the son of Nell and James P. McLaughlin. He attended grammar and high schools at Holy Trinity in Hackensack, NJ. When WWII broke out he enlisted in the US Army Air Corps. He won his pilot's wings and was commissioned a second lieutenant in March, 1944. Stationed with the 15th AF in Italy he completed a tour of 35 combat missions over Germany as a bomber pilot in B-17 aircraft. In recognition of his service he was awarded the Distinguished Flying Cross, and three Air Medals.

At the conclusion of the war he returned to civilian life and enrolled in undergraduate studies at New York University. The outbreak of war in Korea led to his recall to active military duty as a captain in 1953. Having decided to make a career with the newly formed USAF he married the former Wallis P. Angst of Pleasantville, NY. Accompanied by his wife, he underwent further training and was assigned to test pilot duties flying B-36 aircraft in Ft. Worth, TX. He went on to finish his AF sponsored undergraduate studies in Aeronautical Engineering at the University of Colorado and was posted to Japan and Okinawa. Promoted to major, he combined his piloting duties in C-130 Troop Transport Aircraft with a position as aircraft maintenance engineering officer.

Upon return to this country he was assigned to the AF Aircraft Maintenance Depot at Warner-Robins, GA. From there, his earlier experience in the maintenance field led to a tour of duty in Pakistan where he headed an AF team engaged in the training of that country's personnel in the operation and maintenance of the C-130 airplane.

When the first Minuteman ICBM System was deployed in Minot, ND, Maj. McLaughlin was selected to attend combat crew training and became one of the first combat crew commanders to man missiles in the dawn of the nuclear age.

After concurrently completing the academic requirements for a master's degree he was promoted to the rank of lieutenant colonel and served out his military career as a missile squadron commander. Upon retirement he and his wife, Wally, settled in Maine.

RONNE G. MERCER, COL (SEL), served as the deputy commander, 58th Logistics Gp., Kirtland AFB, NM. From Santa Barbara, CA, he entered active duty in 1966 and completed basic and technical training at Amarillo AFB. He served at Norton and McClellan AFBs, Karamursel, Turkey, and Lackland before he entered AECP. He obtained a management degree from Southwest Texas State University in 1974 and completed OTS and both the Titan II Missile Launch Officer Course at Sheppard AFB and ORT Course at Vandenberg AFB as a Distinguished Graduate. He was assigned in 1975 to the 390th SMW and 570th SMS at Davis-Monthan AFB, AZ, where he served as a deputy missile combat crew commander (DMCCC), instructor DMCCC, and MCCC and as wing instructor DMCCC on the upgrade training crew. While a crew commander his crew was selected SAC Titan II Combat Crew of the Month for November 1977. His last missile assignment was chief, Scheduling Branch where he worked integration of females on the crew force.

He attended AFIT acquiring an MS in Logistics Management in 1980; then served as a supply officer at George AFB, CA; with the US Military Training Mission, Dhahran, Saudi Arabia; on the ATC staff at Randolph AFB: and at the ALC, Kelly AFB. He completed ACSC in residence and then became the commander, 341st Supply Sqdn., Malmstrom AFB, MT, where he supported the largest missile wing in the free world. From Montana to sunny Hawaii where he served on the logistics staff at US Pacific Command, then on to Kirtland where he spent two years with the AF Inspection Agency. He then attended Air War College in residence and returned to Kirtland in his current position in 1996. He will be promoted to colonel on Feb. 1, 1998.

He received the DMSM, MSM, JSCM, AFCM, JSAM, AFAM, Combat Readiness Medal, AF Good Conduct Medal, Humanitarian Service Medal, and NDSM. An avid bowler and golfer he participated in MAJCOM-level in each of the past four decades. He lives with his wife, Veronica, and son, Timothy, in the Albuquerque area.

ROOSEVELT (TED) MERCER JR., COL, is commander of the 91st Operations Gp., Minot AFB, ND. In this position he leads 1150 AF men and women to support the nation's deterrent objectives by the operating, maintaining and securing 150 Minuteman III intercontinental ballistic missiles (ICBMS). The group operates 15 missile alert facilities and 150 launch facilities remotely located in North Dakota.

Mercer has spent most of his AF career as a missile combat crew member, instructor, commander, squadron commander and missile maintenance deputy group commander. He has been involved in personnel programs for missile officer assignments, and was assigned to the Pentagon as chief on Congressional Affairs for DCS/Personnel. He served in Europe as chief of the Nuclear Div. of the Plans and Policy Directorte, HQ USEUCOM. While in Europe, he was responsible for the development, coordination and execution of policies and procedures for nuclear weapons and associated systems for the European Theater. Colonel Mercer recently left Patrick AFB, FL where he served as the commander of the 45th Logistics Gp.

He was awarded the Defense Superior Service Medal, Meritorious Service Medal w/OLCs, AF Commendation Medal, Combat Crew Readiness Medal, AF Achievement Medal, National Defense Service Medal.

MAX L. MEYER, born Aug. 2, 1931, Boone IA enlisted in the USAF on Aug. 9, 1949. Major duty stations during his AF career were: Travis AFB, Wahoo, NE Communications Site, Offutt AFB, RAF Station Sturgate, RAF Station Brize Norton, Holloman AFB, Edwards Rocket Site, Vandenberg AFB, Hawes Radio Relay Site and March AFB.

Memorable experiences: he participated, as a launch crew member, in the first THOR ballistic missile launch by SAC Gen. Wade, in his report, listed a number of "firsts" resulting from the launch. (a) The first ballistic missile fired from Vandenberg westward into the Pacific Missile Range. (b) The first known operation where new equipment, a new missile, new personnel and a new organization were integrated into a successful operation. (c) The first ballistic missile to be launched by a SAC operational crew. (d) The first completely automatic launch of an operational ballistic missile using normal launching procedures.

He married L. Joline Willett in 1959 while stationed at Vandenberg AFB. They are now living in San Diego, CA.

JOHN A. MILLER, MAJ (RET), the 54 Lucky Bag predicted Tony would return to Minnesota's sky blue waters and land of 10,000 lakes—and that is just what he did. His AF career saw him served in many fields such as Guided Missile Operations, AFROTC, think tank long-range strategic planning, Command and Control System design and procurement and DOD Graduate Management training.

Before heading home to Minneapolis, he had duty that took him to many locales including Colorado, Florida, Germany, North Africa, Alabama, New Jersey, Washington, DC, New York, and Massachusetts, somewhat in that order. He also was at Cape Kennedy and in the USSR during the Apollo-Soyuz Joint Space Mission.

Major civilian activities have included teaching, coaching, media work, politics plus retail sales at the new Mall of America and marketing work at Caterbury Downs. A favorite hobby, along with his partners, is watching their horses race. He is part of the volunteer sports press box crew at University of Minnesota Golden Gopher hockey, football and basketball games.

Miller tries to stay in touch with classmates and other USNA grads in the area. He and 1954 Classmate, Tom Hooely, attended the 6th and 7th games of the Twins "91" World Series victory over Ted, Jane and the Braves. Proud of his class and classmates, Tony says please let him host you if you are ever in the Twin Cities area of Minneapolis-St. Paul. That's a wrap from Laundry No. 6495.

RALPH A. (ARCHIE) MILLOTTE, MAJ (RET), born May 3, 1929, in Rochelle, IL. After graduation from Northwestern University in May 1951 enlisted in the AF. Spent a short time as a control tower operator at Webb AFB, TX before receiving his commission through Officer Candidate School and serving as an AF officer from June 19, 1953 until Aug. 1, 1971. Duties included detachment commander at 610th Aircraft Control and Warning Sqdn., Itazuke AFB, Japan,

Chief of the Training Devices Branch at Little Rock, AFB, AR, Job Control at 549th SMS, Offutt AFB, NE, staff maintenance officer, Missile Systems Branch at 2AF, Barksdale AFB, LA, missile combat crew commander at 447th SMS, Grand Forks AFB, ND and Missile Operation staff officer at 4751 Air Defense Sqdn., Eglin AFB Aux. Field 9, FL.

Not married, has resided in Rochelle, IL since 1971. Active in community service work, teaching and playing bridge.

ROBERT F. MOORE, LT COL, is an AAFM life member. After graduation from the University of North Carolina at Chapel Hill in 1980, he reported to the 740th SMS, 91st SMW, Minot AFB, ND.

On crew, he served as an instructor and flight commander, and earned the 15th AF Crew Excellence Award. Follow-on assignments include AFROTC instructor, Memphis, TN; deputy chief of Protocol, Andrews AFB, MD; executive officer, 960th AWACS, Keflavik NAS, Iceland; chief of protocol operations and wing executive officer, 89th Airlift Wing, Andrews AFB, MD; executive officer to COMSIXATAF, Izmir, Turkey; and chief of strategic planning, Air Force Communications and Information Center, Pentagon.

His medals include Defense Meritorious Service, Meritorious Service w/2 OLCs, Commendation, Achievement, and Combat Readiness. Upon retirement, Lt. Col. Moore will return to North Carolina and try to forget what "busting" an IG recheck was like.

EDWARD F. MORAN JR., CAPT, after graduation from Georgetown and the University of Pittsburgh, he was assigned to the 390th SMW at Davis Monthan AFB in Tucson, AZ. During his four-year assignment until 1972, he was never rated below highly qualified on any evaluation, he was the first evaluation deputy and commander in the wing's history, and he participated in Titan Launch M1-17 from Vandenberg AFB, CA.

After an initial key turn, Range Safety shutdown the sequence. Launch orders initiated a second key turn hours later. The crew reacted quickly to simultaneous "lift off" and "abort" indication, shutting down the missile and protecting the silo from the explosion as the start cartridge shattered and cut off the turbine exhaust. The bird was recycled and finally launched on June 20, 1971.

He and his wife, Nancy, reside in Tucson where he is a senior partner in a regional CPA firm, Moran, Quick & Yeanoplos. He is currently treasurer of the Arizona Aerospace Foundation and Titan Missile Museum, a national historic landmark.

MARK L. MORGAN, born on Travis AFB, CA and grew up on several SAC bases prior to graduating from the University of New Mexico with a BA in history in May 1976. He has been a member of the association since 1994 and is the son of Lt. Col. Richard H. Morgan (member).

Morgan entered the USN on May 18, 1976, transferred to the Naval Reserve in November 1986 and left the service on Dec. 31, 1988. Duties included administration assistant, Attack Sqdn. 52, NAS Whidbey Island, WA; student naval flight officer, NAS Pensacola, FL; bombardier/navigator, airframes/powerplants branch officer, Attack Sqdn. 42, NAS Oceana, VA; and aviation fuels/flight deck officer, USS Nashville (LPD-13), Naval Station, Norfolk, VA.

In June 1981 he reported to Navy Astronautics Groups, NAS Point Mugu, CA, as satellite launch officer. During his tour he served as command representative to NASA and Western Space & Missile Center and was responsible for scout launch operations from SLC-5 at Vandenberg AFB. In 1983, as acting ops officer for Navy Astronautics Group, he coordinated the successful launch of satellite P83-1 for the Air Force. His subsequent tours included assignment as admin. department head and assistant training officer, Fleet Intelligence Training Center, NTC, San Diego, CA; and air intelligence officer, Helicopter Combat Support Sqdn. 9, NAS North Island, CA.

After leaving the Navy he worked for four years as an aerospace engineer with the Lockheed Skunk Works in Burbank, CA, and General Dynamics in Forth Worth, TX. He then spent five years as a research historian/park ranger with the National Park Service at Vicksburg National Military Park, MS, and Steamtown National Historic Site, Scranton, PA.

He has resided in Oak Harbor, WA since 1996 and is employed as a newspaper reporter, photographer, political cartoonist and consultant on Cold War installations and weapon systems. His first book, on Army Air Defense Command and the Nike Missile system, was published in 1996. He is currently working on a second volume as well as histories of the A-6 Intruder and the ICBM program.

RICHARD H. MORGAN, LT COL (RET), USAF, born in Muskogee, OK and worked briefly as an actor and singer in the Bay Area prior to entering the Air Force Jan. 31, 1952 as an aviation cadet. He left the USAF Aug. 3, 1956, was recalled Sept. 19, 1959 and retired May 31, 1981.

Duties included navigator training with the 3605th Navigator Training Wing, Ellington AFB, TX; bomb/nav in the 31st BS, 5th BW, Travis AFB, CA; bomb/nav, 343rd BS, 98th BW, Lincoln AFB, NE; navigator, 98th Air Refueling Sqdn., 98th BW, Lincoln AFB, NE; navigator, 306th Air Refueling Sqdn., 4147th Strategic Wing and 306th BW, McCoy AFB, FL; Missile Combat Crew Commander, 447th Strategic Msl. Sqdn., 321st Strategic Msl. Wing, Grand Forks AFB, ND; Weapon System Officer, 34th Tactical Ftr. Sqdn., 388th Tactical Ftr. Wing, Takhli RTAFB; Wing Radar Strike Analysis Officer and Chief of Radar Strike, 388th Tactical Ftr. Wing, Takhli RTAB; Weapon System Officer, 8th Tactical Ftr. Sqdn., 49th Tactical Ftr. Wing, Holloman AFB, NM; Wing Chief WSO, 49th Tactical Ftr. Wing, Holloman AFB, NM; Commander, 49th Field Maint. Sqdn., Holloman AFB, NM; Commander, HQ Section, 388th Tactical Ftr. Wing, Hill AFB, UT; and a final assignment as deputy commander for maintenance, 32nd

Tactical Ftr. Sqdn., Camp New Amsterdam, the Netherlands, during which time the squadron won two consecutive Hughes Trophies.

In 1965 Dick earned a BS history degree from Florida Southern College. He earned his MS in industrial management from the University of North Dakota in 1970. During his career he earned the Master Navigator and Senior Missileman ratings.

After leaving the Air Force he worked for six years with McDonnell-Douglas Corporation on the Peace Sun Program, including two years at Khamis Mushayt, Saudi Arabia. He then served as professor of history and drama at Skagit Valley Junior College, prior to retiring a second time. Dick and his wife Dottie have lived in Oak Harbor, WA since 1996.

MICHAEL J. MORNEAU, MAJ (RET), is a da-
tabase analyst in the 10th Comm Sqdn. at the AF Academy (USAFA). After graduation from St. Michael's College, VT in May 1971, he served as an AF officer from June 18, 1971 to Sept. 30, 1992. He was first assigned to the 66th SMS at Ellsworth AFB, SD serving as an "Echo 01" SCP (Squadron Command Post) crew member. Notably, the first SCP  crew deputy to upgrade directly to SCP crew commander, and being more experienced than the instructor, taught his own SCP upgrade class too. In 1977, assigned to AFROTC at the University of Minnesota, MN. He was commandant of cadets, and is Det. 415 Arnold Air Society/Angel Flight (Silver Wings) Honorary Lifetime advisor.

In 1980 he was Distinguished Graduate of his MMM-ILCS IQT Class. Assigned to the 510th SMS at Whiteman AFB, MO, he was an "ERCS" flight commander and received the 8th AF Crew Member Excellence Award. He is further distinguished as one of the few strictly Minuteman combat crew members to approach the 600 Alert mark. He entered the AF Reserves Feb. 6, 1984 as an AF Academy/AFROTC admissions liaison officer (ALO). In January 1986 he joined the 442nd TFW, Richards-Gebaur AFB, MO, as the munitions safety officer, and was in the First Joint Reserves/Active Forces NATO TAC Eval at Sembach AFB, Germany. After 1989, he finished his career as an ALO. A Retired Reservist he is the assistant executive director of the USAFA Colorado Parents Club. In October 1986, Mike entered Civil Service, at HQ Engineering Installation Div., Tinker AFB, OK and August 1988 at the Academy. He has business in International Development and Business-Database Record Keeping and is very involved in his church and community service. He resides in Colorado Springs, CO with his wife, Marlene and his children, Jon and Katie.

DON E. (MOON) MULLINS, MAJ (RET), is a
member of the Association of Air Force Missileers, a life member of the Air Force Association, and a life member of the Retired Officers Association. After graduation from the University of North Texas in 1962, he attended USAF Security Service Communications Intelligence School, Goodfellow AFB, TX and was assigned to the 6950th Security Group, RAF Chicksands, England as a flight commander and exploitation officer from 1964-67. He entered the Minuteman Education Program in 1967 and was a Minuteman I crew commander and wing instructor crew commander in the 91st SMW, Minot AFB, ND from 1967-71. While at Minot he received his MS degree in Industrial Management.

Upon completion of his four year tour at Minot, then Capt. Mullins was returned to the intelligence field and served 1971-72 in SEA with Task Force Alpha, Nakhon Phanom RTAFB, Thailand. Upon returning to the CONUS he served with the 4315th Combat Crew Training Sqdn. as the Minuteman Modernized Senior Instructor Evaluator from 1971-75. He attended Air Command and Staff College, Maxwell AFB, AL and graduated in the class of 1976. Upon graduation, then Maj. Mullins remained on staff at ACSC as an instructor and later the ACSC scheduler. He remained at ACSC until his retirement in April 1983.

Major Mullin's decorations include the Bronze Star, Meritorious Service w/OLC, AF Commendation Medal w/2 OLCs, the Vietnamese Cross of Valor w/palm, Vietnamese Service Medal, Armed Forces Reserve Medal, Outstanding Unit Award w/3 OLCs and a V device for valor and the Air Force Marksmanship Medal.

He and his son, Reade, live in Sherman, TX where he is an economics instructor at Grayson County College.

PAUL F. MURPHY, COL (RET), began Minute-
man duty at Whiteman in 1964. Murph spent more time on standboard without a 3901st check than anyone else in SAC. He was a standboard deputy and instructor on crew S-001. He depostured the last originals before transitioning to force mod. His AMCCC was not qualified as a DMCCC so Murphy was dual-rated and got to run the show from the backseat. Something Jack Rowe always thought he tried to do.

After completing MMEP, Murph got his PhD at Syracuse. He taught at SOS and wrote a missileer poem called "Alert" which was published in 1973 in the AF Times. He had a six-year stint in Washington, DC, culminating at the White House and graduated from National War College in 1980.

From 1980 to 1985, he commanded the 319th SMS, was ADO of the 90th, launched a Minuteman III, was Whiteman's Asst. DCM and became SAC's LGB before returning to Whiteman as vice commander.

He was chairman of the Department of Leadership, Air War College and retired in 1986 to consult in aerospace. Murph returned to missiles as Aerojet's director of quality. He also co-founded a quality consulting firm in Houston and won the International Platform Association's 1994 "Hal Holbrook Cup" as their outstanding speaker. In 1997, he and Helen moved to Punta Gorda, FL and set up their own firm, Murphy Management Innovations.

JOHN T. NAILEN, MAJ (RET), born Sayre, PA,
enlisted in the USAF, Oct. 9, 1956 and retired as major, Aug. 1, 1986.

ICBM assignments: 579th SMS (ATLAS F) Walker AFB, NM, 1962-65; 10th Air Defense Gp., 25th Air Defense Sqdn. (THOR/Program 437), 1969-71; 44th SMW (MINUTEMAN II), Ellsworth AFB, SD, 1972-74; 15th AF, March AFB, CA, 1974-76; SAC, Missile IG Team, Offutt AFB, NE, 1976-79; 351st SMW (Minuteman II) Whiteman AFB, MO, 1979-83; HQ SAC, ICBM Maintenance Directorate, Offutt ABF, NE, 1983-86.

Enlisted NCO for the first 15 years, primarily as maintenance ballistic missile analyst technician (BMAT). While assigned to 25th ADS, as launch console operator, ran booster countdown, turned key and launched a THOR from Johnston Island.

Commissioned Jan. 12, 1972, assigned as missile maintenance officer until retirement. Task Force maintenance officer for GT-143MS-1 in 1982, first successful launch with ARSIP-modified guidance can, following initial hangfire due to ordnance driver failure.

Education: BA in social sciences, LaVerne College, 1971 through Bootstrap Program from Vandenberg, MA, Planning and Public Administration, Pepperdine University, 1976.

He has worked for TRW as systems engineer at ICBM SPO, Ogden ALC, UT, as deployment planner with Rivet MILE (Minuteman Integrated Life Extension), 1986-88 and currently as ICBM Long Ranger Planner.

Living in Ogden, UT; married to Sandra since 1959 and they have two daughters and a son.

ROBERT H. NEBEN, COL, born Jan. 7, 1943 in
Evergreen Park, IL. He has a bachelor of arts degree in economics from William Penn College in Oskaloosa, IA, and a master of arts degree in management from the University of Northern Colorado. He has completed Squadron Officer School, Air Command and Staff College, and National Security Management. He enlisted in the AF in 1966 and was commissioned in  1967 after attending OTS at Lackland AFB, TX. He was a Minuteman I and later a Minuteman III MCCC and an instructor in the 741 SMS at Minot AFB, ND.

He completed the crew that launched the first operational Minuteman III from Vandenberg AFB, CA. He then programmed the Minuteman II launch codes for the 44 SMW at Ellsworth AFB, SD and taught code handling procedures to launch crews, maintenance personnel, and wing staff. He then became an automated systems analyst at the HQ SAC where he led a team that designed and programmed the HQ SAC Missile Early Warning System. As a reservist he was assigned to the Aeronautical Systems Div. at Wright Patterson AFB, OH. When he was promoted to colonel he was assigned to Tinker AFB, OK as an acquisition engineer. In 1995 he assisted the AF Center for Environmental Excellence, Brooks AFB, TX in conducting a critical environmental source selection. In 1997 he was the deputy director for Crisis Action at the HQ US Pacific Command under Operation Pacific Haven. He coordinated the transfer of over 6,600 Kurds from northern Iraq to Guam and then to the Conus to process for US citizenship. He retired from the Reserves in April 1998. He presently works for Science Applications International Corporation in Dayton, OH as a senior systems analyst. He also teaches statistics for the University of Dayton and Acquisition Management for the Air Force Institute of Technology. Colonel Neben is a private pilot and he holds an amateur extra class radio license.

Colonel Neben has been awarded the AF Achievement Medal, Air Force Commendation Medal w/2 OLCs and the AF Meritorious Service Medal.

Colonel Neben and his wife Micky live in Dayton, OH. They have a daughter and three sons. One son has served in the AF, one served in the Army, and one served in the USN.

WILLIAM R. NEWELL, COL (RET), served in
the USAF from Aug. 23, 1960 to Aug. 1, 1989. Following his graduation from the University of Maryland in 1960, his initial assignment was to Whiteman AFB, MO, as the 340th BW's Organizational Maintenance Sqdn. administrative officer. Other Administrative Career Field assignments included: assistant director of Student Affairs at the USAF School for Latin  America, Albrook AFB, CZ; the 4608th Support Sqdn., Ent AFB, CO—serving both HQ North American Aerospace Defense Command and HQ Air Defense Command; and three years at HQ, Pacific Air Forces, Hickam AFB, HI, as assistant executive officer to the deputy chief of staff for Materiel.

His career as an Air Force Missileer began in May 1969 at the 321st SMW, Grand Forks AFB, ND, where he was a missile combat crew commander, the wing senior standardization evaluation crew commander; and chief, Wing Standardization Div.; earned a MS in Industrial Management from the University of North Dakota. He was reassigned to HQ, 15th AF, March AFB, CA, in November 1972 as chief of the directorate of Missile's Standardization Div., serving in that capacity until June 1977. He attended the US Army War College, Carlisle Barracks, PA, from July 1977 to June 1978. The next assignment was to the 90th SMW, F.E. Warren AFB, WY, where he was wing self-inspection manager; chief, Plans and Intelligence Div.; Commander, 321st SMS; and assistant deputy commander for operations. He returned to HQ 15th AF as director of Missile Operations in May 1981. In August 1982, he was reassigned to HQ, USAF in Europe, Ramstein AB, Germany, as the AF's (and HQ USAFE's) first major command director of Ground-Launched Cruise Missile Operations. In August 1985, he was assigned to Sheppard AFB, TX, as inspector general for the Sheppard Technical Training Center. His final base of assignment was Vandenberg AFB, CA, where he served from June 1986 to August 1989, first as the 1st Strategic Aerospace Divs. special assistant for Space Shuttle Operations, and then as chief of the Environmental Management Directorate's Development Div.

He and his wife, Mary Lou, have resided in Colorado Springs, CO, since August 1989. He is currently employed as a senior planner with Higginbotham/Briggs & Associates, an architectural and planning firm that does military master and community planning worldwide, and similar work in the civilian sector. He also enjoys working with his computer, and running for pleasure (competitively also), in the Pikes Peak Region.

JOHN P. NOGUES, MSGT, born Oct. 10, 1936 in Los Angeles, CA, enlisted in the USAF March 11, 1957. After Tech School at Scotts AFB, IL (Radio-Relay) he was assigned to the 6314th Comm Sqdn. (K-55) Osan, Korea. January 1959 he crosstrained into Bomb-NAV Tech on B47E at the 341 A&E Dyess AFB, TX. When the 341st BW was deactivated he attended Atlas-F training at Sheppard AFB, TX. Upon his return to Dyess he served on the Mobile Calibration Team of the 578th he subsequently attended Minuteman I training at Chanute AFB, IL and was reassigned to Whiteman (WG4) on a MMT. While at the 351st he participated in "posturing" "N" Flight (the last to reach Strat Alert) and depostured the first launcher in "A" flight to be turned over to Boeing for Force Mod. Upgrade.

He was the first airman in the 351st to reenlist in a L.F. Workcage while mating a G& C unit. In 1966 he returned to Chanute AFB permanent party and was an instructor in the Officer Course OZR-2825 until 1970. While at Chanute he was the first NCO to reenlist in the classroom with an officer student doing the ceremony.

Upon transferring to 1STRAD Vandenberg AFB, he served with the launch analysis group (LGLA) for SAC and AFLC Minuteman Telemetry Analysis.

Upon transfer in 1974 to Det. 16 AFPRO at Autonetics Anaheim. He served in the deputy for Minuteman-Engineering Office as NCOIC of Missile Software.

He retired April 1978 to Nuevo, CA where he is currently a firearms dealer and instructs in firearm safety for the NRA and California Department of Fish & Game Hunter Safety Course.

JACK A. OAKMAN JR., graduated Clearfield Area High School, Clearfield, PA, June 8, 1966. President Electra Omega Lambda Phi fraternity until graduating from Allegheny Technical Institute, Pittsburgh, PA in 1968. His active AF enlistment began, Dec. 13, 1968. While in basic training at Lackland AFB, he served as dorm chief until graduation. He attended technical training at Chanute AFB, serving as class leader, assistant barracks commander and squadron guidon, completing training as a Minuteman II EMT (316XO). While serving in 91MIMS at Minot AFB, A1C, Oakman became a two stripe EMT team chief and was "dual

trained" for MMI and MMIII. He was rewarded with the SAC Master Team Chief Award then later transferred to 91HQDS as team training instructor. Sergeant Oakman served one term and honorably separated from the AF Dec. 7, 1972. Long a resident of Stockbridge, GA, Oakman has worked for Delta Air Lines, Inc. since Dec. 26, 1972 and is currently regional manager, properties and facilities. Married to Sherry (Campbell) since Aug. 29, 1970, they have two children, Rebecca and Jeffrey. Mr. Oakman is well known for his community involvement where he has served as volunteer fire chief, reserve police captain, and vice chairman of the Henry County Board of Tax Assessor with memberships in the Georgia Association of Assessing Officials, Association of Air Force Missileers, and Post 33 of the American Legion.

A1C Oakman and his entire class of MMII Electro Mechanical Team members were surprised by being assigned to Wing III at Minot AFB, ND which was then a Minuteman I base. It was determined that with Wing III scheduled to receive the new Minutemen III MIRV, and because there were no MMIII trained EMTs, new MMII personnel could be field trained for MMIII.

While servicing a MMIII LF at Wing III, Sgt. Oakman may have been the first team chief to have experienced a REPS fuel leak reading on the new MMIII MIRV. After emergency evacuation and EOD response, it was later determined to have been a fungal growth in the plastic tube leading into the reentry vehicle.

RAY OBERMAN, MSGT, born April 20, 1936, in Andrew, IA to Edwin and Freda Beck Oberman, married Joyce Shipley, Dec. 7, 1957, in Downey, CA. He was a master sergeant in the USAF where he worked in electronics for over 23 years. He served in the radio shop, 9th BW (Medium), Mountain Home AFB, ID and was NCOIC of the guidance shop, 569th SMS, also at MHAFB, involved in maintaining the ground guidance system including the Univac Athena computer for the Titan I ICBM (SM-68). He was subsequently involved in missile maintenance at Vandenberg AFB 1965-1975, retiring as a master sergeant.

He died at the age of 58, July 13, 1994 and is survived by Shirley Jessen, a friend with whom he made his home; a son, Darrell, of Fulton, IL; a daughter, Dori, of Galesburg, IL; his father and stepmother, Edwin and Selma Oberman of Maquoketa; a brother, Leo Oberman of Clinton; a sister, Mrs. George (Dorothy) Donnelly of Seligman, AZ; and two stepsisters, Marge Bott and Helen Bender, both of Clinton. He was proceeded in death by his mother and a son in infancy.

JOHN E. O'CONNOR JR., COL, is director, Space and Missile Systems at the AF Operational Test and Evaluation Center, Kirtland AFB, NM. A 1974 ROTC Distinguished Graduate of Manhattan College in New York, his first assignment was with the 341st SMW, Malmstrom AFB, MT where he was an instructor and evaluator. In 1979 he began a tour at Offutt AFB, NE in the SAC Underground Command Post, first as a missile warning officer and then as chief, SAC Warning and Control Systems Operations Branch. He next went to Hanscom AFB, MA in 1983 in the Missile Warning Systems Directorate, and then as deputy director for Advanced Technology. After graduation from Air Command and Staff College in 1989, Col. O'Connor spent a year at Andrews AFB, MD with the 1875th Computer Systems Sqdn., Pentagon in the Space and Nuclear Deterrence Directorate, assistant secretary for acquisition, as program officer for AF Strategic Defense Initiative programs, and then from 1992 through 1996 as the chief, Mission Support, Congres-

sional and Budget Div. Following a brief assignment as chief, Requirements and Budget Integration at Ogden Air Logistics Center, Hill AFB, UT, he reported to AFOTEC in November 1997.

Colonel O'Connor has been awarded the Meritorious Service Medal w/3 OLCs, the AF Commendation and AF Achievement Medals.

GREGORY W. OGLETREE, MAJ, is commander of the Missile Operations Training Flight, 392d Training Sqdn., 381 TRG, Vandenberg AFB. This flight conducts Initial Qualification Training (IQT) for all future missile crew members. Greg enlisted in 1971, earned a BA in history (cum laude) from the University of Puget Sound in 1980, and later added a BS in Political Science. He was commissioned in 1981 (OTS), sent to IQT (MM/ILCS), and assigned to the 490 SMS, 341 SMW, Malmstrom AFB. In addition to serving as flight commander of the ACP, he performed wing instructor duty and launched a Minuteman missile from Vandenberg (Glory Trip 144-MS). Following his crew tour, Ogletree headed back to Vandenberg as a 4315 CCTS instructor, winding up as a briefer for the PREPS and SMOC courses. In late 1989, he was chosen for airborne missileer duty with the 4 ACCS, 28 BMW, Ellsworth AFB. He qualified in the ALCS weapon system and flew numerous sorties as mission commander aboard EC-135 aircraft. After instructor certification, he also became the executive officer for the Bomb Wing DO. In 1992, he transferred to the 2 ACCS at Offutt AFB and flew aboard the "Looking Glass" ABNCP. He returned to Vandenberg in 1995, serving as EWO chief, and became flight commander in March 1996. Major Ogletree wears the Master Missile Badge with Operations Designator, the Senior Officer Aircrew Member Badge, and the Senior Maintenance Badge.

Ogletree is married to the former Sheri Lewis of St. Joseph, MO. They have two daughters, Melanie and Melissa, and three grandchildren: Kayla, Tyler and Brandi.

FRED HUGH O'HERN JR., CMSGT (RET), born and graduated from high school in Americus, GA. He is a graduate of the MAC NCO Academy at Norton AFB, CA (Distinguished Graduate and Commandant's Award winner), the Community College of the AF, and the Senior NCO Academy. He earned a BSc in management from the University of Maryland and an MSc in Public Administration from Troy State University.

He enlisted Dec. 7, 1961 and attended the Nuclear Weapons (RV) Course at Lowry AFB. After tech school he was assigned to Little Rock AFB where he participated in bed-down and activation of the Titan II missile system, the Cuban Missile Crisis, and bed-down and activation of the B-58 weapons system as a nuclear weapons specialist. From Little Rock he attended the USN Explosive Ordnance Disposal Course.

His assignments include England, Italy, Germany, Thailand, Laos, Little Rock, AFB and Seymour Johnson AFB in the CONUS. He retired from the AF in January 1990 at the 20th TFW, RAF Upper Heyford England where he had served as ammo chief, squadron maintenance chief, and wing maintenance chief. He was a key participant in Operation El Dorado Canyon in April 1986.

In addition to his support of the Titan II system his ...sile experience includes air-to-air and air-to-ground ...m EOD and maintenance viewpoints. He participated in OT&E of the AGM-88 Harm and provided test and configuration support for RAF Vulcan bombers during Operation Peace Rapid in 1982. He oversaw the supply of AGM-45s to the RAF in support of Peace Rapid.

His military decorations and awards include the Bronze Star, Meritorious Service Medal, AF Commendation Medal, AF Achievement Medal, Presidential Unit Citation, AF Outstanding Unit Award w/Combat "V", Navy Meritorious Unit Commendation and a number of foreign awards. He holds the Master Missile Maintenance, EOD Technician, and Master Munitions and Aircraft Maintenance Badges.

Since retirement he has been employed as a senior manager by Royal Mail, currently as operations plans and programme's manager for Royal Mail Northampton, UK. He and his wife, Marion reside in Steele Aston, Oxfordshire, England. During his free time he assists the USAF retired community in the UK as a volunteer worker and deputy director of the RAF Croughton Retired Affairs Office.

BILL ORNE, LT COL (RET) hails from Marblehead, MA, entered the AF in October 1958, as a UPT candidate, time short lived. The majority of his career was then spent in missile operations with brief assignments in personnel and services. He received a BS in economics, Colby College; Industrial Management, Sans Degree, University of North Dakota; MBA, Golden Gate University.

Employment: Homestead AFB, FL; Nouasseur AB, Morocco; 556 SMS, Plattsburgh AFB, NY; 447 SMS Grand Forks AFB, ND; 4315th CCTS, Vandenberg AFB, CA; 91 SMW, Minot AFB, ND; directorate of missiles, HQ SAC Offutt AFB, NE; 363 TRW, Shaw AFB, SC. Central Carolina Technical College.

He met and married his wife Fran, an AF nurse, at Plattsburg AFB, NY. They have a son, Bill Jr., born at Grand Forks AFB. Retiring as a lieutenant colonel in May 1982, he and Fran enjoy retirement in the south and life one day at a time.

Memorable experiences: The North Dakota winters and especially the blizzard of 1966. Turning keys on "Giant Boost"/"Giant Bust" at Grand Forks - all those VIPs waiting in the wheat field for the lift-off that never came. The associations and friendships of the many talented and dedicated people they met and served with during the "missile years" are certainly the most memorable.

EDWARD W. OSBORNE, COL (RET), enlisted in the AF in June of 1952. He completed Aviation Cadets, receiving his wings and commission in July 1954. Following assignments in the B-29, B-36, and B-52, he was assigned to the 532nd SMS, at McConnel AFB, KS, as a Titan II missile combat crew commander. Three years as a Titan II instructor MCCC was followed by Air Command and Staff College, duty as an avionics squadron commander in Southeast Asia, an assignment at SAC HQ as the Avionics Project officer for the SR-71 and U-2 Aircraft. Air War College was next, followed by an assignment as Director of Student Operations at Squadron Officer School. After that it was back to the Missile Career Field at Whiteman AFB, MO, as a squadron commander and then deputy commander for operations. Two years on the SAC ICBM Inspector General Team and two years as the Director of Operations for 4th AD was followed by retirement on Aug. 1, 1984.

He and his wife, Shirlee, reside in Cheyenne, WY, where he works as a financial advisor with American Express Financial Advisors, Inc.

KENNETH E. PADGETT, MSGT, born March 23, 1941, Chicago, IL, entered the AF in July 1959 and spent the next 20 years in SAC bomber and ICBM units, retiring as a master sergeant from Vandenberg's 394th Test Missile Sqdn. He completed Missile Engine Mechanical Technical Training at Chanute AFB, IL and was assigned to the 55th Strategic Recon. Wing at Forbes AFB, KS, as a B-47 jet engine mechanic, until activation of the 548th SMS Sept. 1, 1960. Shortly after participating in an operational test launch of an Atlas ICBM at Vandenberg AFB, he was reassigned to the 90th SMW at F.E. Warren AFB, WY in January 1964. Over the next seven years he worked in Minuteman I target and alignment, became a missile maintenance team chief, team training branch instructor and quality control evaluator, moving on to the 44th SMW Ellsworth AFB, SD in 1971. During this 25 month assignment Tech Sgt. Padgett was an MMT team chief during the switch from Minuteman I to the newer Minuteman II, and was team chief of the 1972 Olympic Arena Missile Maintenance Team.

August 1973 saw MSgt. Padgett become NCOIC of the 3901st SMES Missile Maintenance Team, at Vandenberg AFB, followed in 1978 with a tour as NCOIC of the 394th Missile Maintenance Team Branch until his July 31, 1979 retirement.

CHARLIE F. PALKO, MSGT. (RET), LOO35, enlisted June 15, 1953, cross-trained from Transportation at Vance AFB, OK, to nuclear missiles in October 1958. Starting at Lowry AFB, on TM61C Matador. March 1959, Orlando AFB, TM76A Mace launch crew training. February 1960, live launch of TM76A at Holloman AFB. (Bad Guidance and bad Range Safety package, as they parked that one on Hiway 60 just out of Mountainair, NM). April 1960, 38th TAC Missile Wing Sembach AFB, Germany, assist removal of Matador and set up Mace A from basic launch capability through Rapid Fire Multiple Launch (RFML). April 1963, AFSC, Research and Development at high speed test track Holloman AFB. QA - Track Safety Office, F105 and F111 seat and pod ejection test, Minuteman Guidance Test, weapon impact test, numerous rain test on radar domes. February 1967, 498th TAC Msl. Gp. Kadena AFB, Okinawa, TM76B Mace, launch team February 1968, Nose and Missile Replacement Team NCOIC. September 1969, 91st Msl. Maint. Sqdn. Minot AFB, MMT Team chief MM1. March 1970, first MMT trained for MMIII Force mod., June 1970, postured first MMIII site for Strat. Alert. January 1971 (first) and June

1971 (third), MMIII FOT&E launches at Vandenberg AFB. September 1973, 90th Msl. Maint. Sqdn., F.E. Warren AFB, MMI and MMIII in MMT, Field Sup., First MMIII SELM Test, NCOIC of VECB until retirement March 31, 1976. January 1989, Civil Service, at F.E. Warren AFB on MX (Peacekeeper). TDY to Vandenberg AFB for MX FOT&E Launch, #5 March 1991, #9 March 1992, #12 May 1993 and #15 July 1994. Participated in numerous visits of Russian inspectors in regard to the START Treaty. September 1994 to present, assigned to Missile Stage Processing Facility, Peacekeeper Missile, 90th Missile Sqdn., 90th MW, F.E. Warren AFB. He and his wife, Patricia, have resided in Cheyenne since 1973.

CHESTER D. PALMER, 1LT, born Dec. 28, 1948, Orangeberg, SC, enlisted in the USAF Sept. 15, 1971.

His military locations and stations: 443 MAW, Altos AFB, OK (443rd Supply Sqdn.), 90 SMW, Warren AFB, WY (320 SMSGT); executive officer, 443rd Supply Sqdn. (1971-73); Minuteman Launch Officer, 320 SMSgt. (1974-76) (MMI and MMIII).

His memorable experiences: "The first alert I pulled in Minuteman I (April 1974) and they thought they had a chance of war."; dismantling the last Minuteman I (1974) on alert.

Palmer received the Presidential Unit Citation, four highly qualified crew pins; selected 443d MAW Honor Guard Commander, 1972-73, Altus AFB, OK. Discharged as first lieutenant, April 30, 1976.

He married the former Susan Smith in 1994 and they have no children. He is a 1971 graduate, North Carolina State University with a BA degree. Received an MA degree in 1976 from the University of Northern Colorado. Besides teaching Social Studies he has coached one of the top debate teams in South Carolina.

MICHAEL J. PATTERSON, LT COL, born May 26, 1939 in Oakland, CA, a life member of AAFM, is one of the founders of the Association of the Civil Air Patrol Rocketeers, and has served as the ACAPR executive director since October 1996. He joined the CAP Cadet program at San Francisco in Wing HQ Communications Sqdn. 90 in 1953, rising through all squadron positions to cadet commander. In 1955, he was selected to be Maj. Gen. Lucas V. Beau's cadet aide to the 1955 Air Force Association National Convention in San Francisco. He joined the California Air National Guard's 144th FW in 1956 as an airborne radio operator assigned to the State Governor's C-47. Several years later, he transferred to the 129th Air Commando Sqdn. and rose to RO flight instructor/Stan/Eval examiner and eventually NCOIC of the Airborne Communications Section.

He flew in several operations of Darkcloud/Pinecone and Operation Sidewinder in Arizona, testing one of the two "James Bond" personnel recovery systems. On the civilian side, he was a quality control inspector for Dumont Mfg. Co. in San Rafael, CA, building blast tubes for the Bomarc Interceptor Missile, nose cones for the Polaris and insulation petals for the Minuteman Missile. Following 12 years on flying status and receiving his degree in Fire Science Technology at the College of San Mateo, he transferred to the 129th Combat Support Sqdn. as assistant fire chief for Technical Services. In 1973, he joined the 904th Combat Support Sqdn. at Hamilton AFB as deputy fire chief. In 1974, he was selected as Wing Outstanding Airman of the Year in the 452nd.

In 1975, he became the fire protection superintendent of the 904th Civil Engineering Flight, a 69 man Global Mobile "Prime Beef" organization with four independently deployable 15 man fire fighting teams. In 1976, the organization became the 452nd Air Refueling Wing and transferred to March AFB, CA. Also, in 1976, he won the top award for Aerospace Educators in the six western states of the Pacific Region and the top award in the nation, the Brewer Award. In 1976 he was also selected as Wing Outstanding Airman of the year for the second time.

During his tour at March AFB, he was selected by the AF to go TDY to the Air Force Civil Engineering Center at Tyndal AFB, FL to be the project test conductor for the new Air Force Prototype Rescue Vehicle - MERV. Leaving the 452nd in 1979, he continued to serve in the CAP in the primary areas of Flying Safety and Aerospace Education with the additional specialty of Model Rocketry. The former deputy group commander of North Coast Group 23, he is presently the National Model Rocketry Project Officer and the Northern California area coordinator for Aerospace Education in the California Wing and is the executive director of the Northern California Aerospace Education Teacher Resource Center.

In his civilian occupations, he was a San Francisco policeman, spent five years in the Pacifica Fire Departement before going to the San Francisco Fire Department. In San Francisco, he performed firefighting duties on engines and trucks and became a fire command post technician. On promotion to inspector, he was the project coordinator for Hazardous Materials for four years and then coordinator for the High-Rise Fire Safety Director Program in 475 San Francisco Hi-Rise buildings. On promotion to lieutenant of inspectors, he was assigned as fire marshal, Port of San Francisco, responsible for all fire related activities on 18.2 miles of waterfront, including all bullet hits and explosions in "Dirty Harry" movies. On promotion to captain of inspectors he became the assistant fire marshal of the 49 person Division of Fire Prevention and Investigation.

He retired from SFFD in 1991 and continues his full time activities in Aerospace Education and model rocketry with the CAP and running the NorCal A E Teacher Resource Center. Among his awards and decorations are the USAF Commendation Medal, USAF Outstanding Unit Award, USAF Longevity Service Award, Armed Forces Reserve Medal, Air Reserve Forces Meritorious Service Medal w/3 OLCs and the USAF Small Arms Expert Marksmanship Ribbon. Among his CAP decorations are the Bronze Medal of Valor, the Exceptional Service Medal and 14 other medals and awards. He and his wife Margaret have resided in Forestville, CA since 1981. His son Shaw, a Coast Guard Veteran of 16 years, is stationed at Kodiak, AK with grandson, Tyson and wife, Dana.

ORRIN RICHARD PECK, MSGT (RET), enlisted in the USAF in 1955, retired 1975.

Assignments: LRAF, AR 70 A&E, radio; Presque Isle AFB, ME; 702 Msl. Wing Snark Msl.; GFAFB, ND, 319 AMMS, Hound Dog Msl. Blytheville AFB, AR; 97 AMMS, Hound Dog Msl. Chanute AFB, IL 3364 Inst. Sgt. SRAM Msl. K. I. Sawyer AFB, MI 410 MMS SRAM Msl.

His most memorable experiences included flying maint. status on RB47 - KC97 - B52s; operational launch of SNARK Msl., Cape Canaveral; operational drop of SRAM Msl., White Sands Msl. Range, and assisting Boeing representatives activate SRAM School and WAT, K I Sawyer.

He married Dot in 1957 and adopted sons, Butch

and Mike. They were divorced in 1970. In 1975 he married Mary and they live at 1006, 12th Ave. N.W., Rochester, MN. He has three stepdaughters, and one stepson. After AF he worked at US Post Office until retiring in 1992; he is now employed at Sams Club. He and Mary enjoy their 10 grandchildren, and spending time at their lake cottage.

ELMER L. PETERSON, MAJ (RET), was perhaps the first missile operator and maintainer in the Air Corps/Force. In late 1945/early 1946, he was the only enlisted radar mechanic at Wendover Field, UT, where the initial tests of the German V-1 buzzbombs were made. Working with civilian contractors they installed an SCR-584 (gunlaying radar) with an ancillary plotting board on a hill side overlooking the base. By installing a beacon in the buzzbombs, we could control and track the missiles across the desert floor. Initially, by changing the radar PRF, we had up and down control; by "stepping" the beacon, we could then have right and left control; another "step" put the beacon into arm and dump control. The plotting board followed the missile's track in real time and facilitated course corrections and dump point. Later, 1946, the Boeing Co. started experiments with a ground launch missile called GAPA (Ground-Air Pilotless Aircraft). Again, the SCR-584 was the workhouse for missile tracking and Elmer (a staff sergeant at that time) was the only enlisted person on the project. In March, 1947, the 4145th AAFBU at Wendover was transferred en masse to reopen the base at Alamo-gordo N. Mexico. Once again, the SCR-548 was the workhouse for missile tracking. They also used this radar to help in triangulation of the range. Here, most of the initial Air Corps/Force Missile systems got their start.

In the spring of 1948, Elmer (still a staff sergeant) applied for (1) a direct commission (2) warrant officer, and (3) OCS. He started OCS July 1, 1948. (Then a six-month course) In October, 1948, he received an appointment to WOJG; December 15 (three days before scheduled OCS graduation) be received word that the direct commission was approved, but since OCS was so near completion, the direct commission was waived.

In March, 1957, Elmer was given an appointment into the Regular Air Force. In August, 1958, he entered the University of Oklahoma (AFIT) and earned a BSEE graduating summer, 1960. He was then assigned to Chanute AFB, serving first as commander, 3346th Field Training Sqdn., then as branch chief, GAM-77 Training Branch. Elmer retired May 30, 1962, and went to work for North American Aviation at Downey, CA, where he was the maintainability administrator for the GAM-77, Apollo, and Saturn programs. In 1969, he was given a scholarship to UCLA, where he earned a MSE, Systems Engineering major. In January, 1970, Elmer re-entered the AF as a GS-14, serving at Andrews AFB, the Defense Systems Management College, and at the Pentagon, AF HQ AF/RD as systems engineering staff officer. Here, he was promoted twice, and was personally responsible for the AFR 800-Series of regulations. Elmer retired from the AF the second time in October 1979, then was rehired from North American Aviation, where he was project manager for Interim Contractor Support for the B-1B program, planning and implementing intermediate and depot level support for the B-1B. Since 1987, Elmer has been teaching graduate level classes for National University, short courses for other universities and professional societies, and he and his wife, Marie have been enjoying motorhome travel.

JAMES N. POSEY, COL (RET), is one of the most diversified missileers. After graduation from Drury College in Springfield, MO in 1965, he was assigned to the 308 SMW where he became one of the SAC first lieutenants to command a Titan Missile crew. In 1968 he trained at Chanute and Vandenberg, to become a Minuteman MCCC at the 90 SMW, F.E. Warren AFB, WY. Served as instructor and command post controller and standardization evaluator. In 1972 he joined the 28th BW at Ellsworth AFB, SD as an airborne launch control system commander. He then cross trained to the 44th SMW as a missile maintenance officer where he conducted the first simulated electronic launch minuteman test in the Air Force. In 1974 he became the chief of missile assignments at HQ SAC where he was responsible for the Rivet Save drawdown that resulted in saving 900 manpower spaces. Col. Posey returned to HQ SAC after graduation from Armed Forces Staff College and became the MX project officer for Director of Missiles, and the executive officer to the SAC director of operations and chief of staff.

He attended Naval War College in 1982 and returned to Whiteman AFB, MO to command the 509th SMS, where he won the Golden Missile Award for Best Squadron in 8th AF. In 1983, he returned to HQ SAC to become chief of Missile Tactics Div. for the Joint Strategic Target Planning Staff; also chaired the CINC's Missile and Space Panel. In 1985 he became the deputy commander of the 351st Combat Support Gp., and the assistant deputy commander for OPS in 1986. In 1987 he became the Ground Launch Cruise Missile (GLCM) deputy commander for operations at the 486th TMW, Woensdrecht AB, Netherlands. In 1988 he became the Space Command, CO. He then became the 9th Space Div. DO responsible for launch operations at the Eastern & Western Test Range. His last assignment was as the commander of 45 Operations Gps. at Patrick AFB, FL where he retired from active duty February 1992.

His education includes an MBA from the University of Northern Colorado. His PME includes Squadron Officers School, Armed Forces Staff College, and Naval War College in residence and Air Command & Staff College, and National Security Management by correspondence. He has the Master Missile Badge & Senior Space Badge. He and his wife, Maggie, children, Jeffrey and Jolene reside in a beautiful lake community in Winchester, TN. Leisure time finds them on the lake, in the mountains, or on their Harley's touring. He is currently employed by ACS - a joint venture of CSC, General Physics & Dyne Corp.

CHARLES S. (CHARLEY) PUGSLEY, COL (RET), is currently employed by Aero Thermo Technology in Arlington, VA. After graduation from LSA in 1970, he attended OTS. His first assignment was to Vance AFB, OK as the wing safety officer.

In 1973, he was reassigned to the 351st SMW, Whiteman AFB, MO serving as DMCCC, MCCC, instructor, and evaluator.

Assigned to the AF Inspection and Safety Center, Norton AFB, CA in 1977 as a safety action officer. In

1982, he attended ACSC. Following graduation, reported to 2nd ACCS, Offutt AFB as chief SAC Abn. Battle Staff and HQ SAC/DOCA.

He assumed command of the 320th SMS 90th SMW, F.E. Warren AFB, WY in 1987 and also served as ADO before attending National War College in 1990.

Pugsley was assigned to the Pentagon as chief of ICBM Modernization Div. He and his wife, Nancy, reside in Fairfax Station, VA.

JOHNSON V. (RANDY) RANDOLPH, TSGT (RET), an early member of the Association of Air Force Missileers, enlisted in July 1954. Upon completion of basic training he was assigned to Lake Charles AFB, LA as a finance clerk. In August of 1955 he was transferred to RAF Brize Norton as an aerospace ground equipment specialist. Upon return from the UK he served with the 303rd BW Davis-Monthan AFB, AZ. In November 1960, he was assigned to the 395th SMS Vandenberg AFB, CA in the Titan I program, until February 1961 when he was assigned to the Titan II Operation, duties included contractor launch support and refurbishment. He was assigned to the 390th SMW in November 1964 as a maintenance team chief and later a maintenance scheduler. In March 1969 he was assigned to 570th SMS as a launch crew BMFT. In July 1970 he was assigned as wind instructor crew member. On June 21, 1971 as a member of crew S-212 he participated in a launch of a Titan II from 395-C at Vandenberg AFB, CA. Upon return to Davis-Monthan he assumed duty as senior instructor crew BMFT until he was assigned to Sheppard AFB, TX, as BMFT course supervisor in February 1979. He retired August 1976.

He and his wife Jean live in Abilene, TX where he retired from the local electric utility company. They are active in church and the utility retirees club. They spend a lot of time traveling in their travel trailer and he fishes the local lakes.

DAVID R. RAWSON, born Dec. 13, 1949 in Queens, NY, enlisted in the service April 27, 1970. He received basic training at Lackland AFB, TX; MFT Tech School, Sheppard AFB, TX; ORT, Vandenberg AFB, CA; Post, ORT, McConnell AFB, KS; ORT (Operational Readiness Training); Assigned 381 SMW, 533 SMS; 1976-77, ACP (Alternate Command Post); 1977-79, 532 SMS instructor crew (DOT), 1979-81, Standardization/Evaluation Crew (DOV).

Chosen for the crew of the month with Crew S-181 (533SMS); chosen for NCO of the Quarter Fall 1977; chosen to be on the first crew in Titan to have a female MCCC; served under that commander on line crew and evaluator crew; commander was Capt. Pat Fornes and was one of the best commanders around; earned AF Commendation Medal for crew work; had 17 highly qualified readiness checks and over 700 alerts. He was discharged April 27, 1981 as technical sergeant (E-6).

Hired by FlightSafety International and is now the third shift lead technician. (17 years in April). He is married to Dona L. Rawson and they had a son, Darian,

born while he was in USAF. Darian is senior in high school and going to Kansas State University in 1998. Their second son, Eron, was born in 1984 and is in the 7th grade (middle school).

ROSS D. REED, SSGT, enlisted in November 1984 and graduated as an honor student from basic training. After training at Chanute he was assigned to the 44th SMW Ellsworth AFB, and initially served as a PMT team member. In 1989 he earned NCO status and held various positions in FMT, TTB, and QA through the remainder of his time in service. In 1993 he was selected to represent part of the Ellsworth FMT OA team, and won the FMT trophy.

He earned several local and SAC level awards, and was involved in the D01/D09 site turnovers to the National Park Service. One of the last people to leave in July 1994, Ross helped to set the rededication monument near the PRIDE hangar prior to the closing ceremony. In 1994 he earned an AS degree and held the rank of staff sergeant prior to being honorably discharged.

Most memorable: the beautiful country and good people up in South Dakota, Col. Payne's retirement ceremony, the MMII in spotlights near the PRIDE hangar at night, winter survival kits, the pride of being a missileer from the 44th, and just about every day spent on duty in general.

HAROLD J. RENNINGER, MSGT, born April 24, 1937, enlisted in the USAF, April 5, 1954, master sergeant.

His military locations and stations: F.E. Warren AFB, 389th SMW, Hahn AFB, Germany, 38th TMW. He is a member of Launch Crew 3, 389th SMW (Atlas - D).

His memorable experiences: participant in operation: "Quickstart" April 22, 1960. The first launch of an ICBM by an all AF launch crew. Atlas-D from Vandenberg. He was awarded the Meritorious Service and Air Medal w/OLC.

Renninger was discharged June 1, 1974 and achieved the rank of master sergeant. He is married to Ruane E. and they have four daughters: Diana, Nancy, Barbara, Cheryl. He restores vintage aircraft.

RICHARD R. RENTERIA, born July 4, 1928, San Marcos, TX, served in the USAF Nov. 14, 1947-Aug. 31, 1975.

His military locations and stations: Scott Field, IL; Goose Bay Labrador; 97th BW Biggs AFB (Airborne Radio Tech B-29) El Paso, TX, October 1950; Lowry AFB, Denver, CO; 95th BW Biggs AFB (Bomb Nav Tech. B-36) El Paso, TX, March 1954; 93rd BW (team chief bomb Nav System B-52) Castle AFB, Merced, CA November 1955; 576 SMS (Missile Sys. Anly Tech Minuteman) VAFB, CA, March 1960; 341st SMW WS-133A Minuteman Malmstrom AFB, Great Falls, MT, August 1965; January 1970, 355 TFW PACAF Thakli RTAFB Thailand (NCOIC Missile Shop F-105) reassigned to 341 SMW October 1970.

TDY Upper Hayford RAFB England 1997, ARS B29; TDY to Midway 97 ARS KB-29 (Korean conflict); SAC Bomb Comp Walker AFB, NM B-36; TDY to Anderson AFB Guam (Bomb Nav Team Chief B-36) included 36 hour non-stop B-36 flight from Biggs to Guam; TDY from Guam to Fairchild AFB, WA for SAC Bomb Comp; Biggs AFB, TX back to Guam to complete TDY; SAC 15AF NCO Academy as youngest master sergeant with DOR Dec. 1, 1957 (10 years); Primary launch/MOCAM crew category II Testing at OSTF-1 Atlas E; Combat Ready Missile Crew duty during the Cuban Crisis; Senior Standardization Evaluation and instructor Crew R-14 (SM-65 Atlas E); Launched Atlas E at OSTF-1 (launch manually sequenced in support building by BMAT and facilities tech prior to LOX load); escorted and briefed in Spanish during VAFB visit of Inter-American AF Chiefs; Team chief OSTF-1 Deactivation (assigned to site during construction and locked gate when site was deactivated); NCOIC Technical Engineering & Analysis Div., 341st SMW Malmstrom AFB, Mt. (troubleshooting team on 24 hour seven days a week); recommended circuit change which enabled Voice Reporting System (VRSA) to alert LCC when standby power diesel was left in the off position by maintenance teams design change was accepted and installed at all WS-133A missile sites by Depot Teams.

He is married to Gloria and they have children: Roger, Daniel, Rita and Rebecca. They attended two years of college at Great Falls, MT. Employed Air Force Quality Assurance (AFQA) with SATAF the Boeing Co. Malmstrom AFB, Missile Site update 1977-1980. Air Force Plant Representative Office (AFPRO) TRW Redondo Beach, CA 1980-81; selected to activate AFPRO at Rockwell International El Sequendo, CA for the revived B1-B bomber program 1981-83. Configuration manager for the Ground Airborne Integrated Terminal (GAIT) Texas Instrument, Control Segment IBM/LORAL and the Nudet Detection System Sandia AFB, NM for the Ground Positioning System (GPS) Joint Program Office Space and Missile Center, LAAFB, CA, 1983-94. He resides in Torrance, CA (senior complex with golf course). Enjoys golf and grandkids.

HOWARD L. RICE, LT COL (RET), graduated from Arizona State University in 1954 and served as AF officer from August 1954 to December 1977. Duties included navigator, instructor and flight examiner with the 47th ATS, Hickam AFB, HI, and 50th ATS, Travis AFB, CA. He entered ICBMs in 1962 with the Atlas F, 576th SMS, Vandenberg AFB, CA as DMCCC and MCCC, and the 394th SMS (Minuteman) as training officer. In 1966 he moved to Grand Forks AFB, ND as MCCC and instructor in the 448th SMS. He was called back to flying in 1970 as C-130E navigator, instructor and flight examiner with the 345 Tactical Airlift Sqdn., CCK AB, Taiwan. He flew 200+ mission throughout Vietnam, plus 65 combat missions (369 flying hours) over hostile territory. In 1972 he moved to Forbes AFB, KS as chief navigator and ended active flying with 7300 hours. In 1973 he returned to 321st SMW, earned an MS in Industrial Management from the University of North Dakota while serving as chief, Standardization and Evaluation, commander 448th SMS and assistant deputy commander, operations.

In civilian life he joined ITT Corp., working three years as quality assurance manager, BMEWS site, Clear, AK. He moved to Vandenberg AFB as ITT program manager for a variety of test programs conducted there. He retired from ITT in 1993.

He and his wife, Genevieve, reside in Monument, CO. They started business, G&H Antiques and Collectibles, doing antique shows in Wyoming, Colorado and Kansas.

RANDY L. RIVERA, CAPT, born Sept. 20, 1965, Ramey AFB, Puerto Rico. Attended New Mexico State University for his BBA in management with minor in economics, 1988 and MBA in business administration, Univ. of South Dakota, 1992.

Entered active duty May 18, 1989; 4315th Combat Crew Trng. Sqdn., Vandenberg AFB; 1989-1993, 66 Msl. Sqdn., 44 Msl. Wing, Ellsworth AFB, SD (SAC), deputy combat crew commander and instructor; 1993-94, student, Wright Patterson AFB, OH; assigned to Vought Aircraft Co., Dallas, TX; 1994-96, Wright Patterson AFB, aircraft systems cost analyst and airframe contracts manager, C-17 Systems Program Office; 1997-present, USAF Logistics Career Broadner, Oklahoma City Air Log. Center, Tinker AFB.

Medals include AFCM w/OLC and AFCRM plus numerous awards including Distinguished Graduate, Undergraduate Msl. Trng. He participated in first ever Olympic Flag (1992) and wrote first scrip depicting missile ops.

Married Manuelita "Lita" in 1988 and they have two children, Luis and Joseph.

PHILIP A. RIZZO, COL, born April 19, 1945, Johnstown, PA, enlisted in the USAF. His military locations and stations: Davis Monthan, Vandenberg; Grand Forks AFB, ND; Offutt AFB, NE; Rapid City, SD; March AFB, CA.

He was awarded the Legion of Merit, MSM w/3 OLCs, AFCM w/4 OLCs. Rizzo was discharged as colonel, April 30, 1992.

Married to Ethel and they have five children: Stephanie, Yvonne, Andrea, Philip and Joseph, two grandchildren: Annessa, Alexandra. Employed as director, Governmental Relations, Riverside County Office of Education.

STANLEY B. ROADARMEL, MAJ (RET), was launched May 5, 1937 from Albion, NY, married Carole Hayes in 1959, and graduated from Syracuse University with commission in 1962. Understandably, after initial flight training, communications security, SOS, and recruiting assignments, he volunteered for Titan II in 1969. Thereafter (except for a short 1980 Incirlik Turkey tour), he became a "SAC Asset" exclusively.

Sheppard AFB Titan II training led directly to 390th SMW missile maintenance at Davis Monthan. One of the first comany graders to lead a field maintenance organization, with earned propellant transfer/missile recycle certifications, Capt. Roadarmel was tapped in 1971 as Titan II Maintenance Evaluation Team Chief

for HQ SACs 3901st SMES. Subsequent Vandenberg assignments from 1974, within the 394th TMS/1st STRAD, directly supported both Titan II and Minuteman launch operations, and staff assisted key logistics functions as well. Concurrently, Maj. Roadarmel completed ACSC and ICAF, then transitioned into SAC and USAFE contracting for the balance of his military career.

Post 1982 retired at Vandenberg meant Space Shuttle SLC-6 and Titan IV SLC-4E construction administration for Martin Marietta and Lockheed until 1992. Thereafter, AF civil service commenced as a 30th Space Wing Contract Negotiator/Administrator for ongoing AFSPC mission support.

Off duty? Among other things, Stan and Carole still watch missile launches, facilitate their church's "Pre Cana" marriage preparation seminars, get conspicuously involved with Pro-Life issues, and so fare are thrice blessed grandparents.

WILLIAM B. RODIE, LT COL, is the director of the Plans, Programs and Requirements Div. The division is responsible for all wing plans and coordinating military initiatives and exercise between the US Army, US Marines, and US Navy in Panama. Before that, he was commander of the 24th Services Sqdn., a position he held from January 1993 until September 1995. Under his leadership, the squadron was rated first for over a year among the 22 squadrons in Air Combat Command as determined by Quality Performance Measures which track customer satisfaction and financial performance in 25 key areas. He was born Feb. 22, 1949 in Ridgewood, NJ. He graduated from Ridgewood High School in 1967 and from Ohio Wesleyan University in Delaware, OH, in 1971 with a degree in Economics. He was commissioned a second lieutenant after completing the four-year AFROTC program.

Lt. Col. Rodie reported to Vandenberg AFB, CA, in August 1971 for Minuteman II missile crew member training. He graduated in November 1971 as a deputy missile combat crew commander. In December 1975, he participated in a successful operational launch of a Minuteman III missile from Vandenberg AFB to Kwajalein Atoll in the Marshall Islands of the Western Pacific. He later upgraded to missile combat crew commander and was awarded the 15th AFs Crew Member Excellence Award.

Lt. Col. Rodie was assigned to the 390th SMW at Davis-Monthan AFB, AZ in July 1976, as the wing executive officer. The wing was responsible for 18 Titan II ICBMs. In October 1978, Lt. Col. Rodie was assigned to the 20th TFW, RAF Upper Heyford, Oxfordshire, England, as the wing executive officer for the F-111 wing, and later served as chief of the Services Div.

In October 1981, he was assigned to Loring AFB, ME, and commanded the 42nd Services Sqdn. In November 1982, the squadron was recognized as the best services squadron in the 8th AF for 1982. In March 1984, Lt. Col. Rodie was part of the initial cadre selected to open the first ground launched cruise missile base in Europe at Florennes AB, Belgium. In April 1985, he was assigned as chief of the Services Branch, Office of the Air Training Command Inspector General. In that capacity, he was responsible for inspecting all Air Training Command Services operations and those of Air University and AF Academy.

Lt. Col. Rodie assumed command of the 12th Services squadron at Randolph AFB, TX, in August 1988. Under his command, the squadron won the ATC Hennessy Award for both 1989 and 1990 and was named AFs Outstanding Services Unit of the Year runner-up for 1990. He was then assigned as deputy director, Housing and Services, DCS/Engineering and Services, HQ Air Training Command. He was named director, business operations, DCS/Morale, Welfare, Recreation and Services, when the two careers combined.

Lt. Col. Rodie's awards and decorations include the Meritorious Service Medal w/3 OLCs, AF Commendation Medal, AF Achievement Medal, Joint Meritorious Unit Award, AF Humanitarian Service Medal, Outstanding Unit Award w/5 OLCs, AF Organized Excellence

Award w/2 OLCs, Combat Readiness Medal and the National Defense Service Medal w/star device. He completed Squadron Officer School in residence and graduated from the Air Command and Staff Course and Air War College by seminar. He also holds a master of business administration degree from the University of Montana and a master of arts degree in Public Administration from the University of Northern Colorado.

Lt. Col. Rodie is married to Lt. Col. (ret) Sandra M. Rodie, USAF Nurse Corps, of Pardeeville, WI. They have one son, Christopher (11). Lt. Col. Rodie and his family will make their home in San Antonio and invite their 24th Wing friends visiting or locating to Texas to contact them.

CHARLES E. ROGERS, COL (RET), enlisted in the USAF on July 25, 1951 after graduation from Morton High School, Cicero, IL. After basic training at Lackland AFB, TX and Sheppard AFB, TX and graduation from the Weather Observer Course at Chanute AFB, IL, he was assigned to Det 2, 7th Weather Gp., Thornbrough AFB Cold Bay, AK in May 1952 serving as a weather observer and chief observer. He was then assigned to Det 11, 25th Weather Sqdn. as chief observer, in June 1953 at Ardmore AFB, OK and was separated from the USAF July 24, 1955.

After completing three years at the University of Chicago, he reenlisted in the USAF as a weather observer in May 1958, assigned to Det 17, 8th Weather Gp., Reese AFB, TX. He was selected for OCS in June 1960 and was commissioned a second lieutenant in March 1961. After attending the Personnel Officer Course at Greenville AFB, MS, he was assigned to the 818th AD, Lincoln, NE as chief of personnel records and assignments. He was then reassigned to the 10th SMS, Malmstrom AFB, MT in January 1964, serving as a missile deputy combat crew commander, missile combat crew commander and instructor.

In August 1968, he went to the Air Force Institute of Technology, Wright Patterson AFB, OH receiving his ESEE in 1970 and MSEE in 1971. He was then assigned to the 42nd Civil Engineering Sqdn. as chief of operations and maintenance, Loring AFB ME until May 1973 when he was assigned to the 388th Civil Engineering Sqdn., Korat Royal Thai AB as chief of programs.

He was re-assigned as a civil engineer programmer in the Directorate of Engineering, HQ Command, Bolling AFB, DC in July 1974 and moved to Andrews AFB, MD as chief of programs, 1185th Civil Engineering Gp., on July 1, 1976. He then served as Director of Programs, Directorate of Civil Engineering at HQ Electronic Systems Command from January 1980 to June 1982 when he became commander, 819th Civil Engineering Sqdn. Heavy Repair (Red Horse) at RAF Wethersfield UK from June 1982 to June 1986.

Then he commanded the 1776 Civil Engineering Sqdn., Andrews AFB, until retiring April 1, 1991. He is the recipient of the National Defense Service Medal w/two stars, Air Force Good Conduct Medal w/2 Loops, the AF Commendation Medal, Meritorious Service Medal w/5 OLCs, Bronze Star and Legion of Merit. He and his wife Joan reside in Austin, TX and they do a lot of traveling.

GORDON C. ROLSTON, SMSGT (RET), born at Walworth, NY, April 9, 1928 and entered the Army Air Corps in March 1946. He received basic training at Keesler AF, MS and went to Radio Mechanics School Airborne Equipment at Scott Field, IL; then deployed to the Far East AF at Haneda AF, Tokyo, Japan. Worked the flight line as radio mechanic maintaining electronic equipment on training aircraft, and staff aircraft of General's MacArthur, Whitehead and Walker. He also maintained electronics aboard the Russian Embassy aircraft and all transients. He returned to civilian life early in 1950 and went to work for the Ford Motor Company.

Was recalled to active duty for the Korean Conflict and served with the 47th BG Communications Sqdn. and the 3rd Communications Sqdn. at Sculthorpe, England. Her returned to the states to the 2nd AF HQ at Barksdale AB, LA as communications programmer for SAC bases and was also NCOIC of the new receiver site. He was selected to attend Nuclear Weapons School at Lowry AB, CO and on to the 806th AD (SAC) with the 45th Munitions Maintenance Sqdn. at Lake Charles AB, LA where he helped convert the nuclear weapons from IFI to sealed units, and also attended Air Sampler School at Sandia Base, NM.

He then attended the SM-62 SNARK missile arming and fusing school at Lowry AB, CO, then off to Cape Canaveral, FL with the 6555th Guided Missile Sqdn. at Patrick AB, FL, for launch crew training. Then to the 702D SMW (SAC) (ICM-SNARK) at Presque Isle AB, Presque Isle, ME. When it closed he was assigned to the famed 305th BW (SAC) at Bunker Hill AFB in Indiana with the B58 Hustler Bomber which carried the MB-1 and/or BLU-2 TCP and a cluster of MK-45 nuclear weapons.

In December 1962 he received an assignment for Turkey as quality control inspector for the DET 10, TUSLOG (USAFE) munitions maintenance and storage area at Incirlik AB. When he returned to the states he was assigned to the 52nd FW, Suffolk County AB, Long Island, NY as quality control NCOIC of the AIR-2A missile maintenance area. His next assignment to Paine Field, WA he organized and trained the munitions branch of the 57th Combat Support Squadron (ADC) where he rounded out his active military career and retired. In amongst his many assignments he managed to attend the 2nd AF NCO Academy at Barksdale AB, Louisiana graduating in the top quarter.

He has graduated from many technical and management courses through the extension course institute and attended college on base and off in the evening. He continued his education after retirement and received a BA in Human Resources from the University of Washington, Officer Candidate Diploma from the US Army in 1986 and served, after active duty retirement, in the California and New Mexico State Military reserve rising to the rank of lieutenant colonel, acting as Plans, Operations and training officer culminating in battalion commander. He is a 32nd degree master mason and a member of Deming Lodge Number 12 in Deming, NM.

He had many memorable experiences during his career but two stand out: When a B47 on alert strip at Lake Charles AB caught fire with a nuclear weapon on board and three crew members perished; while at Sculthorpe AB in England a tidal wave swept down the English channel and back up into the Hunstanton Wash inundating the beach housing at Heacham and Hunstanton drowning 35 American families. Some of the rescuers were awarded the English George Medal, highest English foreign award given to enlisted men. Gordon holds 18 awards and decorations from active, reserve and state military reserve but treasures his Guided Missile Badge the most.

He and his wife, the former C. Mona Steuber of Holly, NY, reside in Deming, NM. They have three married daughters, Cheryl, Roxanne and Jeri who reside in the Pacific Northwest.

JOSEPH B. RONCHETTO, LT COL, born Aug. 24, 1940, Butte, MT. Commissioned Aug. 17, 1962, USAF. Education includes MS degree, aerospace engineering, AF Institute of Technology; BS degree in electrical engineering, Montana State University; Air Command and Staff College and Squadron Officers School, Air University, Maxwell AFB, AL; numerous training courses at various Air Force Bases.

Assignments include 12th Strategic Msl. Sqdn., 341st Strategic Msl. Wing, SAC, Malmstrom AFB, 1962-66; AF Aeronautical Systems Div., AF Systems Cmd., project officer and test engineer, Wright Patterson AFB, 1966-67; AF Avionics Laboratory, AF Systems Command, deputy program manager Wright-Patterson AFB, 1967-71; Site Alteration Task Force, Space and Missile Systems Organization, AF Systems Cmd., program engineer, chief of engineering then SATAF Commander, Ellsworth AFB 1971-76; Ballistic Missile Office, AF Systems Command, electronics division branch chief, program manager then deputy division chief, Norton AFB, 1976-81; AF Inspection and Safety Center, Inspector General Office, HQ USAF, Norton AFB, systems acquisition management and inspection team chief, 1981-84.

Awards and medals include AFMSM w/2 OLCs, AFCM w/OLC, AF Organizational Excellence Award w/4 OLCs, Basic, Senior and Master Missileman Badge, plus ribbons and numerous Letters of Commendation.

He and wife Diane live in Redlands, CA. They have two sons, Keith and Greg, and two grandchildren, Kyle and Kayla. He is active in the Boy Scouts and received the Silver Beaver Award.

JAMES D. RUDY, COL, is the deputy commander for maintenance, 91st SMW, SAC, Minot AFB, ND.

Born Sept. 29, 1939, in Huntingtown, PA, and graduated from Bald Eagle Area Joint High School in Wingate, PA, in 1957. He received his bachelor of science degree in history from Lycoming College, Williamsport, PA, in 1961. Earned his master's degree in education administration in 1973 from South Dakota State University, Brookings, SD. Colonel Rudy's military education includes Squadron Officer School and Air Command and Staff College through correspondence in 1965 and 1970, respectively.

The colonel was commissioned in the AF through Officer Training School in November 1962. He was initially assigned to the 763rd Radar Sqdn. from November 1962 through May 1966, at Lockport, NY, as personnel officer. He completed personnel officer training in May 1963 at Greenville AFB, MS. Colonel Rudy served as flight training officer from May 1966 through May 1968 at Officer Training School, Lackland AFB, TX.

In July 1968 the colonel was assigned to the 3rd Combat Support Gp., Bien Hoa AB, Vietnam, as personnel officer through July 1969. He was then assigned to the Minuteman Missile System at Ellsworth AFB, SD. He served as crew commander, instructor and evaluator for the Minuteman Missile System and was the last senior Minuteman I standardization evaluation crew commander at Ellsworth.

Following his four and a half years at Ellsworth AFB, SC, he was assigned to the 3502nd USAF Recruiting Gp., McGuire AFB, NJ. In August 1975 he returned to the missile career field and served as officer in charge, vehicle and equipment control branch, and maintenance supervisor in the 44th Field Missile Maintenance Sqdn. He also served as chief of the Training Control Div. for the deputy commander for maintenance.

In August 1978 Col. Rudy was assigned to Whiteman AFB, MO, as chief, Training Control Div.

He commanded the 351st Organization Missile Maintenance Sqdn. from August 1979 to February 1980, and the 351st FMMS until October 1980. In November 1980 he was assigned as assistant deputy commander for resource management, 351st SMW.

In September 1982 he served as assistant deputy commander of maintenance, 91st SMW, Minot AFB, ND. In May 1984 he assumed his present position.

Colonel Rudy wears the Master Missileman Badge. His awards and decorations include the Legion of Merit, Bronze Star Medal, Meritorious Service Medal w/OLC, AF Commendation Medal w/2 OLCs, Combat Readiness Medal, National Defense Service Medal and the Republic of Vietnam Service Medal w/4 Service Stars. His date of rank is Aug. 1, 1983.

He is married to the former Kitty Lou Stiffler of Boalsburg, PA. They have two children, Roben and James. The colonel calls Port Matilda, PA home. Employed in family owned and operated business, Rudy Frame Shop and Gallery.

ARLYN H. SAGE, MAJ (RET), born at Dover, KS, Feb. 25, 1939, attended Washburn University at Topeka, KS and received a BBA in 1962. He received his commission in February 1963 and retired in October 1982.

Assigned to Offutt AFB, NE as the AFW supply officer and materiel control officer for the 3902nd ABW until June 1964.

Upon completion of Titan II missile training, he was assigned to the 381st SMW, McConnell AFB, KS as DMCCC, MCCC, and MPT operator until June 1970.

From July 1970 until August 1975, he was assigned to the 4515th CCTS as an ORT instructor, chief of the Titan II ORT Section, and Plans and programs officer, MPT Management Branch.

Returned to supply career field in 1975 with assignments at 43rd SW, Guam, 96th BW, TX and 8th TFW in Korea.

Sage participated in an operational test of the Titan II at Vandenberg AFB. He served as MPT operator during the 1971 SAC Missile Combat Competition. Helped validate the exercise scripts from 1972 to 1974.

His awards/medals include the AFCM w/OLC and the AF Outstanding Unit Award w/4 OLC. Badges include the Senior Missileman and Master Missileman.

Major Sage and his wife, Sun Ye, have resided in Rossville, KS since 1983. Employed by Roach True Value, Topeka, KS. E-mail address is asage@cjnetworks.com.

LOUIS L. SAMPLES, LT COL, born Dec. 28, 1929, Hastings, NE. Attended Hastings College for BBA degree in 1951 and University of Missouri for MBA in 1968.

Enlisted July 2, 1951 in the USAF with basic training at Lackland AFB; flight line maint., Reese AFB; basic pilot trng., Class 53-A, Hondo AB; advanced pilot trng., Wings and commission, Feb. 2, 1953, Reese AFB.

Pilot at Walker AFB, 1953-54; Waco AFB, Lake Charles AFB, 1955-56; Loring AFB, 196-63; Whiteman AFB, 1964-68; Pleiku AB, Vietnam, 1968-69; March AFB, pilot, director of training, asst. DCO SAC 22nd BW, 1969-76.

Participated in Cuban Missile Crisis, Suez Crisis, ARC Light and Bullet Shot. Memorable experiences include primary launch crew, Ivy Tower, 46th launch of Minuteman Missile, Vandenberg AFB.

Retired in May 1976. Awards include Air Medal w/ 11 OLCs, AFCM, DFC, AFLSA w/5 OLCs, GCM, NDSM w/BSS, AFOUA w/4 OLCs, CRM, VSM w/3 BSS and RVNCM.

Married 41 years to Marjorie and they live in Canyon Lake, CA. They have three children (middle daughter is a doctor and Lt. Col at Travis AFB, CA) and four grandchildren.

He worked for General Dynamics until retirement in 1990. Currently working at Repeat Aircraft where they build reproductions of 1930 ERA racing planes

OTTO B. SCHACHT, LT COL, joined the USAF in March 1971. During his career as a "SAC Trained Killer" he performed duties as a 316X0G-1, EMT specialist, at Whiteman AFB with the 351st SMW. He spent a year each as a EMT member, EMT chief, and as a quality control inspector evaluator. He was honorably discharged in March 1975.

Between dispatches he worked on the Needmore Ranch, Dutch and Mildred Jaegar owners. The MS in Agri-technology at CMSU was completed prior to being honorably discharged in March 1975. PhD in agri-engineering at TAMU was earned in August 1975.

He and his wife, Madonna, reside in Levelland, TX where he is Dean of Arts and Sciences at SPC. Otto Jackie-Don, eldest son, is an adult probation officer in Bryan, TX. Byron Fritz-Nelson, youngest son, is a student at SPC. As a USAR officer he commands the 5thBN(CA/PO)/95thREGT/3rdBDE/95thDIV(IT) in Lubbock, TX.

Memorable experience: Schacht and Sgt. Rick Elkins were to be evaluated by the 3901st during one of their visits to Whiteman. At the time both Elkins and Schacht were members of the 351st EMT Quality Control and Evaluation Branch. They knew what they were doing? After being briefed at the LF by the 3901st personnel it became apparent that they had left a very important piece of equipment for the task they were to perform, LF Start-up. You guessed it. The start-up gear was still in the dispatch area at Whiteman. Think the worst, double it, and you can visualize what little of their back sides was left after their one-way talk with the chief.

GARY C. SCHAFF, LT COL (RET), graduated Rutgers University 1968. Served as a Minuteman deputy missile combat crew commander and missile combat crew commander/instructor at the 321 SMW, Grand Forks AFB, ND from 1969-74. Received MBA from University of North Dakota while at Grand Forks. Selected for Project Top Hand, he was a charter member of the 1st Strategic Aerospace Divs. Test & Evaluation Deputate, where he served as a launch director, test manager and chief of the Launch Data Div. for Minuteman operational test flights at Vandenberg AFB, CA (1974-79). At SAC HQ (Offutt AFB, NE), he was an ICBM requirements officer, working on improvements to the Minuteman weapon system and serving as SACs representative to the Army's Ballistic Missile Defense Initiatives (1980-83). From 1983-85, he was operations officer of the 564 SMS and chief of the Wing Command Post at 341 SMW, Malmstrom AFB, MT. Switching to USAFE and the Ground Launch Cruise Missile, he served as the assistant deputy commander for Operations, 38 TMW and commander, 89 TMS, Wueschheim AS, Germany, from

1986-89. His final assignment was as chief, Plans and Operations at the Joint Communications Center, Ft. Ritchie, MD (1989-90).

He and his wife, Bonnie (they married at Grand Forks, 1972), now live in Connecticut, where she owns/operates a book store and he is employed as a firefighter for the city of West Hartford. Their two children, Holly and Jared, are heading for careers as musician and naval officer respectively.

RENE' LAMARA SCHEUERMAN, A/2C, RIF'd, entered active duty Feb. 10, 1955 (with a nomination to the AF Academy), taking basic training in the 3280th BMTS, Flt. 97, Parks AFB, CA, after completing two years at San Diego Junior College, but missing the GI Bill by 10 days. He was disqualified from entering the AF Academy and from OCS because of poor eyesight. After receiving a Top Secret Security Clearance,  he entered 3430th STURON, Special Weapons training at Lowry AFB, CO, in May 1955, receiving certificates as nuclear specialist, Dec. 16, 1955, and weapons fuzing systems specialist (electronics), Feb. 26, 1956. He was assigned to the 17th TMS at Orlando AFB, FL, in May 1956, with duties as a clerk in the orderly room.

Because of a surplus of personnel with Special Weapons training, on Jan. 21, 1957, he was assigned to the 819th AC&W Sqdn., Ellington AFB, TX, and cross-trained as an FM Radio-Telephone repairman, helping man a relay site on Matagordo Bay. During the reduction in force of 1958, it was determined that his training was not needed in the US, and with insufficient time left in his enlistment to go overseas, was discharged six months early on July 11, 1958. Returning to San Diego, he was hired by Convair-Astronautics as an electrician on the Atlas, working on the first Atlas to orbit the earth. Returning to college in 1960, he received a BA from the University of San Diego in 1964, and after becoming eligible for the GI Bill in 1974, completed course work for an MPA at San Diego State College in 1976.

Retired from the county of San Diego as a vocational counselor in 1985, he works alternately on temporary assignments with the county, and with a local railroad. He has lived with his wife, Cynthia, in Spring Valley, CA since 1975. They gave two sons, Geoff and Gerald. He served as a trustee on the board of the San Diego Railroad Museum, and volunteers in various job categories on their demonstration train. Hobbies include reading, writing, photography, and model trains.

BERNARD A. SCHRIEVER, GEN (RET), born in Bremen, Germany, Sept. 14, 1910, immigrated to the US as a seven year old in 1917 with his parents. He became a naturalized citizen at age 13 and finished his early schooling in San Antonio, TX. In 1931, he received a bachelor of science degree from Texas A&M, and a reserve appointment in the FA. He earned his wings as a second lieutenant in the Army Air Corps Reserve in June 1933.

After obtaining his master's degree in Aeronautical Engineering from Stanford University, 1942, he gained rapid promotions and positions of increasing responsibility during WWII. He was chief of staff of the 5th AF Service Command and later commander of the Advanced Headquarters for the Far Eastern AF Service Command. After the war he became the chief of the Scientific Liaison Section at HQ USAF and held other scientific evaluation jobs as they pertained to military weaponry.

Beginning in 1954 when he assumed command of the AF Ballistic Missile Div. and later the Air Research and Development Command, Gen. Schriever pushed forward research and development on all technical phases of the Atlas, Titan, Thor and Minuteman ballistic missiles, while concurrently providing the launching sites and equipment, tracking facilities, and ground support equipment necessary to the deployment of these systems.

With expansion of the Air Research and Development Command, he became commander of the newly created AF Systems Command (AFSC). Among the many creative programs he conceived and directed at AFSC was Project Forecast I, completed in 1964, which enlisted the best scientific and technological minds of that period in the projection of the aerospace world for the future.

After retiring from the AF, Aug. 31, 1966, with more than 33 years of active military service, Gen. Schriever became a consultant to government and industry where he could most effectively use his knowledge and experience pursuing technology and its management into military operational capabilities.

General Schriever has had several important government advisory assignments since his retirement in 1966, including: by Executive Order, chairman, President's Advisory Commission on Management Improvement (PACMI); member, National Commission on Space; member, President's Foreign Intelligence Advisory Board; member, Strategic Defense Initiative (SDI) Technical Advisory Committee; Chairman, SDI Institute and various ad hoc advisory committees and panels involving national security (DOD) and space (NASA).

General Schriever has been awarded four honorary Doctor of Science degrees, one honorary Doctor of Aeronautical Science degree, one honorary Doctor of Engineering degree, and one honorary Doctor of Laws degree, by various colleges and universities. Inducted into Aviation Hall of Fame in 1980. Elected Honorary Fellow AIAA, recipient of James Forrestal Award 1986. Member of NAE. He received the National Air and Space Museum Trophy for Lifetime Achievement in November 1996.

KARY RANDALL SCHRAMM, born April 28, 1966, Olivia, MN. Commissioned June 16, 1989, through ROTC, University of Minnesota, Missile Combat Crew. Military locations, stations were: 1989 - Vandenberg AFB, CA, 4315th Combat Crew Training Sqdn., student; 1989-93 - F.E. Warren AFB, CA, 320th SMS, Deputy Missile Combat Crew commander; 90th SMW, Minuteman III Evaluator Deputy; 320th  Missile Sqdn., Squadron Command Post Flight Commander and 90th Operations Support Sqdn., Minuteman III Instructor commander; 1993-present - Vandenberg AFB, CA, 392nd Training Sqdn., Missile Operations Training Instructor and EWO Instructor. Currently ranked as captain.

Memorable experience: GEN Chain's annual Christmas address over the Primary Alerting System.

Awarded the Air Force Commendation Medal, Air Force Outstanding Unit Award, Combat Readiness Medal and National Defense Service Medal.

Married the former Mary Beth Dawes on Sept. 23, 1995, and resides at Vandenberg AFB, CA where he is chief, ICBM EWO Training (392nd TRS).

ROBERT R. SCOTT, BGEN, born in Long Beach, CA, Dec. 14, 1923, graduated from Miami Senior High, Miami, FL, and attended the University of Tennessee, entering the Army Air Corps aviation cadet program in February 1943. Commissioned a second lieutenant, he received his pilot wings at Moody Field, Valdosta, GA, June 1944.

During WWII, he went overseas and flew combat missions in B-17 aircraft in the European Theater of

Operations as a member of the 92nd BG stationed at Podington Field, England. His flight crew led the 326th BS on day missions against targets in Germany.

General Scott joined the SAC during its activation in March 1946 when he was transferred to MacDill AAF, FL and remained with the command for the duration of his career.

In 1947 he was transferred to the 509th BG at Walker AFB, NM. In 1948 he became a B-29 combat crew commander and later was squadron operations officer for the 830th BS. In 1953 he went to Loring AFB, ME as operations officer for the 70th BS, was later operations and training officer, then chief of training for the 42nd BW. Next, he was a squadron commander of the 69th BS, which was equipped with B-36 aircraft. During 1956 he attended B-52 transition training and converted the 69th BS to the B-52 bomber. His squadron became the first fully operational combat ready B-52 Sqdn. in the USAF.

He was transferred to Westover AFB, MA, in 1957, where he served with HQ 8th AF as chief, Standardization Div., then chief, Training Div. in 1960 he attended the "Snark" (LRM) training at Cape Canaveral and was the 8 AF HQ Operations "Snark" staff officer for the Presque Isle, ME, "Snark" MW. In 1961 he was assigned as deputy commander for Operations, 72nd BW, and in 1962 he became base commander of Ramey AFB, Puerto Rico.

In 1964 Gen. Scott was assigned to HQ SAC at Offutt AFB, NE, where he served in the office of deputy chief of staff, personnel, as chief, officer Manning Branch and, in 1965, became the HQ deputy chief of staff. In 1966 he was assigned as commander, 390th SMW, with Titan II ICBM missiles at Davis-Monthan AFB, AZ; in 1968 he became commander of the 90th SMW with Minuteman ICBM missiles at F.E. Warren AFB, WY.

General Scott was assigned as commander, 17th Strategic Aerospace Div., Whiteman AFB, MO, in 1969, with responsibility for one Minuteman ICBM Wing, two Titan II ICBM wings (Little Rock and McConnell), one B-52 BW, and two KC-135 Tanker Sqdns.

He was the first commander of the 4th Strategic Missile Div., activated at F.E. Warren AFB, WY, on July 1, 1971, with responsibility for the 1,000 Minuteman ICBM force of the USAF (all six Minuteman Wings and bases). with this assignment Gen. Scott had been commander (Wg. or Div.) of all 9 ICBM Wings (Titan II and Minuteman) during the period of 1966 to 1972. In 1972 Gen. Scott assumed duties at deputy chief of staff for personnel, HQ SAC, Offutt AFB, NE.

General Scott is married to the former Terry Lou Ratliff of Cheyenne, WY. He has a son, Robert R. (Randy), who was a command pilot in the USAF and retired as a lieutenant colonel with 20 years service in fighter aircraft.

General Scott was promoted to the grade of brigadier general in January 1970. A command pilot and master missileer, he was the first SAC missile wing commander to be promoted to general. He retired from the USAF in November 1973 after 31 years service.

After retirement he established his home in Cheyenne, WY, in February 1974. Starting a second career, he joined the staff of the American National Bank for 14 years, retiring as senior vice president of operations. He now lives in and enjoys Wyoming.

DAVID F. SEARES, COL (RET), currently is program manager of Northrop Grumman Non-Contractual Technical Activities in Melbourne, FL. After graduation from the University of California at Los Angeles in 1966, he worked for Univac Corporation in San Diego,

CA as a computer programmer before serving as an AF officer from May 27, 1967 until April 1, 1993. His first assignment was at Malmstrom AFB, MT where he was a line crew member, weapon system instructor and evaluator. He earned his master's degree in systems management from the University of Southern California during this time. Other assignments were a missile combat crew instructor at Vandenberg AFB, CA; force application officer on the JSTPS at HQ SAC, developing SIOP plans for ICBMs and SLBMs; nuclear policy officer for the Under Secretary of Defense, Washington, DC; Senate Legislative officer for the Secretary of the AF, Washington, DC.

After graduating from Air War College in 1984 he became chief, Force Applications Div. in the JSTPS and then special assistant to CINCSAC at HQ SAC. Seares assumed command of the 91st SMW at Minot AFB, ND in June 1988 and returned to Washington, DC in 1989 to become deputy director for program evaluation on the Pentagon's Air Staff and was a senior fellow, Foreign Politics and National Interest, MIT in 1990. His final assignment was the assistant director for plans, Operations and Security, Office of Emergency Operations in the White House Military Office.

He and his wife Nancy have resided in Melbourne, FL since 1995. He enjoys playing golf and playing the piano in his free time.

CHARLES D. SENIAWSKI, COL, RET, born March 18, 1943, Brockport, NY. AFROTC and BS Industrial Management, MIT, Cambridge, MA, 1965. M.S. Logistics Management, AFIT, 1973. Greatest part of career spent in SAC with Minuteman I, and II and III and Peacekeeper missiles. Other responsibilities through the years included SAC reconnaissance aircraft (SR-71-/U-2/RC-135) logistics planning, conventional and nuclear weapons maintenance, explosive ordnance disposal, transient aircraft and AGE maintenance activities, and joint services logistics planning. Assigned to Malmstrom AFB, MT (twice), Vandenberg AFB, CA (twice), HQ SAC Logistics Plans, FCDNA at Johnston Atoll, and F.E. Warren AFB, WY. Initial assignment to Malmstrom AFB, MT, as a Minuteman Targeting and Alignment (T&A) officer. Then to the 3901st SMES as a T&A evaluator and overall maintenance team inspection team chief.

Commanded 341FMMS, 394th ICBM Test Maintenance Sqdn. and Detachment 60, Ogden ALC. Retired Aug. 31, 1989 and his decorations include the Legion of Merit.

PME included Squadron Officers School, Armed Forces Staff College, Air Command and Staff College and National Security Management.

Resides in Cheyenne, WY, with Sue, his wife of 28 years. Their two children, Barbara and David, graduated from Dartmouth College and Stanford University, respectively, and are now living on their own.

DR. JOHN (PAPPY) SHAFER, PhD, MAJ (RET), has Titan II maintenance experience and operations experience in the Minuteman and Titan III systems. He enlisted in 1959, working ten years as a weather observer. After a break in service, he reenlisted in 1971, serving in maintenance, 390 MIMS, Davis-Monthan. There, he also taught enlightening classes on prevention of the silent enemy - corrosion. Commissioned in late 1972 he held a variety of crew jobs at the 90th. Mostly, he was a MMI and MMM/CDB instructor. Following that assignment, he had crew and staff jobs at the 390th. Primarily, he was an EWO instructor. Leaving a hectic Titan II plans/intelligence/codes division, he returned to the relatively calm Minuteman world at

Malmstrom. He was the ILCS senior instructor and an EWO instructor. After Malmstrom, he spent his final five years as a command-level evaluator/inspector with the 3901st and the SAC IG, positions where he effectively maintained the CHIME program. Faced with a situation where he had two children in a Nebraska university and an impending assignment to Minot, he retired in 1987. He took and taught courses, completing a doctorate in sociology from the University of Nebraska-Lincoln. After almost four decades of the military and academics, he finally figured out where the money is and is a management/training/accreditation consultant to the nuclear power industry. He and his lovely and patient spouse, Dee, live on a small acreage in the Arizona desert and periodically travel for business and pleasure.

MICHAEL N. SHAPIRO, MAJ (RET), graduated from Tufts University in June 1965 and reported in to the 321 SMW at Grand Forks, AFB, ND in July. He is one of the signatories to the activation roster of the 448 SMS, but was reassigned to the 446 SMS for his initial missile combat crew assignment.

He completed his MS in Industrial Management from the University of North Dakota before going pcs in June 1970 to the 90 SMW at F.E. Warren AFB, WY to become a missile maintenance officer.

In June 1973 he was reassigned to HQ SAC Office of Logistics Analysis, where he served as the USAF OPR for the Ballistic Missile Status Report, and later as deputy chief, logistics analyst.

In June 1977 he was transferred to Titan II with the 390 SWM at Davis Monthan AFB, AZ where he served in various maintenance assignments and as chief of safety. Following the 1984 successful deactivation of the 390 SMW, he moved across the base to the 868 TMTS to manage safety for ground launch cruise missile training. He retired Aug. 1, 1985.

He and his wife, Karen, moved to Denver where he joined Martin Marietta's Logistics Support Analysis for Peacekeeper. Later he worked payload integration for the space shuttle.

He is a certified fellow in Production and Inventory Management, is certified in Integrated Resource Management, is adjunct faculty at University of Phoenix (Colorado Campus) and University of Denver, and works as senior consultant for IBM Global Services' Enterprise Resource Planning Practice.

RICHARD E. SHARP, CAPT, born May 11, 1944, Pittsburgh, PA, commissioned into USAF, SAC, June 1966, received training September 1966-November 1966, Chanute AFB, IL and Vandenburg AFB, CA. Stationed at Minot AFB, ND, deputy missile combat crew commander, instructor Starboard then missile combat crew commander, Minuteman I. Separated from service February 1970. Served in inactive reserves until November 1977 when he was discharged as captain.

He was the first commander on alert when the first Minuteman I was taken off alert for conversion to Minuteman III.

Sharp was employed in industrial sales from August 1970 until December 1981. He is now self-employed as a landlord.

CHARLES W. SHAW, LT COL (RET), was certified as deputy crew commander in March 1970 (Minuteman I), 91st SMW, 742nd SMS, Minot AFB, ND. Awarded 91 SMW Crew of the Quarter, and SAC Crew of the Month, December 1970 for response to a dual emergency situation (LCC fire and an LF Lockout). Transitioned to Minuteman III system, 1971. Certified as ACP/SCP crew commander; flight leader for Mike Flight (SCP); and Wing Select, instructor crew commander. After 268 alert tours, transferred to 4315 CCTS, Vandenberg AFB, CA in 1974 as instructor, then training program manager for Minuteman III. Transferred to "Top Hand" Program, 1st STRAD, VAFB, CA in 1976 as a Minuteman launch director on Western Test Range. Transferred to NASA, Houston, TX, 1980. Certified as shuttle flight controller, shuttle systems instructor, and simulation supervisor. Selected as first non-NASA flight director for Manned Spaceflight in 1983. Retired from USAF, 1989 and joined NASA.

Shaw and his wife Connie still live in Houston, where he is a senior flight director for the Space Shuttle and International Space Station Programs.

JOHN W. SHORTER, LT COL, native Virginian, commissioned a second lieutenant, USAF after graduating AFROTC from Virginia Tech & State University with a BS in 1954. Distinguished Military Graduate of both the Air Command and Staff College in 1963 and the Air War College in 1974. Received a master of science degree in systems engineering from the University of Southern California in 1978. The majority of his career was served in the SAC in Strategic Missile (ICBM) operations. He was combat qualified in the Titan II and Minuteman Weapon Systems and operationally qualified in the Emergency Rocket Communications System and the Airborne Launch Control System. Assisted in the development of Positive Control System for the USAF.

He retired as deputy director, Strategic Missiles, HQ 15th AF in 1976 as a lieutenant colonel. His awards include the Master Missileman's, Senior Air Operations and Combat Crew Badges, decorated with Meritorious Service Medal, AF Commendation Medal, Combat Readiness Medal, National Defense Service Medal, Armed Forces Expeditionary Medal, Air Reserve Medal w/Bronze Hour Glass, Presidential Unit Citation w/OLC and the AF Outstanding Unit Award w/2 OLCs.

During the next 12 years, served as a senior systems engineer for the USN Standard Missile Weapon System at the Naval Warfare Assessment Center. He and his wife, Cynthia Ann, reside in Moreno Valley, CA.

CHARLES G. SIMPSON, COL (RET), is one of the founders of the Association of Air Force Missileers, and has served as the AAFM executive director since January 1993. After graduation from the University of Miami, in August 1959, he spent a short time as a de-

sign engineer with Pratt Whitney Aircraft before serving as an AF officer from Nov. 25, 1959 until July 31, 1989.

Duties included aircraft maintenance officer at Hanscom Field, MA, Titan I Missile maintenance officer in the 569th SMS, Mountain Home AFB, ID and Missile Combat Crew commander, instructor and evaluator, at the 321st SMW, Grand Forks AFB, ND. He was a member of the Blanchard Trophy winning team in the SAC Missile Combat Competition, Olympic Arena 1969. He competed again in 1970, and earned an MS Industrial Management at University of North Dakota.

In June 1970, he reported to the 3901st Strategic Missile Evaluation Sqdn., Vandenburg AFB, CA, serving as a Minuteman II evaluator and Minuteman operations team chief. He became chief, accuracy and then chief, evaluation, Ballistic Missile Evaluation, HQ SAC, Offutt AFB, NE, developing SIOP planning factors for ICBMs. He attended Air War College in the 1978 class, and then became commander, 68th SMS and then assistant deputy commander, operations in the 44th SMW, Ellsworth AFB, SD. He commanded the 44th Combat Support Gp., Ellsworth AFB, SD, the new 487th Combat Support Gp., Comiso AB, Turkey, and the 406th Combat Support Gp., Zaragoza AB, Spain before his final assignment as chief of staff, 57th AD, Minot AFB, ND.

He and his wife, Carol, have resided in Breckenridge, CO since 1989. He serves as a director on the boards of the Breckenridge Music Institute and the National Repertory Orchestra, and spends most days either skiing or playing golf.

WILLIAM F. SIMS, CAPT, born June 3, 1957 in Windsor, Ontario, graduated from Eisenhower High School in Saginaw, MI in 1975, and joined the USAF within the month. After basic training at Lackland AFB he attended the Armed Forces School of Applied Cryptological Sciences at Goodfellow AFB, TX. His initial training as a cryptanalyst was immediately followed by an assignment to the 6912th Security Sqdn. Tempelhof Central Airport, Occupied West Berlin, Germany.

After returning from Germany he was assigned to the AF Electronic Warfare Center, Kelly AFB, TX, where he served as a liaison analyst. Captain Sims managed two computer systems in the EW center, and was liaison between the center and several operating locations worldwide.

The San Antonio assignment was followed by a second overseas posting with "Skivvy Nine," the 6903rd Electronic Security Sqdn., at Osan AB, Korea. He conducted preliminary intelligence analysis and reporting while at Osan, and operated several intelligence collection systems. While still in Skivvy Nine, he was promoted to the rank of staff sergeant.

A second stateside tour followed Osan, when Capt. Sims was assigned to the 6940th Electronic Security Wing at Ft. Meade, MD. At Ft. Meade, he worked as a signals intelligence analyst and briefer in the National Security Agency. He received an associates and a bachelor's degree at Ft. Meade, and obtained his seven-

level as a radio communications analyst. His tour with NSA was completed on selection to attend Officer Training School at Medina Annex, San Antonio, TX.

Capt. Sims completed OTS in July 1984, and moved to Vandenburg AFB, CA, for initial qualification training as a missile combat crewmember in the Minuteman III Command Data Buffer Intercontinental Ballistic Missile System. Initial combat crew training was followed by a three-and-a-half year tour at Minot AFB, ND, in the 91st SMW. At Minot, he became a deputy combat crew commander at the alternate command post, and spent his last year on missile crew as a combat crew commander. He completed a master's degree at Minot, as well as Squadron Officer's School.

In 1988, he was selected as one of the first missileers to retrain into the newly-created operations management career field. He attended the operations management officer course at Keesler AFB, MS, and moved to Suwon AB, South Korea on course completion.

Capt. Sims became chief of the AB Operability Div. at Suwon, managing the 6170th Combat Support Grps. Disaster Preparedness and Explosive Ordnance Disposal branches.

Suwon was followed by a fourth stateside tour; Ellsworth AFB, SD. He was assigned to the 28th BW's Command Post as an emergency actions controller in December 1989. At Ellsworth, Capt. Sims completed a doctorate in Operations Management, with emphasis in Military Science.

Captain Sims was assigned to the 603rd Airlift Support Group's Air Mobility Command Center, Kadena AB, Okinawa, Japan, on leaving Ellsworth in 1992. This was his fourth overseas tour, and eighth assignment in the AF career.

His Kadena tour was interrupted by a 91-day stint as chief of the Mogadishu International Airport AMC Command Post, and concurrent duty as vice commander, USAF Somalia. Somalia service introduced him to combat. He experienced several mortar, rocket propelled grenade, and small-arms attacks at the airport and during convoys to the embassy/university compound in Mogadishu. One month to the day after his arrival in Somalia, Capt. Sims was a participant in "Bloody Sunday," the longest episode of urban combat since the Vietnam War. He managed Air Mobility Command and United Nations airflow following Bloody Sunday, including the post-battle buildup of forces and armament, medical evacuation and repatriation of well over 100 combat casualties, and the return of CWO Michael Durant, the conflict's only POW.

Among Capt. Sim's accomplishments during his 19 year career are selection as Airman of the Quarter at Kelly AFB, Unit Career Advisor of the Year at Ft. Meade, Suggester of the Quarter at Minot AFB, Junior Officer of the Quarter at Ellsworth AFB, and three-time selection as one of the SAC's most professional combat crew-members. He is also proud of the fact that during his time as vice commander in Mogadishu, no AF personnel were wounded or killed.

Captain Sims has 40 separate awards and decorations, including the Army of Occupation Medal, Armed Forces Expeditionary Medal, United Nations Medal, and Combat Readiness Medal. He wears the Senior Missile Combat Crew Badge, Basic Command and Control Badge, and the Senior Intelligence Badge.

Captain Sims is married to the former So Sun Ki of Pusan, South Korea. He has two children, Kimberly and George. On retiring from the AF in 1994, Capt. Sims and his family moved to San Antonio, where he is now a sales representative for USAA automobile insurance.

WALT SKRAINY JR., COL, born Jan. 29, 1943 in Swoyersville, PA, enlisted in the USAF, Nov. 13, 1964, SAC, USAFE.

His military locations and stations: Malmstrom AFB, MT; Vandenberg AFB, CA; Maxwell AFB, AL; Offutt AFB, NE; Pentagon, Minot AFB, ND; RAF Greenham Common, UK.

His memorable experiences: Started up 564th SMS, Malmstrom AFB, 341st SMW, pulled first alert in Squadron LCCS. Competitor SAC Missile Competition

(1969); 15th AF Missile Crew of the Year; JCS, advisor US, Soviet Union START Team.

Skrainy was awarded the Meritorious Service Medal and was discharged Dec. 1, 1990 as colonel. He married Phyllis Jan. 15, 1966 at Malmstrom AFB Chapel and they have daughters: Karen, Kristin and Kelly. He is now employed in the insurance business and a member of Knights of Columbus.

EUGENE W. SLEGEL, MAJ (RET), is one of the pioneer missileers. He was born in Reading, PA, Oct. 7, 1920, enlisted in the Army Air Corps May 22, 1939 at Chanute Field. Graduated from Radio School and was assigned as an instructor in the Training Command. He was commissioned second lieutenant, Oct. 28, 1943, after completing radar school was assigned to the Experimental Guided Missile Sq. (later 550th GM Det) at Hollomon AFB on Dec. 4, 1949.

The MX771 was a secret weapon, therefore he does not have any photo's of the Mid wing streamlined missile slightly smaller than a fighter, that was launched from a zero length launcher assisted by a JATO bottle. This missile later named the Matador as operations were shifted to Patrick AFB, and a improved model was produced. His speciality was guidance and control officer and beeper pilot. Pilotless bomber squadrons. After duty tours with: HQ TAC, PAFB Comm. Sq., 1935th AACS Sq., ACIC, and 1374th Mapping and Charting Sq. He returned to the 4504th Msl. Tng. Wg. at Orlando AFB, May 13, 1959 serving as missile maintenance officer until his retirement, June 30, 1960. One of the little mentioned, but outstanding events of his last duty assignment was launching six TM-61c's at Patrick AFB in one day.

After retirement from the AF he was employed by Martin Co., as a test engineer, he participated in tests of early Pershing and Patriot missiles. He and his bride of 55 years reside in Winter Park, FL.

B. DEAN SMITH, COL, born Nov. 20, 1929, Bryn Athyn, PA, USAF, enlisted in the USN, 1947. Graduated Naval Academy, June 2, 1953. Took commission in the Air Force.

His military locations and stations: Charleston AFB, SC; Johnson AB, Japan; Patrick AFB, FL; Pentagon, Andrews AFB, Maxwell AFB, Pentagon.

He participated in airborne telemetry of all missile systems tested ATL Missile Range.

His memorable experiences: member of Reentry Society (charter); recovered first nose cone from ballistic missile; Atlas South Atlantic. Had first space system fire ever experienced by NASA.

Smith was discharged July 1, 1974 as colonel and was awarded the Bronze Star Medal, Meritorious Ser-

vice Medal, Air Medal, AFOUA w/OLC, Guided Missile Insignia, VSM w/Silver Star, AFCM w/OLC, SOG, NDSM w/Bronze Service Star, RVCM, PUC, AFLSA w/4 OLC, AFM.

He has four sons, Stewart (civ.), Christopher (Army), pilot; Matthew (civ), Aaron (Army) pilot.

Started his own company (safety) 1977. He is retired as past commander of Military Order of the World Wars. He now runs youth leadership conferences.

DOUGLAS G. SMITH, born Nov. 26, 1948 in North Conway, NJ, received a BA in psychology, 1970 Colby College, Waterville, ME; Montana State University, 1974-75; Doctor of Optometry, Pacific University 1979.

His professional activity: clinic staff, Pacific University College of Optometry, 1979; private practice, Medford, OR, 1979 to present; chairman, Oregon Laser Eye Center, 1996.

His academic appointments: adjunct associate professor, Pacific University College of Optometry, 1996; guest lecturer, Pacific University, 1984-1995; guest lecturer, Southern Oregon State College 1981-present.

State appointments: Oregon Commission for the Blind, 1985-1991; Children's Services Division Citizens Advisory Committee, 1984-1990; Oregon Board of Optometry, 1991 to present.

National appointments: National Board of Examiners, 1984 to present; senior examiner NBEO, 1990 to present.

Community activities: chairman, Southern Oregon State College Learning Disabilities Clinic; Chairman, Jackson County Juvenile Services Commission; Board of Directors, Rogue Valley Alcohol Rehabilitation Center; Board of Directors, Willaway Ranch for the Handicapped; Jackson County Task Force for Pre-School Handicapped; Jackson County Head-Start Medical Advisory Board; Oregon Teacher Standards & Practices Committee Task Force, Handicapped Learner, SOSC; Medford Rotary Club 1984 to present.

Professional Associations: American Academy for the Advancement of Science, American Public Health Association, American Optometric Association/Oregon Optometric Association, Chairman, Oregon Optometric Association Children's Vision Section, College of Optometrists in Vision Development, Optometric Extension Program, Optometric Extension Program, Beta Sigma Kappa, Phi Theta Upsilon, AOA Low Vision Section, AOA Contact Lens Section.

Awards: Who's Who in American Colleges and Universities, 1978; Outstanding Young Men of America, 1978-1984; William M. Feinbloom Low Vision Award, 1979; Oregon Optometrist of the Year, 1982; AOA Optometric Continuing Education Recognition Award, 1984, 1990, 1991, 1992, 1993, 1994, 1995, 1996; Who's Who in the West, 1992, 1993, 1994, 1995, 1996.

ROGER C. SMITH, BGEN (RET), born March 8, 1937 in Orange, NJ. Commissioned in 1959 through AFROTC at Kenyon College, OH. Master's degree, 1973, University of Oklahoma. Retired in 1988. Master Missileman.

Senior military representtive to the US-USSR Defense and Space Talks, 1986-1988; Command Director, US Space Command, 1985-1986; Assistant Chief of Staff, HQ SAC 1984-1985; Commander, 351st SMW, Whiteman AFB, 1982-1984; Base Commander, deputy commander for maintenance and assistant DCM 1979-1982. Member of first SAC combat-ready launch crew for Titan II missile system, Davis-Monthan AFB, 1962. Standardization/Evaluation crew member, 1962. ICBM Test and Evaluation officer, HQ SAC, 1962-1968. SAC

liaison officer for Minuteman III development, Norton AFB, 1969-71. Served in HQ USAF, 1972-1975, in Strategic Forces Div., Directorate of Plans, and later became chief of Policy Analysis for three secretaries of the AF, 1975-1978.

Graduate of National War College, 1979; Armed Forces Staff College, 1971; Distinguished Graduate of Squadron Officer School, 1965. Awarded Defense Distinguished Service Medal, Defense Superior Service Medal, Legion of Merit w/OLC, and numerous other decorations.

Upon retirement, established Roger Smith Associates, a security and intelligence program development consulting firm, of which he is president.

Married in 1966 to Sybil Marie Leonard of Portsmouth, VA. They have three children, and reside in Fairfax Station, VA. Active in the Antique and Classic Boat Society; Division Chairman, National Defense Industrial Association; Subcommittee Chair, Intelligence Committee, AFCEA.

CHARLES A. SNYDER, LT COL (RET), is professor of management, Auburn University, AL. After graduation from the University of Georgia in 1956, he entered pilot training at Moore AB, TX. Assigned to Dover AFB, DE, then to Guam; later to Yokota, AB, Japan as Detachment Commander. After McGuire AFB and SOS, he was assigned to the 68th SMS, Ellsworth AFB as minuteman crew member and instructor. He was one of the initial ALCS crewmembers and instructors in 1967. Chief, ALCS Operations, 4th ACCS, 28th BW. He received the MBA from Ohio State University in 1967. In 1971, he was chief, Long Range Plans, HQ MACV-EA in Saigon. In 1972, he received the MS in economics from South Dakota State and completed the Industrial College of the Armed Forces.

Snyder returned from Vietnam when his home in Rapid City was destroyed by flood. He next served as ALCS operations officer, 2nd ACCS, 55th SRW, Offutt, AFB and chief of ALCS for SAC and DOCA. He was selected Symbolic Graduate of the Air War College Associate programs, 1975, then assigned to Ellsworth AFB as assistant deputy commander, Operations, 44th SMW and commander, 68th SMS.

He retired in 1977 and completed PhD at the University of Nebraska. His decorations included the Bronze Star, two Meritorious Service, Joint Services Commendation, five AF Commendation Medals.

Snyder became a professor at Auburn University in 1978. He is the author of a textbook, over 100 refereed scientific articles and papers. He has been a consultant to such firms as AT&T, Bellsouth, TRW, Coors, etc. He is active as director of a software firm.

He and his wife, Margrit, split their time between their homes in Auburn, on Lake Martin, and in Breckenridge, CO.

FRANK SPIZZIRI, E-4, born Oct. 22, 1943 in Paterson, NJ, entered the USAF/SAC 33150 Special Weapons.

His military locations and stations: Lowry, F.E. Warren, Westover, Atlas D, Atlas E, B-52D, B-52E.

Spizziri was discharged Jan. 23, 1967 as E-4. He was awarded the Good Conduct Medal and the NDSR.

He was married and has two children, both just married in 1997. Employed with Xerox Corp. as corporate instructor.

HUBERT O. SPRABERRY, LT COL (RET), born on April 5, 1931 near Post, TX, grew up on farms near Lamesa and Whitharral, TX. Graduated from Whitharral High School in 1948 and immediately began attending

Texas Technological College in Lubbock, TX. After graduation he served as an AF officer from May 14, 1952 until March 31, 1976. Duties included adjutant, administrative officer, assistant professor of Air Science at Northeastern A&M College, Miami, OK and chief, Dependent Schools Branch, RAF Lakenheath, England. In 1963 he reported to Minot AFB, ND. He became a missile combat crew commander of a select instructor crew in the 741 SMS. In 1967 Spraberry reported to HQ SAC where served as a missile operations staff officer and eventually chief, Test Operations Branch, DCS/Plans. In 1972 he reported to the 321st SMW where served as commander, 446 SMS; commander, 448 SMS, and deputy commander, 321st Security Police Gp. The final assignment was at U-Tapao, Thailand where he served as commander of an 1,000 person Combat Defense Force. A very harrowing experience occurred during this tour. Spraberry's Criminal Investigations Div. broke up a drug traffic ring which was operating on the AB and the local village. As a result the criminals decided to assassinate the base commander. Extra security precautions were taken so the attempt did not succeed. Then, they decided to assassinate Spraberry. Needless to say this was a very stressful time until the threat went away.

Awards and decorations received include the Meritorious Service Medal, AF Commendation Medal w/OLC, Korean Service Medal, and the UN Service Medal.

Spraberry has a BA degree from University of Maryland; BS from Texas Technological College; MA from the University of North Dakota; MS from Pittsburg State University; PhD from Texas Tech University.

Since retirement he has served as assistant professor and chairman, Business Div. at Wayland Baptist University, Plainview, TX; Assistant professor of Economics at Texas Tech University, Lubbock, TX; and professor and dean, School of Business Administration, Howard Payne University, Brownwood, TX. He now serves as professor of Economics and Finance at Howard Payne.

RALPH E. SPRAKER, M/GEN (RET), entered the AF in March 1956 after graduating from Utah State University. He was a navigator instructor and standardization evaluator in B-47s at Schilling, Forbes and Pease until 1965. He was a Minuteman missile crewmember, chief of standardization, commander of two SMS, Missile Evaluation Sqdn., a combat support group and two missile wings at Malmstrom, F.E. Warren, Vandenberg, Ellsworth and Whiteman AFBs.

He held positions of chief, Missile Div., asst. DCS, Operations and Plans and DCS, Space Warning and Surveillance at SAC HQ. After activation of AF Space Command, became the first commander of the 1st Space Wing. Subsquently, served as chief of staff for NORAD and AF Space Command and assistant chief of staff of U.S. Space Command.

Spraker's final assignment was as vice commander of AF Space Command. He and Sandra reside in Colorado Springs where he started a finance company.

DANIEL S. STACHOWIAK, TSGT (RET), born Aug. 15, 1924, Buffalo, NY, enlisted in the USN, Dec. 7, 1942. He served in the USN, Dec. 7, 1942 to April 5, 1946; USN, Jan. 10, 1949 to June 24, 1957; USAF, July 19, 1957 to Nov. 1, 1965.

His military locations and stations: NTC Newport, RI; NATC, Norman, OK; NATC, Memphis, TN, NAS San Diego, CA; NAS Niagara Falls, NY, VAH-7 Patuxent River, MD; VAH-7 Port Lyautey, French Morroco; VAH-7 USS Randolph (CVA-15); VAH-7 USS Intrepid (CVA-11); NAAS Sanford, FL; ATU-212 NAAS Kingsville, TX; 6555 GMS Patrick AFB, FL; 4751 ADSM Eglin AFB Field, #9, Florida; 10 ADS Vandenberg AFB, CA; Johnston Island.

His memorable experiences include the end of WWII VE and VJ Days; blackouts in WWII; flight operations of his squadron VAH-7 aboard carriers USS Randolph (CVA-15) and USS Intrepid (CVA-11) in the Mediterranean Sea; member first all AF crew to process and launch Bomarc Missile at Cape Canaveral 1958.

Retired Nov. 1, 1965 as technical sergeant and awarded the USAF Commendation, USAF Good Conduct, Navy Good Conduct, Occupation (Europe), American Campaign, National Defense, Master Missileman Badge, WWII Victory.

Stachowiak is divorced and has three married sons, seven grandchildren and two great-grandchildren. He was employed at LTV Aircraft Corp. for 9 1/2 years and Dallas Ft. Worth International Airport for 20 years. He currently resides in N. Richland Hills, TX. Now enjoying retirement, woodworking is his hobby. He has a small work shop at his home.

SHERRY L. STEARNS, CAPT, is an Intercontinental Ballistic Missile (ICBM) Operations Plans and Program Officer, Twentieth AF, Plans and Program Branch, F. E. Warren AFB, WY. She assumed her current rank on Jan. 13, 1992.

Stearns was born May 23, 1966, in Colorado Springs, CO. She graduated from Holland Hall College Preparatory High School, Tulsa, OK, in 1983, and earned a bachelor of arts degree in Human Biology from Standford University in 1988. She also earned a master's in aeronautical sciences degree from Embry-Riddle Aeronautical University in 1993.

Commissioned through the AFROTC program at San Jose State University in California, Capt. Stearns entered active duty in May 1988. She attended the ICBM Initial Qualification Training Course and graduated as a "Top Performer" in September 1988.

Captain Stearns was first assigned to the 10th SMS at Malmstrom AFB, MT. She served as a deputy missile combat crew commander, from September 1988 to September 1989. In October of 1989, she earned an instructor position in the 341st SMW, Operations Training Instructor Branch. While there, she won instructor of the Quarter award. In September 1990, she upgraded to the position of missile combat crew commander. While in this position, she acquired three "Top of the Line Commander" awards. In April of 1991, she was hired by the 341st Operations Gp., Standardization Evaluation Div., as an evaluator. In June of 1992, she

was promoted to be chief of the Procedures AFB, CA in November 1992. While at Vandenberg, she performed multiple duties as a simulator instructor, simulator courseware developer, and academic classroom instructor. She advanced to the position of chief, Academic Operations Element in June 1994. In February of 1995, she was hired as an emergency war order instructor. While at Vandenberg, she accumulated several awards to include Instructor of the Quarter, company grade officer of the Quarter/Year, and Volunteer of the Quarter. In June of 1995, Capt. Stearns was selected as an evaluator for the 20th AF, Standardization Evaluation Div., F.E. Warren AFB, WY. She performed duties as chief, Standardization Policy, and Chief, Technical Order Programs. During this period, she was selected as 20 Air Force Company Grade Officer of the Quarter. She has served as an ICBM operations plans and programs officer there from May 1997 until current. While holding this position, she was selected as 20 Air Force Company Grade Officer of the Year for 1997.

Captain Stearns' awards and decorations include the Air Force Commendation Medal w/OLC, the AF Achievement Medal, the AF Outstanding Unit Award w/4 OLCs, the Combat Readiness Medal w/OLC, and the National Defense Service Medal. She also earned Air Education and Training Command's Master Instructor Badge.

She completed Squadron Officer School by correspondence in June 1994, and in residence in April 1995. She received her regular commission in the AF in March 1995.

Her professional affiliations include being a member of the following: Air Force Association, Association of Air Force Missileers, Officer's Club, Delta Sigma Theta Sorority, Inc., National Association of Female Executives, and Cheyenne Young Professionals.

Outside her job, she heads several positions in the community. She serves as president of F. E. Warren's Chapel Protestant Parish Council and the Wyoming Buffalo Soldiers Association. She also serves as vice-president of Warren's Protestant Women of the Chapel and leads Sunday School classes. She is a member of the Chapel Gospel Choir and the Cheyenne Community Gospel Choir.

CARLTON A. STIDSEN, born Nov. 20, 1943, was commissioned from the AFROTC program at the University of Massachusetts on June 13, 1965. He was assigned to the 571st SMS, 390th SMW, at Davis-Monthan AFB, AZ. He upgraded to deputy missile combat crew commander on June 18, 1966. Eighteen months later, on Dec. 15, 1967, he completed upgrade training to missile combat crew commander.

He left USAF active duty on March 31, 1970, as captain. Since then he has been a safety consultant, working with a number of insurance companies. He earned the designation of "Certified Safety Professional" in 1979.

In 1974, entered the USAF Active Reserve and was assigned to the 9004th Air Reserve Sqdn. as a liaison officer with the Connecticut Wing, CAP. He served in that capacity until he entered the USAF Retired Reserve as a major, Jan. 6, 1990.

Stidsen married Doris M. Anderson, Oct. 21, 1967, in Worcester, MA. They have lived in Tolland, CT since 1973. He holds a commercial pilot certificate. He is active in CAP as the inspector general for Connecticut Wing, and also as a Senior Emergency Services Search & Rescue Pilot, Counter-Narcotics Pilot, and Cadet Flight Orientation Pilot.

HAROLD A. STRACK, BGEN (RET), born 1923 in San Francisco, graduated 1943 from Selman Field Navigation School. Flew 52 combat missions as navigator in 15th AF B-24s; squadron navigator, chief, navigation ground training at Mather Field. Flew 29 missions as radar operator in B-29s over Korea; Wing Staff Radar and Observer, flying B-47s and YB-52, chief, Radar Bombing Branch, SAC HQ, commander, 1st Radar Bomb Scoring Gp. As chief, missile requirements, HQ SAC, staff manager for Snark and all Atlas, Titan and Minuteman series (1959-62). Signed AF acceptance of Thor and operational readiness documents for all but Minuteman before ICAF (1962). DCO and vice commander, 90th MW, F.E. Warren from activation (1963) until operational readiness (1965). Chief, chairman, JCS, Special Studies Gp. Strategic Branch; Deputy assistant to JCS chairman for Strategic Arms Negotiations (1968). Commander, 90th MW (1969), promoted to brigadier general, 1970; flew "Looking Glass" and launched one of his Minuteman Missiles from Vandenburg with ALCS.

Director, JCS Studies Analysis and Gaming Agency, retiring 1974. His decorations include the Distinguished Service Medal, Legion of Merit, Distinguished Flying Cross, Air Medal, Purple Heart and Army Air Force, Joint Service Commendation Medals, Master Navigator; Master Missileer.

Employed with Northrop Electronics Div., as vice president and program manager for MX/Peacekeeper AIRS inertial guidance during advanced development phase, then UP for strategic planning 1988 retirement. Air Force Association Doolittle Chapter and California State President and national official. Residing in Incline Village, NE, with wife Margaret; still active musician, writer and speaker.

THOMAS R. SWANSON, LT COL (RET), enlisted in the USAAF in 1943, served as a radio/radar technician, with the 458th BG, 8th AF, England. He was discharged as a buck sergeant.

Received a bachelor of arts degree and was commissioned a second lieutenant USAF in 1950.

Called to active duty and sent to Arabic Language studies at Georgetown University and assigned to the 580th Air Resupply and Communications Wing, Tripoli, Libya as language officer. After a tour in Germany, he returned to Libya as wing liaison officer to the Libyan Government. He was assigned to the directorate of intelligence, HQS SAC in 1956.

In 1960 he joined the 566th (549th) Atlas D Missile Sqdn., Offutt AFB, NE. His crew participated in the 1962 Tall Tree Exercise. This was the launch of an Atlas D, removed from alert status, air-lifted to Vandenberg AFB and launched on March 2, 1962. Legend states it looked like "The second coming" as it lit the evening skies; exploding at 4,000 feet. Atlas 102 D almost made a complete loop before its unfortunate demise. After this success Maj. Swanson, became a standboard officer. He was on duty in November 1962 when President Kennedy ordered the Soviet ships approaching Cuba to turn back. Lieutenant Col. Swanson remained with the Atlas D until it was phased out in 1965.

After attending the Armed Forces Staffs Staff College, he was assigned to HQ Military Assistance Command, Vietnam. Lt. Col. Swanson served with the Defense Intelligence Agency and then was assigned as assistant deputy chief of staff for Intelligence, 8th AF, Guam 1970 to 1972. As a senior staff officer he flew B-52 missions over South Vietnam. His tour included the December B-52, Eleven Day War, Linebacker 11.

He and his wife, Anne, and family retired in 1976 to Clearwater, FL.

GERALD R. TAFT, MAJ (RET), commissioned via AFROTC at the University of Minnesota in 1964. Following duty at Hanscom Field, MA; Ching Chuan Kang AB, Taiwan and Bien Hoa, Vietnam. He arrived at Whiteman AFB, MO in June 1968. Training at Chanute and Vandenberg preceded local "upgrade" at Whiteman and Taft was certified as combat ready MCCC in November 1968 with the 508 SMS. He served as a flight commander, ACP crew and earned seven  "highly qualified" evaluation ratings. He participated in the Modified Operational Missile (MOMS) test and a Follow On Test (FOT) launch task force at Vandenberg. In 1972 he became a standardization evaluator advancing to "select" wing crew status.

In March 1974 he became an instructor with the Minuteman Modernized Section of the 4315 CCTS at Vandenberg AFB, CA. Here Taft progressed from team chief, scheduler, instructor evaluator, Academic chief and finally section chief. He was a key player in the development and testing of the Mission Ready training program at Vandenberg and use of the Instructional Development concepts.

His next two assignments, Elemendorf AFB, AK and Wurtsmith AFB, MI were as chief of public affairs. Taft retired in 1984 and has since resided in Anchorage, AK with his wife Helen. He completed 12 years work with the State of Alaska's Division of Public Assistance in 1997 and is currently serving as a VISTA with the Anchorage Literacy Project.

CLIFFORD V. TAYLOR, COL, USAF (RET), designated as the first of field ballistic missile assistants.

In 1958 he was assigned to the Rochester, NY, air material field office to assist in the start-up of Atlas Missile and Classified Surveillance Programs.

He next saw duty with the HQ USAF Ballistic Missile Div. in Inglewood where he participated in Atlas and Titan Nite Exploration and Site Activation Task Force (SATAF) Activities.

At completion of the Air University Staff and Command College, he saw duty on the Air Staff at HQ USAF in the Pentagon with the directorate of procurement and production involving Minuteman Readiness and Deployment.

His military decorations/awards include the Legion of Merit, Distinguished Flying Cross, Air Medal, AF Commendation Medal. He also holds the Master Missile Man Badge.

Taylor currently resides with his wife, Beverly, in an adult golf retirement community. He was a member of the Board of Directors of the Progress Aerospace Corp. and is now active as a VFW trustee and tutors in an adult literacy program.

One of his most memorable experiences was as the HQ directorate of Systems and Logistics duty officer under Gen. Thomas Gerrity when the Titan silo exploded at Vandenberg AFB.

JAMES PATRICK THOMAS, MAJ (RET), born Sept. 24, 1946 in Chicago, IL, enlisted in the USN on June 11, 1964. He served aboard the USS Savage (DER 386) on which he did a 1966 tour in Vietnam. Next, he served in Naval Special Warfare Gp. as a radarman on Patrol Craft, Fast (PCF) Swift Boats in Vietnam from November 1967-August 1969.

After returning from Vietnam, Maj. Thomas attended Simpson College, Indianola, IA and graduated from Drake University, Des Moines, IA in 1973. He was commissioned in the USAF in Dec. 6, 1973. In the AF, Maj. Thomas served as a missile launch officer assigned to the 91st SMW, Minot AAFB, ND. After serving on missile crew and as a codes controller, Maj. Thomas changed his AFSC and became a signal intelligence officer. In this career field, Maj. Thomas served with the 6916th Electronic Security Sqdn., Athens, Greece as an RC-135 crewman, chief of operations production, and squadron executive officer. Other assignments for Maj. Thomas include, two tours at the National Security Agency, two tours at the Pentagon, and a tour at HQ USUCOM assigned to J-2. Maj. Thomas also attended in residence and graduated from the Army Command and General Staff College at Fort Leavenworth, KS.

He retired from the USAF in October 1990 with the rank of major and his award and decorations include: Purple Heart, Defense Meritorious Service Medal w/ OLC, Meritorious Service Medal, Air Medal, AF Commendation Medal, AF Achievement Medal, Navy Combat Action Ribbon and the Vietnamese Gallantry Cross.

He was selected Who's Who in the World, 13th edition, Who's Who in America, 52nd edition; Who's Who in Education; Who's Who in the East, Colleges and Universities, 1995; president of the Swiftboat Sailors Association, Inc. 1995-present; charter member of the Swiftboat Sailors Association, Inc.; member of Phi Delta Gamma National Graduate Honor Fraternity, Gamma Chapter vice president 1994-1996.

Major Thomas' education includes: certificate of Advanced Graduate Studies in Special Education; Johns Hopkins University, MD; master of science, Special Education (Severe and Profound, Mild to Moderate Handicapping Conditions); Johns Hopkins University, MD; master of science, Public Administration, Troy State University, AL; BA Political Science and History, Drake University; Des Moines, IA; graduate, US Naval Instructor Course and the SAC instructor course.

Major Thomas is currently employed as a special educator in charge of a 3.6 behavior disorder program at Hiatt Middle School in Des Moines, IA. He resides in Iowa and they have one son, Nicholas J. Thomas, who lives with his mother in Ellicott City, MD. Major Thomas enjoys, golf, sailing and flying.

RECTOR A. THOMPSON (AKA REX TOMMY), born May 18, 1931, Faulkton, SD, enlisted in the service Dec. 28, 1949. He served with USAF, promoted to E-6, selected for OCS. Commissioned Sept. 28, 1956.

His military locations: Lowry AFB, CO; Hunter AFB, GA; Larson AFB, WA; McConnell AFB, KS; Vandenberg AFB, CA; Ubon Thailand; Fairchild AFB,

WA; RAF Bentwaters, England; Davis-Monthan AFB, HI.

His mission/operations: Project Fire-out (B-36); Reflex Action (B-47); Titan I update; Titan II Launch "Longlight"; Missile competition; Project Constant Guard II.

His memorable experiences: his first command, which was 8th AMS in Ubon, Thailand. A 800 man squadron which 13 special systems supporting combat missions in SEA; Titan II Launch "Longlight"; taking command of the 81st AMS RAF, Bentwaters, England.

He retired May 1, 1979 as lieutenant colonel (0-5) and was awarded the Bronze Star, Meritorious Service Medal, AFCM w/2 OLCs, AFOUA w/4 Dev., Good Conduct Medal w/2 Loops, National Defense Medal w/ 2 Devices; RVN Camp Medal/ AF Loyal Service Ribbon w/6 Devices.

Married Beverly J. Eavenson, Dec. 19, 1951 at Lowry AFB, CO. They have three daughters. Presently they are playing golf and are completely retired.

ROBERT ELSTON THOMPSON, MAJ (RET), Course OBR 1821G-1, missile launch officer, WS133A, class CO66AM-8, Chanute AFB, IL, Oct. 24-Nov. 19, 1966. ORT, Class MMM-9 Vandenburg AFB, CA Nov. 28, 1966-Jan. 20, 1967. Assigned to 509 SMS, Whiteman AFB, MO as MCCC, line crew S-103 from May 22, 1967 to July 31, 1971.

Operations participated in: Project Glory Trip 19M, March 26-April 16, 1969, Vandenburg AFB, CA. First operational test of an operational missile selected from the Whiteman alert missile force.

His memorable experiences: dropping a simulated 8,600 pound atomic bomb with a B-36 Norden bombsight in April 1956.

Being tossed out of bed by B-52s dropping 10K bombs near Dong Ha, RVN July 1965. Launching a Minuteman April 16, 1969.

He was awarded the Meritorious Service Medal w/ OLC, AF Commendation Medal w/2 OLC, CRM, AF Good Conduct Medal w/OLC, Good Conduct Medal, NDSM w/OLC, AFEM, VSM (2 Bronze Stars), AFRM (1 HG), RVCM, Navigator Wings, Senior Missileman Badge, Miniature Minuteman Missile Launch Pin, Highly Qualified Standboards "10" Pin, 15th AF Crewmember Excellence Award.

Thompson is married to Dorothea E. Thompson and they have children, Dennis R. and Kenneth R. Thompson. They reside in Syosett, NY. Currently retired and engaged in family genealogy.

RANDY P. THREET, MAJ, is the chief, Airborne Launch Control System (ALCS) Standardization/Evaluation, USSTRATCOM, Offutt AFB, NE. He graduated from Memphis State University, August 1983 and attended IQT for the "Deuce" weapon system at Vandenberg AFB, CA August 1984. His first assignment was the 321st SMW at Grand Forks AFB, ND. In July 1992 he reported to the 90 MW at F.E. Warren AFB,  WY. In October 1995, he moved to USSTRATCOM as the chief ALCS Recurring Training. Some of his most memorable experiences include becoming an Olympic Arena Crew Commander and the youngest assistant operations officer in SAC history; implementing the training and evaluation flight concept and the first REACT weapon system in AFSPC; ushering in the first replacement to the EC-135 "Looking Glass" in 37 years - the Navy E6-B. He currently resides in Bellevue, NE with his wife Mary Anne and sons Bailey and Brandon.

RICHARD G. TOYE, COL (RET), was deputy commander for operations, 321st SMW in 1980-81.

He was commissioned from West Point in 1959. Georgetown University awarded him a master of arts and juris doctor. He graduated from Squadron Officers School, Air Command and Staff College, Industrial

College of the Armed Forces, and National War College.

He served with the AF Technical Applications Center; USAF in Europe; US Forces/UN Command, Korea; Air Staff and US State Department; and Organization of the Joint Chiefs of Staff. He was senior military advisor to the US Strategic Nuclear Arms Reductions Talks Delegation.

President Reagan appointed him director of Policy Planning at the Arms Control and Disarmament Agency. Afterwards, he became vice president of Pacific-Sierra Research Corporation.

He now practices law before the US Court of Veterans Appeals.

PHILIP W. TRENT JR., COL, Mission Realignment for the 321st Missile Gp., Grand Forks AFB, ND, enlisted in the AF in April 1968. He attended Technical School at Keesler AFB, MS and was stationed with the 662d Radar Sqdn., Oakdale, PA until March 1972. He then attended Officer Training School and was commissioned in June 1972. From 1972 to 1978, he served in various missile positions at the 91st SMW, Minot AFB, ND, including: missile combat crew commander, instructor, evaluator, and Emergency War Order Instructor. From 1979 to 1983, he was assigned to the 1st Strategic Aerospace Div., Vandenberg AFB, CA, in the ICBM Test and Evaluation Program where he was chief, ICBM Test Management Div., participating in 48 test launches. From 1983 to 1985 he was the Minuteman II program manager at HQ SAC, Offutt AFB, NE, and later deputy ICBM test director for the command. In 1985, Col. Trent was assigned as chief, plans and later as chief, Plans and Intelligence Div. for the 487th TMW, Comiso AS, Sicily. In 1986 he returned to Offutt AFB to the Joint Strategic Target Planning Staff as chief, Nuclear Force Timing and Document Production Div. In 1990, he was assigned to the AF Space Command as chief, Special Access Programs and then as chief, Plans Div. Colonel Trent assumed his current position in July 1995.

His decorations include the Defense Meritorious Service Medal, Meritorious Service Medal w/4 OLCs, AF Commendation Medal, Combat Readiness Medal, AF Good Conduct Medal and the Legion of Merit.

He is married to the former Sondra Thompson and they have a married daughter, Amy Elizabeth Harrison of Omaha, NE and one grandson.

MICHAEL L. VERES, served on Minuteman crew in the 564th SMS/341st SMW at Malmstrom AFB, MT from June 1976 to March 1981. He was the Tango flight commander and was twice honored as SAC Crew of the Month. He participated in the first Global Shield exercise, and has 251 missions beneath Montana. After leaving the crew force, he was a project manager at Wright-Patterson AFB, OH, developing a training system for the FB-111A/Short Range Attack Missile.

Veres currently works at NASA's Johnson Space Center in Houston, TX. He is the flight lead for two space station assembly missions. Prior to his participation in the Space Station Program, he was a Space Shuttle rendezvous flight controller. He provided real time support in NASA's Misson Control Center for many shuttle flights, including some of the dramatic satellite repair missions. He also wrote astronaut crew checklists for several missions.

He holds a BS in Natural Sciences from the University of Akron, OH and an MS in Systems Management from the University of Southern California. He grew up in the small northeast Ohio town of Macedonia. Mike and Susan Veres have been married for 23 years, and

have three teenagers. The whole family is active in scouting. Veres has written two novels based on his Minuteman experiences.

THOMAS E. VITITO, COL (RET), served 24 years primarily in the ICBM maintenance career field. After graduating from Florida Southern College he was commissioned in 1961, and reported to Malmstrom AFB, MT, the first operational Minuteman I SMW.

He received a master's degree in Logistic Management from AFIT in 1967 and was assigned to the 498th Tactical Missile Gp. in Okinawa. He transferred to HQ SAC in 1970, serving in the Directorate of Missile Maintenance. In 1974, he was assigned to the AF Inspection and Safety Center.

In 1976, he arrived at Grand Forks AFB, ND, where he served as the Organizational Missile Maintenance Squadron commander prior to becoming the deputy commander and later the commander of the 321 Combat Support Gp. After attending Air War College, he served as the ROTC Detachment commander at South Dakota State University.

Since retiring in 1985, he and his wife, Elaine, have resided in Texas, The Woodlands, Dallas and McKinney, where Tom's favorite activity is enjoying his children and grandchildren.

GERALD L. WAKEFIELD, LT COL (RET), born June 26, 1983 in Gainesville, TX. After graduating from Texas A&M University (1960), he entered the AF on April 24, 1961 as a supply officer at Andrews AFB, MD. Here as a second lieutenant he maintained flight clothing records on 86 general officers stationed at the Pentagon. These records included a pair of "flying overalls" for Gen. Curtis Lemay. In 1963 he was assigned as a Minuteman I, Missile Combat Crew member to the 741st SMS at Minot AFB, ND. There he eventually served as the 455 SMW senior instructor crew commander.

After receiving his master of science degree in Industrial Management from the University of North Dakota, he reported to the 366th TFW (Gunfighters) as the munition supply officer at DaNang AB, South Vietnam. He returned to the USA in February 1969, and began his data processing career at HQ SAC, Offutt AFB, NE. There he was one of three officers who started the design and development of the SAC Warning and Control System (SWCS). After a short stint at the Armed Forces Staff College, he returned to HQ SAC to manage the SWCS and assist in turning it into the Command Center Processing and Display System (CCPDS) which was used at SAC, NMCC, ANMCC, and NORAD. His last assignment was to provide support to the CCPDS executive manager as the Warning Division Chief, Director of Command Control, HQ SAC. After retiring on May 31, 1981, Wakefield became a manager of Computer Data Services at Mutual of Omaha. In July 1987, he left Mutual of Omaha for a better opportunity at Central States Health & Life Co. of Omaha where he became the vice president of Information Services. Wakefield retired from that position on Aug. 31, 1997.

He and his wife Clara live in Bellevue, NE where

he now spends his time working part time for the new College of Information Science & Technology at the University of Nebraska at Omaha, working on family genealogy, feeding wild Canada geese, and enjoying his two granddaughters.

ROBERT E. WALKER, COL (RET), born Oct. 17, 1930 in Detroit, MI raised in Mt. Lebanon, a suburb of Pittsburg, PA. Entered navigator training at Harlingen AFB, TX, Nov. 28, 1952. Following commissioning and graduation in 1954 received flying assignments to England AFB, LA; James Connally AFB, TX; and Naha AB, Japan. In March 1963 reassigned to Minuteman Maintenance Officer Training. Following training served in Missile Maintenance positions at the 90th SMW, 15th AF and the Directorate of Aerospace Safety. In 1974 reassigned to Minot AFB as commander 91 MIMS. In 1976 became deputy commander for maintenance. While serving in that capacity the 91st SMW received the Riverside Trophy (1977) and won the Blanchard Trophy (1978). Returned to 15th AF in 1978 and served as assistant DCS Logistics until retirement in December 1981 as a colonel.

He married Louise Fisackerly, June 18, 1955. They have two children and two grandchildren. They currently reside in Ingram, TX.

CLARK W. WARD, COL (RET), is a career missileer. A graduate of Oklahoma State University, where he commanded the AFROTC Cadet Air Div., received his commission, Jan. 21, 1961. His active service began April 23, 1961, Amarillo AFB, TX and continued until his retirement Oct. 31, 1987, Offutt AFB, NE. Duties included: wing supply officer, 4108th ARW, Plattsburg AFB, NY, launch control officer and material control officer, Emergency Rocket Communications System (ERCS), Scribner AFS, NE, Wing Code Custodian, 351st SMW, Whiteman AFB, MO where he also completed a master of arts degree in history from Central Missouri.

In July 1970, reported to HQ, SAC as a staff officer responsible for Minuteman II and Titan II targeting programs. He attended Air Command and Staff College in 1972 where he earned "distinguished graduate." Upon departing ACSC, he began a long association with the Minuteman Weapon System by becoming a missile combat crew commander, squadron operations officer, chief of operations training, 90th SMW, F.E. Warren AFB, WY; squadron commander, 564th SMS and deputy commander, 341st Combat Support Gp., 341st SMW, Malmstrom AFB, MT; chief, Space and Missile Future Concepts and Chief, ICBM Operations Inspections, HQSAC, Offutt AFB, NE; and, deputy commander for operations and wing vice commander, 44th SMW, Ellsworth AFB, SD. He also served a remote tour at TUSLOG Det 184, Balikesir, Turkey. He was director, Force Applications, HQSAC/Joint Strategic Targeting and Planning Staff, Offutt AFB, NE before he retired in 1987.

He and his wife, Patricia, have resided in McAlester, OK since his retirement. He is a cattle rancher and directs the Chancel Choir at First Presbyterian Church and a community choir, "The McAlester Singers." Hob-

bies include golf, fishing, following son Robby's sports and music activities and playing with his five grandchildren belonging to daughters, Cheryl and Linda.

GLENN E. WASHINGTON, CMS, is currently assigned as superintendent propulsion directorate, AF Research Laboratory, Edwards AFB, CA.

After graduating from high school in 1969, he enlisted in the USAF. His duties included: Titan II Missile crew chief, missile maintenance technician and quality control inspector/evaluator at Davis-Monthan AFB, AZ; Titan II Missile Handling Shop Chief at Little Rock AFB, AR; superintendent, Titan Unit, 3901st SMES, Vandenberg AFB, CA; superintendent, Maintenance Management Div., Woensdrecht AB, Netherlands; Superintendent Missile Maintenance Div., Comiso AS, Italy. He was a member of the best missile alignment team in SAC during the 1975 Missile Combat Competition, Olympic Arena.

Chief Washington is a graduate of the Senior NCO Academy and has a master of business administration. He is married to the former Heather L. Martin and has four children: Dawnyeala, Adrian, Andre and Alicia. They reside at Edwards AFB, CA.

KENNETH D. WEBB, MSGT (RET), born Dec. 9, 1935, Stanton, NE, graduated from Stanson High School and enlisted USAF May 1954. After completing basic training at Lackland AFB, TX and Aircraft Electrical School, Chanute AFB, IL, he was assigned to 6091 SRS & 6067 A&E SQ. (PACAF) Yokota AB, Japan until April 1959. He was assigned 43 FMS (SAC) Davis-Monthan AFB, AZ until being selected for ICBM. Training and assigned to 565 SMS (SAC) F.E. Warren AFB, WY October 1959.

After completing Liquid Fuels Maintenance School at Chanute AFB, IL he was assigned to Launch Crew 12 and considered combat ready after completing IWST Vandenberg AFB, CA July 1960. He was "Missileman of the Month" January 1961 and rewarded with a ten day R&R trip to Hawaii. He was assigned to 389 MMS at Warren AFB September 1961 and became Pad chief as well as NCOIC Pneudraulic shop before assignment to 576 SMS (SAC) Vandenberg AFB, CA November 1964.

At Vandenberg he participated in a number of Atlas E & F launches to test new types of warheads. November 1966 found him assigned to 35 CES (PACAF) Phan Rang AB, RVN where he was NCOIC of the Liquid Fuel Maintenance Shop. He received the Air Force Commendation Medal for his service in Vietnam and was assigned to 3750 C.E.G. (ATC) Sheppard AFB, TX November 1967. He completed Petroleum Tank Cleaning Supervisor School at Chanute AFB, IL before being assigned as NCOIC Liquid Fuel Maint. Shop 824 C.E.S. (PACAF) Kadena AB, Okinawa (February 1969).

Webb was NCOIC Mechanical Section at Kadena AB and Ellington AFB, TX until he retired August 1974. He obtained employment as a refrigeration and air conditioning mechanic at the University of Houston in July 1974 and was serving as assistant mechanic maintenance ship foreman when he again retired January 191. He and his wife Chieko currently own Chi's Florist and Gifts, Houston, TX. They have daughters, June Hodges and Teresa Webb; grandchildren: Krissa, Jason, Kim and Lisa and great-grandchildren: Gregory and Victoria.

GUY F. WELCH, LT COL (RET), was a member of one of the initial Titan II Missile Combat Crews to activate the 571st SMS at Davis Monthan AFB, AZ. Colonel Welch was born in Galesburg, IL, in 1940 and attended high school in Aledo, IL. After graduation from Monmouth College with a bachelor of art's degree in physics and mathematics, he joined the AF and was commissioned a second lieutenant through Officer Training School at Lackland AFB, TX, in 1962. Upon completion of missile launch officer training at Sheppard AFB, TX, he was assigned as a Titan II deputy combat crew commander on crew MCC-117 with the 571st SMS, 390th SMW, Davis Monthan AFB, AZ.

Primary tour of duty was at the SAC launch site 571-5 in Madera Canyon. In 1965 he was selected by the AF Institute of Technology to attend Rochester Institute of Technology where he received a master of science degree in electro-optics in 1967. Upon graduation, he reported to the Office of the Secretary of the Air Force, Special Projects, Space and Missile Systems Organization, Los Angeles AFB, CA, as research and development engineer for a military space systems. In 1971 he was selected by Air University to serve as an AF ROTC instructor at Norwich University, Daytona Beach, FL. Welch returned to the OSAFSP in 1974 where he served as deputy for payloads, SP-7, for two nationally important satellite programs. He ended his career on special assignment to the AF chief of staff responsible for development of ICBM treaty inspection and verification methods.

His military decorations include the Legion of Merit, Meritorious Service Medal, and the Air Force Commendation Medal.

He and his wife Danae, have resided in Los Angeles, CA, since his retirement in 1982. Colonel Welch is a satellite development program manager for TRW Space & Electronics Group in Redondo Beach, CA.

THOMAS D. WELCH, CAPT (RET), career included assignments to AFIT, Space and Missile Systems Organization (SAMSO), Logistics Command, 351st SMS and Rome Air Development Center. At SAMSO he managed instrumentation of high explosive tests and development testing of the Minuteman missile suspension system. In the logistics command he provided service engineering for components common to all services. After this, he attended the first 'Mission Ready' class for launch control officers and was assigned to the 351st where he became Juliet capsule flight commander and chief of the Technical Engineering Branch. His final assignment was to the Rome Air Development Center where he managed Signal Intelligence research efforts.

Awards include Air Force Commendation w/OLC and Meritorious Service Medal. Captain Welch currently provides computer and software development support to businesses in rural Maine where he resides with his wife Jane.

DANIEL G. WELLS, LT COL, KYANG, has served both the USAF and the Air National Guard during his varied career. After graduation from the University of Louisville and receiving an AFROTC commission in

May of 1970, he jumped into the Titan II program with schools at Sheppard AFB, TX and Vandenberg AFB. McConnell AFB in Wichita, KS with his first assignment starting as a deputy and moving on to missile combat crew commander in the 533rd Sqdn. from 1971 to 1972. The Kentucky Air National Guard kept him occupied with many assignments and he is currently special projects officer for the 123 Airlift Wing. In 1988 he was awarded the USAF Distinguished Rifleman Badge.

He is married to Marian Wells and has two children, Daniel III and Sarah. His family lives in Danville, KY and he manages engineering information systems for an electronic lock company.

FRANCIS (FRANK) J. WELSH, LT COL

(RET), began his missile career in July 1962 after graduating from St. Joseph's College in Philadelphia, PA and receiving his AFROTC commission. Assigned to the 532nd SMS, 381st SMW at McConnell AFB, he initially worked with the Site Activation Task Force in coordinating the installation of the 18 Titan II missiles around Wichita, KS. On one of the first crews to man these sites in 1963, he served as a deputy missile combat crew commander. (DMCCC). In 1965 he was made DMCCC instructor and was eventually assigned to the senior instructor crew in 1966. That year the wing received the Col. Lee R. Williams Memorial Trophy. He was a member of the team that was named "Best Titan II Wing" at the first SAC Missile Combat Competition, "Curtain Raiser," in April 1967. He was appointed in the Regular Air Force earlier that year.

He was selected to work with NASA at the Manned Spacecraft Center in Houston, TX on the Apollo Program and supported the first lunar landing (Apollo 11) in July 1969. In September 1971, he reported to Norad Cheyenne Mountain Complex to serve as lead orbital analyst. After spending a year in the Ballistic Missile Early Warning System (BMEWS) at Thule, Greenland as a space surveillance officer and assistant operations officer, he reported to the 2nd Communications Sqdn. at Buckley ANGB, CO in September 1973. Here he served as chief of operations training and disaster preparedness. In July 1976, he moved to the Pentagon serving four years on the Air Staff. He as then selected as an exchange officer to work with the Royal AF at HQ Strike Command in High Wycombe, England on the British segment fn the BMEWS. His final assignment was a chief of the C3 Branch in the Plans Div. of HQ Space Command in Colorado Springs, CO.

He and his wife, Jeri, have resided in Berwyn, PA since retirement in 1985. He now works as a consultant for Lockheed Martin in King of Prussia, PA on a part-time basis.

JAMES A. WIDLAR, A1C/E-4, born Nov. 25, 1940, Cleveland, OH, enlisted in the USAF Dec. 15, 1960, as missile mechanic.

Assignments: December 1960-January 1961, AFSC 00010, basic military training, 3725 BMTS, Lackland AFB, TX; January 1961-June 1961, AFSC 43010 Student Missile Mechanic (ballistic) 3764 School Sqdn. (ATC), Sheppard AFB, TX; June 1961-July 1961, AFSC D43330 Missile Mechanic with launch maintenance,

706th MMS, 706th SMW (SAC), F.E. Warren AFB, WY; July 1961-October 1964, AFSC D43350A missile mechanic with launch maintenance, 389th MMS; 389th SMW (SAC), F.E. Warren AFB, WY.

He was honorably discharged (convenience of the government), 389th MMS (SAC), F.E. Warren AFB, WY, Oct. 22, 1964, A1C/E-4.

Married to Christie and they reside in Breckenridge, CO. Widlar is owner of Breckenridge Electric. He enjoys life, traveling and skiing.

EUGENE C. WILLIAMS, LT COL (RET), graduated from the University of Alabama in 1971 and was assigned to the 448th SMS Grand Forks AFB, ND in 1973. He was an evaluator and instructor deputy crew commander and crew commander. Competed in the 1975, 1976 and 1977 SAC Missile Combat Competitions. In 1980, he earned a MBA from the University of North Dakota. In July 1980, reported to the 2nd ACCS, Offutt AFB, NE. As chief, Abn. SIOP planning team he was an instructor and integrated Computerized Data Management into the airborne planner's capabilities. In July 1982, assigned to HQ SAC directorate of ICBM requirements, he initiated the Minuteman/Peacekeeper ALCS development and was responsible for Unauthorized Launch Studies. He graduated with Armed Forces Staff College 78 in January 1986. In January 1986, assigned to the National Security Agency Nuclear Weapons Systems Div. where he led the development of a prototype encryption device. He was assigned to the 446th SMS as operations officer in July 1990. In 1991, assumed command of the 446th SMS. His final assignment was NSA to the Network System Security Office.

Medals include the Defense Superior Service Medal, Meritorious Service Medal w/2 OLCs, Air Force Commendation Medal, National Defense Service Medal w/ one device and the Humanitarian Service Medal.

Williams and his wife, Jackie, and their sons, Stephen and Robert reside in Madison, AL. He is a principal engineer for SPARTA, Inc. providing technical engineering support to the Department of Defense.

DAVID M. WILLIAMSON, COL (RET), born in Dayton, OH, served in the USAF.

His military locations and stations: Andrews AFB, MD; Bien Hoa AB, RVN; Malmstrom AFB, MT; Ellsworth AFB, SD; Offutt AFB, NE; DNA, Washington, DC; Pentagon, Grand Forks AFB, ND; Scott AFB, IL.

He participated in forward air control, ballistic missile operations (Ground and Airborne), PACCS and Arms Control, Strategic Plans.

His memorable experiences include: night flights and OPS in RVN; Standboard for 16 (twice); Glory trip missile launches; start talks, US Team chief for US/USSR Missile Verification Process; AF (SAC/TAC/MAC) Reorganization Plans; AMC Master Plan.

Williamson was discharged Jan. 1, 1994 as colonel. He was awarded the Legion of Merit, Bronze Star, Air Medal, DMSM, MSM, AFCM, and Combat Ready Medal.

He is married to Judy and they have a son, Mark (a captain USAF, RC-135 pilot); Rick (a TV sports anchor and reporter). Employed as training and development specialist and executive director, State of Ohio Independent Board.

JAMES R. (RON) WILLIS, COL (RET), is employed by Aero Thermo Technology in Huntsville, AL.

He began his career in 1967 on Titan II crews with the 308th SMW in Little Rock where he served as an instructor and evaluator. He was a member of the Olympic Arena team that won Best Titan Wing in 1971. After graduating from AECP in 1973, he served as a DMCCC, MCCC, and senior evaluator with the 351st SMW, Whiteman AFB, MO. He served with the 3901st SMES as a minuteman operations evaluator.

After Armed Forces Staff College, he served as chief, Missile Tactics Branch, JSTPS, Offutt AFB. In August 1989, he assumed command of the 740th SMS, Minot AFB, ND. While an alert qualified squadron commander, he became the first person to perform at least one alert in four separate decades.

After Air War College, he was assigned to the Pentagon's ICBM Modernization Div. in the Office of the Assistant Secretary of the AF for Acquisition.

He and his wife, Marsha, have resided in Huntsville since 1996.

JAMES MURRAY WINGATE, MAJ (RET), born July 3, 1929 outside Davidson, NC, enlisted AAF, June 18, 1947. USAF Transfer Sept. 18, 1947. Served Lackland, FL; Lowry Field, Harmon Field; Harmon AFB, Guam; Pope AFB, NC; Eglin AFB, FL; Pentagon; Eniwetok Atoll; Scott AFB, IL; Fairchild AFB, WA; Amarillo AFB, TX; Patrick AFB, FL; Presque Isle AFB, ME; F.E. Warren AFB, WY; Sheppard AFB, TX; Vandenburg AFB, CA; Barksdale AFB, LA.

Missile assignments: Presque Isle, ME; F.E. Warren, WY. Units: 702 SMW; 566 SMS. Served as MCCC at both bases.

Operations: JTF3 1951, supported unmanned flight of T-33 and B17 flights over nuclear explosions: JTF7 1958-still classified activities.

Wingate retired at Barksdale AFB, LA, Feb. 1, 1968, and now enjoying retirement on the SE Gulf Coast of Texas. He is active in the Military Coalition working to preserve benefits of retired members of our Armed Forces. He and his wife, Mary Alice, enjoy travel.

KENNETH W. WOLFE, MAJ (RET), born Jan. 26, 1949, Roxboro, NC, enlisted in the service in 1968 and was commissioned in 1978. Retired as a major in 1992. He was an electro-mechanical technician (EMT), Grand Forks, 1969-72; EMT and job controller Vandenberg 1972-78. After OTS, was missile maintenance officer at Ellsworth 1978-82; GLCM safety officer at HQ USAFE, 1982-85; missile maintenance officer at Davis-Monthan 1985-90; squadron commander at Myrtle Beach, 1990-92.

Returned to civilian life, built his home, sold real estate, and worked as a supervisor until 1996 when his wife was promoted to her corporate headquarters in Dallas. He quit his job to support her as she had done for him for 24 years.

Married in 1972 to Janine Serrero, they now reside in Hurdle Mills, NC. Ken works for the county school system and Janine is in marketing. They have two grown children, Scott and Adam.

JOHN H. WOMACK JR., MAJ (RET), BA, Liberal Arts, Southern Illinois University, 1956. Graduated Harlingen AFB, TX, December 1957, and K-system school at Mather AFB, CA, June 1958. 1958-63: Langley AFB, VA, 427 AREFSQ, KB-50, navigator, instructor and standboard; commissioned Regular AF, 1959. 1963-68: Little Rock AFB, AR, 308 SMS, Titan II, instructor MCCC at ACP. 1968-71: Minot AFB, ND, 906 AREFSQ, KC-135, including Young Tiger to CCK, Taiwan & Utapao, Thailand, instructor navigator. 1971-

73: Statesboro AFS, GA, 3901 CEG, served two six-month Skyspot tours at senior mission director and operations officer in Ubon, Udorn, Bien Hoa, Son Tra, Da Nang, Tan Son Nhut, MADV HQ. 1973-76: Rickenbacker AFB, OH, 301 ARS, KC-135, instructor navigator, then financial management control officer for NAF. Retired in 1976 as master navigator and senior missileer.

Major Womack was MCCC of a crew chosen as 2AF crew of the Month, SAC crew of the month, served as senior instructor crew commander on the West Coast at ACP; launched Titan II, Giant Train, July 1966. He was a member of SAC Speakers Bureau, speaking to civic organizations and participating in university seminars on Titan II; published articles in Combat Crew; completed Squadron Officers' School, Command and Staff College, Air War College; received MBA in 1976. He has taught management, finance, accounting and economics courses at colleges in Ohio, Georgia, Florida, and North Carolina. He worked for state of Florida as budget officer, auditor supervisor, and supervisor of accounting services in comptroller's office.

John and his wife JoAnn, parents of five children, now live in the western North Carolina mountains. John is a naturalist, professional photographer, writer and publisher. He has photographic art gallery, teaches nature photography, and has published several books through his company, Sililoquy Press, in Franklin, NC. One of his books, Titan Tales, is a diary of two years of his missile crew duties including descriptions of operational alert tours, missile maintenance, a seven-week trip to Vandenberg AFB, and the launch of a Titan II.

ROBERT A. WYCKOFF, born in Newark, NJ, graduated from Colgate University, Hamilton, NY with a BA in English Literature and was commissioned in September of 1962. Subsequently he served as an administrative officer at Lackland and with the Armed Forces Courier Service based at Kelly. In 1965 he was assigned to missile operations duty at Malmstrom as part of the initial cadre of the 564th SMS "Odd Squad".

After training at Chanute and Vandenberg he served as a deputy, alternate and crew commander. In 1971 he became the 341st SMW WS-133B EWO instructor. During this period he earned an MS in Systems Management from USC. He moved to the 1st Strategic Aerospace Div. at Vandenberg in 1972 as an operations plans officer and later joined the Top Hand Program as a Minuteman II test manager. Subsequently he became the chief of Test Management and chief of Advanced Systems. In 1972 he wrote the poem Missileer. From 1978 to 1982 he was the command acquisition officer for Peacekeeper and DOD shuttle training support programs for Air Training Command. He retired in 1982 to work for Martin Marietta as a field engineer on the West Coast Shuttle Program. He moved to the Western Range Technical Services contractor (ITT/FEC) as a program manager for MX DT&E and Titan IV. He is now a program director for the Western Commercial Space Center.

Wyckoff is an Associate Fellow of the American Institute of Aeronautics and Astronautics and is past chairman of the Vandenberg Section.

He and his wife Eileen have raised a son, Steven, and a daughter, Lynne, and operate a Beechcraft Bonanza named "Chaquita" from the Lompoc, CA airport.

CHARLES E. WYNNE, LT COL, pulled 292 alerts under Kansas as a Titan II launch control officer for the 381st SMW, McConnell AFB, KS (1981-1985) and served as a ground launch cruise missile launch control officer for the 11th Tactical Missile Sqdn. ("Mr. Jigs"),

501st TMW, RAF Greenham Common, United Kingdom (1986-1988). He took two sorties permanently off alert during Titan II deactivation at McConnell and was the only launch control officer at Greenham to lead a dispersed flight back to base from field training at Salisbury Plain.

In May 1988, he became the public affairs officer at Greenham. He would go on to win prestigious USAFE and TAC public affairs awards and be named AETC Public Affairs Field Grade Officer of the Year twice. Promoted to lieutenant colonel Feb. 1, 1998. He continues to serve. He and Vicki have one son, Nicklaus.

ANDREW YASENOVSKY JR., CAPT (RET), born May 27, 1937 in Yonkers, NY, enlisted in USAF, Oct. 19, 1955. Received a commission as a second lieutenant September 1970. Served as an air policeman from 1955 to 1961 at Loring AFB, ME; Wheelus Field, Tripoli, Libya; and Plattsburg AFB, NY.

Cross trained into the Missile Field, November 1961 at Plattsburgh AFB, NY, into the Atlas "F" 556th SMS as a missile safety specialist. In 1963, assigned to HQ 8th AF Safety Directorate at Westover AFB, MA as Numbered AF missile safety technician for all Atlas "D", "E", and "F", Titan I and II, Minutemen (Whiteman AFB, MO locations) Weapon Systems, and the Houndog & Quail Air-Launched Missile Systems. In 1968, assigned to 44th SMW, Ellsworth AFB, SD, as safety superintendent with the rank of master sergeant. In July 1970, he was assigned to Officers Training School, Lackland AFB, TX, and was commissioned a second lieutenant, September, 1970. October 1970, assigned to the University of Southern California to attend the USAF Missile Safety Officers Course. Upon completion in December 1970, he was assigned to the 351st SMW at Whiteman AFB, MO as the wing missile safety officer. In April 1974, assigned to HQ 15th AF Safety Directorate at March AFB, CA, as the numbered AF missile safety officer. He was then selected to become chief of safety at the 351st SMW, Whiteman AFB, MO. He stayed in this position until his retirement in November, 1980, as a captain with 25 years of service.

Yasenovsky received his associates degree at American International College in Springfield, MA; bachelors degree in Social Sciences at Black Hills State University, Spearfish, SD; and masters degree in safety at Central Missouri State University, Warrensburg, MO.

Yasenovsky attended the Atlas "F" Launch Control Officers Course, Sheppard AFB, TX; Missile Safety Technicians Course, Chanute AFB, IL; Officers Training School (honor graduate), Lackland AFB, TX; Missile Safety Course, University of Southern California, Los Angeles, CA; Squadron Officers Course, Maxwell AFB, AL.

He and his wife, Heli, reside in Riverside, CA. They have four children and six grandchildren.

KENNETH W. YOUNG, LT COL, born Feb. 29, 1956, San Mateo, CA, is the Chief of Plans, 71st Flying Training Wing, Vance AFB, OK. Under his auspices, his division provides planning, exercise evaluation, and

deployment support to the Joint Specialized Undergraduate Pilot Training Mission. The wing consists of over 2,200 personnel and flies over 41,000 sorties in over 200 T-37, T-38, and T-1 aircraft graduating over 200 US and allied pilots each year.

A native of San Mateo, CA, Col. Young entered the AF in 1978. He was commissioned a second lieutenant through the AF Reserve Officer Training Corps at Baylor University. Colonel Young has served in the missile operations, nonrated operations management, international political military affairs and personnel career fields.

His awards include the Defense Meritorious Service Medal, Meritorious Service Medal w/2 OLCs and Air Force Commendation Medal.

Colonel Young is married to the former Karen Ferrell of Ponca City, OK. They have two children: Sara and Michael.

JOHN J. YUHAS, born June 28, 1926, Cleveland, OH, enlisted in the USAF, July 12, 1944 and served through Oct. 1, 1964.

From 1944 to 1964 he served at twelve stateside AF bases as well as the European theatre to include England, France, Germany and other overseas bases.

He flew as a ball gunner on B-24 aircraft. Also flew as crew chief and flight engineer on C-45, C-47, C-180, B-25, B-29 and KC-97 aircraft, and held various other aircraft maintenance positions as a aircraft maintenance superintendent. He worked as an instrument specialist and B-29 crew member while working on the development of the TM-61 Matador in 1950 and 1951.

As a pioneer in the missile field, he worked with Glenn L. Martin Co. and AF personnel at Hollomon AFB, NM and Patrick AFB, FL in the operational and maintenance phases of the TM-61 Matador, the first operational guided missile in the US inventory (1950-1951). A very exciting and rewarding experience.

He was the first enlisted man at Lockbourne AFB, OH to receive the missile badge for work on the Matador missile in 1950 and 1951.

Yuhas retired Oct. 1, 1964 as SMSGT (E-8). He married Marty and they have a daughter, Karen; son, Ray and granddaughter, Alli. Resides in Delaware, OH and is restoring classic cars and collecting military memorabilia.

WAYNE ZITZKA, CAPT, born June 21, 1951, Chicago, IL, enlisted in the USAF, Sept. 14, 1977. His military locations and stations: Kessler AFB, MS (training); Malmstrom AFB, MT (2153 CS); Vandenberg AFB, CA (training); Ellsworth AFB, SD (66, 67, 68, SMS, 44 SMW), Wright-Patterson AFB, OH.

His memorable experiences on project manager on B1-B after first operational crash. Returned aircraft to flight status, member of 66 SMS, when they received best 15 SAC 80, 81 and 81 (three out of four years).

Awards include Meritorious Service Medal w/OLC, AF Commendation Medal, Achievement Medal, Outstanding Unit, Organization Unit, Good Conduct Medal, National Defense Medal.

He was discharged Aug. 31, 1993 as captain. Married to Pamela J. and they have two daughters, Rhonda J. and Tamara R. Employed as a budget officer with Bernallilo County, NM.

AAFM Members Roster

A

ABBEY, DENNIS, O.
ADAMS, CHARLES, J.
ADAMS, GORDON, S.
ADAMS, RICHARD, E.
ADAMS JR, CHRISTOPHER, S.
ADAMS JR, CLARK, N.
ADAMSKI, RICHARD, H.
AGRONT JR, ABRAHAM
AGUE, KENNETH, R.
AKINS, JERRY, G.
ALBERT, DAVID, H.
ALBERT, JOHN, G.
ALBERT, GEORGE, E.
ALBERTI, GORDON, E.
ALBRO, WILLIAM, A.
ALDRICH, LINDA, S.
ALDRIDGE, DONALD, O.
ALDRIDGE, ROY, E.
ALLEGER, ARTHUR
ALLEN, STEVE
ALLEN, JAMES, R.
ALLEN, CRAIG, E.
ALLEY, RON
ALMEDA, LESTER, E.
ANARDE, RUSSELL, J.
ANDERS, BERNARD, C.
ANDERSON, RICHARD, L.
ANDERSON, GORDON, L.
ANDERSON, JACK, L.
ANDERSON, JIMMY
ANDREW, JOSEPH, D.
ANDRUS, THOMAS, C.
ANGEL, RALPH, E.
APRIL, PAUL, K.
ARMENTA, RODOLFO, R.
ARMSTRONG, THOMAS, F.
ARNOLD, ROBERT, O.
ARNOLD, DAVID, C.
ARNOLD, JOHN, K.
ASBILL JR, MORRIS, W.
ASHENBERGER, RICHARD, H.
ASHY, JOSEPH, W.
AST, DANIEL, A.
ATKINS, ALLAN, W.
AUDETTE, JEAN, MR,
AUST, STEPHEN, H.
AVANT JR, ARTHUR, W.
AXELSEN, MAX, M.
AYARS, DICK

B

BABBIDGE, MIKE
BAILEY, EDWARD, L.
BAILEY, ROBERT, M.
BAILEY, PAUL, L.
BAIR, JEFFREY, A.
BAKER, JAMES, R.
BAKER, GARY, G.
BALDWIN, GLENN, R.
BALLANTINE, TERRY, R.
BALMER, NEAL, F.
BANE, LENNIE, M.
BARAN, JOHN, A.
BARBER, CHARLES, J.
BARBER, RODERICK, D.
BARKHURST, P. D.
BARLOW, N. HENRY
BARNA JR, FRANK, J.
BARNARD, JAMES, W.
BARNARD, ROYAL, J.
BARNETT, WILLARD
BARR, CHARLES, M.
BARRETT, RUSSELL, W.
BARRIERE, JORGE, M.
BATES, DAVID, B.
BATES, NEIL, G.
BAUM, IRVING
BAYLOR, KEITH, R.
BEATON, KENNETH, L.
BEAUDRO, BOB, O.
BEDKER, VONARD, R.
BEDSOLE, HERSHEL, K.
BELL, PAUL, E.
BELT, CHARLES, D.
BENDER, HENRY, J.
BENDER, EDUARD
BENDER, JON, C.
BENJAMIN, PHILIP, G.
BENNETT, THOMAS, A.
BENNINGTON, RAYMOND, O.

BENTON, CLAY
BERG, DAVID, H.
BERGMAN, SHARON, M.
BERNSTEIN, WILLIAM, E.
BERRY, RAYMOND, W.
BETTIS, ALVIN, R.
BETTS, JOHN, M.
BEYER, LAWRENCE
BIDDLE, DREXEL, R.
BIEBER, FRED
BIELESKI JR, STANLEY, T.
BIERMAN, MERVIN
BIFFORD, WILLIAM, E.
BIGGS, RICHARD, J.
BIKKER, ARTHUR, W.
BILLAR, WILLIAM, L.
BILLE, MATTHEW, A.
BILOTTA, JOHN, A.
BINA, ROBERT, E.
BINDER, MICHAEL, S.
BIRD JR, JAMES, R.
BISHOP, LEE, R.
BISHOP JR, RONALD, J.
BITNER, CHARLES, B.
BLACK, RONALD, E.
BLACK, RICHARD, W.
BLACKBURN, CHARLES, E.
BLACKMORE, G. D.
BLAISDELL, FRANKLIN, J.
BLANCHARD, PAUL, D.
BLEAKLEY, BRUCE, A.
BLEDSOE, GARY, L.
BLYTH, DAVID, A.
BODENHAMER, KENNETH, D.
BODOVINAC, JOHN, A.
BOE, RAYMOND, W.
BOELLING, DONALD, L.
BOHNSACK, ROBERT, H.
BOINEST, JAMES, W.
BOLTON, ROBERT, L.
BOONE, JOHN, S.
BORDEN, WALLACE
BORNHOFT, KEVIN, A.
BORON, JOHN, W.
BOTTORFF, GARALD, L.
BOURCIER, LUCIEN, E.
BOVERIE, RICHARD, T.
BOYD, STEPHEN, B.
BOYKINS, JAMES, H.
BOZEMAN, JERALD, D.
BRASWELL, GEORGE, L.
BRENDLE, GEORGE, R.
BREWER, RICHARD, A.
BRIGHT, BURTON, K.
BRINER, JOHN, G,
BRISTOL, WILLIAM, E.
BRITT, BRIAN, L.
BROHAMMER, RONALD, G.
BROOKS, ELMER, T.
BROOKS, STEVEN
BROOKS, JERRY
BROOKS, CHERYL, P.
BROOKSHER, WILLIAM, R.
BROWN, RICHARD, I.
BROWN, LELAND, T.
BROWN, JERRY, E.
BROWN, MANNING, C.
BROWN, HOBERT JOE
BRUCE, WILLIAM, E.
BRUCH, JOSEPH, M.
BRUMFIELD, JERRY, P.
BRYANT, CHARLES, E.
BUCHERT, RONALD, V.
BUCHOLZ, STEVEN, J.
BULLOCK, HEIDI, H.
BULLOCK, JERRY, M.
BURBA, JAMES, G.
BURCHFIELD, EDWARD, L.
BURDAN, JOHN, W.
BURDICK, MARTIN, M.
BURDULIS, PAUL, V.
BURG, ROGER, W.
BURKE, WILLIAM, J.
BURNETT, DANNY, A.
BURNETT, PAUL, J.
BURNS, ROBERT, B.
BURRESS JR, EARL, W.
BURRILL, WILSON, E.
BURTON, ROBERT, L.
BUSH, ROBERT, V.
BUSH, FITZHUGH, G.
BUSH, WILLIAM, H.
BUTLER, DONALD, R.
BUTSKO, J. E.

C

CAISSE, JAMES, T.
CALLAHAN, GEORGE, T.
CAMPBELL, JOHN, R.
CAMPBELL, RICHARD, H.
CAMPBELL JR, HENRY, G.
CANAVAN, WILLIAM, J.
CANCELLIER, MICHAEL, G.
CANCELLIERI, ROBERT
CAPOZELLA, WILLIAM, P.
CARLISLE, RONALD, G.
CARLTON, JAMES, A.
CARLTON-WIPPERN, KITT, C.
CARMENA JR, JOSEPH, W.
CARMODY, MELVIN, D.
CARUANA, PATRICK, P.
CARVER, JOSEPH, H.
CASE, RUSSELL, C.
CASEY, ALOYSIUS, G.
CASTRO, STEVEN, M.
CASTRO, DAYNA, L.
CERNY, JOSEPH, P.
CERVIN, BENNETT, W.
CHAGNON, DAVID, H.
CHAMBERS, JOHN, E.
CHAMBERS JR, JOSEPH
CHANDLER, JACK, D.
CHAPMAN, PHILIP, W.
CHAPMAN, PRESTON, J.
CHAPPELL, VAN
CHAPPELL, BRIAN, K.
CHARAMUT, CHARLES
CHASTEEN, CALVIN, L.
CHOCK, STEPHEN, K. W.
CHRISTENSEN, GEORGE, M.
CHUVALA, RAYMOND, D.
CLARK, MARK, W.
CLARK, DANIEL, R.
CLAYTON, HAROLD, I.
CLEARY, MARK, C.
CLEMMER, DEAN, L.
CLEVELAND, RAYMOND, H.
CLICK, ROBERT, T.
CLOYD, FRANK, R.
COBB, HAROLD, E.
COCHRAN, DANNY, E.
COGLITORE, SEBASTIAN, F.
COGSWELL, BRENT, E.
COLEMAN, JAMES, P.
COLSTON, MARVIN, A.
COMITZ, RONALD, W.
CONDER JR, CHARLES, E.
CONDIT, CHARLES, O.
CONNER, PAUL, H.
CONRAD, JOSEPH, P.
COOK, OWEN, D.
COOK, TED, L.
COOK, DONALD, G.
COOMES, JAMES, W.
COONEY, LARRY, G.
COOPER, LESTER, D.
COPELAND, JACK, L.
CORK, HERBERT, L.
CORLEY, MONTFORD, J.
CORNELLA, ALTON
CORWAY, CHARLES, E.
COSTAS, JOSE, A.
COTNER, JOHN, F.
COURTNEY, HAROLD, D.
COX, WILLIAM
COX, ROBERT, M.
COX, LEE-VOLKER
COX, WILLIAM LEE
COXE, ALFRED, C.
CRAFT, TERRENCE, W.
CRAIG II, JOHN, E.
CRANKSHAW III, HENRY, E.
CROCCO, KEVIN, R.
CROUCH, JAMES, L.
CRUZ, HECTOR LUIS
CRYTZER, WILLIAM, F.
CSABAI, STEVEN, R.
CUBICCIOTTI, LEROY, L.
CULBERTSON, GRADY
CURTIN, GARY, L.
CWIKOWSKI, RAYMOND, T,

D

DANNER, BRYAN, L.
DANSRO, DANIEL, A.

DARR, JOHN, W.
DAUGHERTY, J. C.
DAVIDSON, RONALD, L.
DAVIS, BUSTER, J.
DAVIS, JOHN EVAN
DAVIS, RANDALL, R.
DAVIS SR, BENJAMIN, J.
DAVY, ALBERT, W.
DAVY, JAMES, J.
DAY, TERRELL, G.
DEAL, DUANE, W.
DEARNESS, JOHN, L.
DEDE, VERNON, L.
DEHART, LOUIS, E.
DEICH, KENTON, P.
DENINGTON, MICHAEL, R.
DENMAN, JAMES, E.
DEPPE, THOMAS, F.
DEREU, WAYNE, E.
DEWAR, DUDLEY, R.
DEWOLF, HOWARD, G.
DIAL, CHARLES, E.
DIAZ, MILTON, E.
DIEKMANN, PAUL, J.
DIISHCER, JOHN, B.
DISHONG, RAYMOND, C.
DIX, LOUIS, O.
DOBBINS, CLIFFORD, R.
DOBBS, DAVID, L.
DOBSON, JESSIE, J.
DOCKUM, ROBERT, R.
DODGE JR, JOHN, I.
DODSON JR, BERNARD
DOELKER, PAUL, T.
DOLAN JR, DANIEL, H.
DOLL, JAMES, F.
DONALD, ROBERT, K.
DONNELLY, RICHARD
DONOGHUE, VINCENT, J.
DORFMAN, STEVEN, N.
DOUGHTY, ORVILLE, L.
DOUGLAS, GEORGE, M.
DOUGLAS, RICHARD, W.
DOWELL JR, RALPH, H.
DOWNING, DARRELL, A.
DOWNS, FRED, W.
DRAPER, JENNIFER, A.
DRENNAN, JERRY, M.
DREW, DENNIS, M.
DREYLING, ROBERT, A.
DRYE, KENNETH, J.
DUNCAN, JOHN, A.
DUNKER, DOUGLAS, H.
DUNKIN, A. O.
DUNLAVEY, WALTER, F.
DUNN, ROBERT, E.
DUNNET, CHARLES
DUQUETTE, JOSEPH, A.
DWYER, CHARLES, F.
DYER, WARREN, M.
DZIEWULSKI, KENNETH, J.

E

EARLS SR, JOHN, M.
EASTWOOD, BRIAN
ECKERT, JON, S.
EDINGER, MICHAEL, R.
EDSON, MICHAEL, M.
EDWARDS, JOHN, W.
EDWARDS, HAROLD, P.
EDWARDS, ROBERT, M.
EICHEL, ROBERT, F.
EIGENRAUCH, DAVID, W.
EIRIKSSON, JACK, R.
EISSNER, RONALD, L.
ELLEN, JOHN
ELLERSICK, GORDON, J.
ELLIOTT, ARDEN, D.
ELLIS, STEPHEN, H.
ELLIS, RICHARD, W.
ELSNER JR, JAMES, E.
EMDIN, LIONEL, M.
EMERSON, LARRY, A.
EMERSON, JOAN, F.
ENGLEHART, JAMES, C.
ENGLISH, WALLACE, E.
EPLER, FREDERICK, J.
EPTING, JAMES, C.
ERLENBUSCH, WILLIAM, C.
ESELBY, RICHARD, H.
ESPINO, ERIC
ESTES III, HOWELL, M.
EVA, WILLIAM, D.

EVANOFF, JON
EVANS, KIT, D.
EVANS, DAVID, J.
EVANS, RICHARD, T.
EVANS III, ALBERT, L.
EVERHART, DAVID, E.

F

FAGAN, JERRY, D.
FARFOUR, GEORGE, R.
FARLEY, PAT
FARNEY, EARL, J.
FAY, LELAND, G.
FEDIE, JAMES, R.
FEDOR, WILLIAM, P.
FELKER, T. RENE'
FEUERSTEIN, LEWIS, E.
FICKLIN JR, JOHN, B.
FIGEL JR, WALTER
FILLEY, OLIVER, D.
FINAN, SANDRA, E.
FISHER, GENE
FISHER, THOMAS, W.
FITZGERALD, ANDREW, M.
FITZPATRICK, THOMAS, F.
FLOOD, JOHN, F.
FLORES, PETER, J.
FORAND, JAMES, M.
FORBES, STEPHEN, T.
FORD, JAMES, E.
FORGETTE, JOHN, F.
FORNES, GLENN, L.
FORNES, PATRICIA, M.
FORSELL, JOHN, R.
FORSTER, ARTHUR, C.
FORSYTH IV, JOHN, F.
FOSS, DAVID, W.
FOSSEN, CHADWICK, M.
FOSTER, THOMPSON, H.
FOUGHTY, MICHAEL, A.
FOWLER, STEVEN, L.
FOX, GARY, J.
FRANCIS, JOHN, E.
FRANK, DONALD, G.
FRANK, STEPHEN, R.
FRANTZ, JAMES, R.
FRANZETTI, JOSEPH, N.
FRAUMANN, ROGER, L.
FREDELL, MICHAEL, J.
FREER JR, ALBERT, P.
FREWERT, KEVIN, K.
FRIEDMAN, JOSEPH, C.
FUCHS, DELMER, H.
FUNICELLO, ROBERT
FUNK, JOEL, B.
FUTRELL, CHARLES, A.

G

GANGER, TODD, A.
GARDNER, AUTREY, T.
GARDNER, EAUL, L.
GARFINKEL, BERNARD
GARLAND, ROBERT, E.
GARNER, WILLIAM, S.
GARRETT, KENNETH, L.
GARVIN, RAYMOND
GASHO, JAMES, B.
GATLIN, JAY, P.
GAU, DAVID, J.
GAVITT JR, WILLIAM, F.
GAWELKO, JOSEPH, J.
GAYNOS, NICOLAUS
GECOWETS, JANET F. S.
GECOWETS, GREGORY A. S.
GEIST, LAWRENCE, W.
GEORGE, GETTY, J.
GERVAIS, FREDERICK, B.
GEZELIUS, JOHN
GIBBS, DONALD, J.
GIBSON, STEPHEN, T.
GILBERT, THOMAS, M.
GILBERT, WILLIAM
GILL, EDGAR, A.
GILLIKIN, STEVEN, V.
GILLIS, ROBERT, W.
GILMORE, JOHN, E.
GINN, ROBERT, W.
GIOCONDA, THOMAS, F.
GLASCOCK, STEVEN, M.
GLUSH, SCOTT, W.
GOERING, ROBERT, G.
GOETZ, WILLIAM, L.

GONZALES PSY D,NORM, D.
GOODWIN JR. REGINALD, S.
GORDON, JOHN, A.
GOSLING, THOMAS, J.
GOSS, KENNETH
GOUIN, RICHARD
GOULD, DOUGLAS, R.
GRADY, EDWIN, L.
GRANT, WILLIAM, J.
GRAVES, CURTIS
GRAY, RONALD, D.
GRAY, WILLIAM, R.
GRAYDON, MICHAEL, T.
GREEN, JAMES, D.
GREEN, BONITA, D.
GREENE, WALTER, E.
GREENE, LEROY, V.
GREENLAW, JAMES, A.
GREENLEY, CHARLES, H.
GREER, DONALD, T.
GREGORY, CHARLES, F.
GRIECO, MICHAEL, C.
GRIMMNITZ, LOUIS, C.
GROSSART, CHARLES, K.
GROSSHOLZ, TED, G.
GROSSMILLER III,WILLIAM, J.
GROVES, JAMES, S.
GUETTLER, JEFFERY, W.
GUIDRY, LEE
GUNN, DUANE, D.
GUNST, RICHARD, A.

H

HAAS, VICTOR, J.
HABENICK, HENRY, W.
HAFNER, THOMAS, H.
HAHN, ROLAND BUTCH
HALE, JAMES, E.
HALEY III, JOHN, W.
HALL, PRESTON, H.
HALL III, ALONZO
HALLCZUK, MICHAEL
HAMEL, RAYMOND, F.
HAMILTON, ROBERT, L.
HAMILTON, KEN
HAMILTON, PAUL, S.
HAMLIN, NANCY, C.
HAMMILL, DONALD, W.
HANAM, RUSSELL, T.
HANDSHY, WAYNE, H.
HANSEN, WAYNE, N.
HANSON, BRIAN, A.
HARALSON, BOBBY, J.
HARAZDA, CASIMIR, J.
HARDY, THOMAS, J.
HARGER, BRUCE, S.
HARLAN, RAYMOND, C.
HARRELL, WAYNE, K.
HARRIS, RICHARD, P.
HARRISON, MARTIN, T.
HARTMAN, NATHAN
HARTMAN, HAROLD, E.
HARTMANN, MICHAEL, P.
HASBROUCK, LAWRENCE
HAUSWALD, ARNOLD, C.
HAUTALA, BERNARD, M.
HAWKINS, EDWARD, D.
HAYLES, LESLIE, W.
HEALY, ANDREW, D.
HEIN, DAVID, R.
HEITKAMP, DENNIS, M.
HELLINGA, JERRY
HELMS, THOMAS
HELMS, BRUCE, A.
HELT, HAROLD, C.
HELTERBRIDLE, ERNEST, R.
HENDRICKSON, DEAN, P.
HENDRICKSON III,SIR
BENJAMIN,S.
HENNESS, GLEN, E.
HENRY, JAMES, P.
HENSCHEN, CRAIG, C.
HENSHAW, K. MICHAEL
HERBST, DANIEL, L.
HERNANDEZ, MICHAEL, W.
HERNANDEZ, JOSE, R.
HESKE, WILLIAM, J.
HESS, DONALD, R.
HESTER, TOMMY, H.
HETTERLY, MARK
HICKEY, MICHAEL, L.
HICKMAN, WARREN, W.
HICKS, GREGORY, W.
HIGLEY, LELAND, C.
HILDEN, JACK, G.
HILLEBRAND, LAWRENCE, J.
HILLIARD, JOHN, R.
HINDS, HUBERT, T.
HINES, BENJAMIN, W.
HINKLEY, RONALD, E.
HIPSHER, JOHN

HIRTLE, RONALD, W.
HODGE, WILLIAM, R.
HOFREITER, ALLEN, L.
HOGAN, ANTHONY, T.
HOGANCAMP, PAUL, L.
HOGLER, JOE, L.
HOIHJELLE, ERIK, E.
HOKE, CHARLES, S.
HOKEL, THOMAS, A.
HOLLINGA, KENNETH, L.
HONEYCUTT JR. JOHNNY
HOOKER, ROGER, D.
HORN, ARTHUR, F.
HOSELTON, GARY, A.
HOUCHIN, LLOYD, K.
HOWARD, TIMOTHY, H.
HOWARD III, WILLIAM
HOWE, JOHN, P.
HOWES, LAWRENCE, P.
HUBBARD, RON
HUDELSON, GARY, L.
HUDSON, LEE, F.
HUEY, WILLIAM, B.
HUFF, RONALD, D.
HUGHES, LARRY, B.
HULSEY, THOMAS, C.
HUMMER, WILLIAM, R.
HUNT, WELLS, E.
HURD, CALVIN, W.
HURSH, BURTON, J.
HURST, ROBERT, L.
HURST JR, ARNOLD, E.
HUTCHINS, RICHARD, M.
HYDE, KURT
HYNDMAN, BRAD, J.

I

ILLINGER, DEAN, F.
INTIHAR, WILLIAM, R.
IRVIN, RONALD, P.
IRVINE, ROBERT, B.
ISAACS, TERRY

J

JACKSON, MICHAEL, D.
JACQUE JR, MITCHELL, A.
JAMES, RICK
JAMESON, ARLEN, D.
JANUSHKOWSKY, VICTOR
JAQUES, RICHARD, P.
JAYNE, RONALD, W.
JENKINS, JAMES, M.
JENSEN, LOYD, E.
JOHNIGAN, IRENE
JOHNSON, MICHAEL, R.
JOHNSON, ARTHUR, E.
JOHNSON, RICHARD, M.
JOHNSON, JAMES, E.
JOHNSON, J. MICHAEL
JOHNSON, THOMAS, H.
JOHNSON, AUSTIN, L.
JOHNSON, ERIC, A.
JOHNSON, BYRON, G.
JOHNSON, RICHARD, T.
JOHNSON, RICHARD, L.
JOHNSTON, CHARLES, V.
JONES, MICHAEL, T.
JONES, GARY, A.
JONES, DAVID, B.
JONES, JAMES, W.
JONES, JAY, P.
JONES, DORIS, J.
JONES, JERRIE, E.
JONES JR. SAMUEL, A.
JORDAN, EDGAR, W.

K

KACZOR, CHARLES, S.
KALASKIE, WILLIAM, J.
KAMMEYER, LAMAR, D.
KARLAGE, JOSEPH, F.
KARNOUPAKIS, VICTOR, L.
KASS, NICHOLAS, E.
KEEFER, ROD
KEEN, RICHARD, O.
KEEN, RONALD, L.
KEHLER, C. ROBERT
KELCHNER, ROBERT, H.
KELLEY, JAY, W.
KELLEY, DOUGLAS, E.
KELLY, MICHAEL, J.
KELLY, THOMAS, J.
KELSAY, RAYMOND, E.
KELTNER, DONALD, E.
KEM, ROBERT, S.
KENDERES, MICHAEL
KENNEDY, GEORGE, B.
KENT, ROMAN, H.
KERMES, WILLIAM, J.

KERR, FRANK, W.
KESSLER, STEPHEN, J.
KIDD, WILLIAM, A.
KIME, WILLIAM
KING, CARL, L.
KING, RODNEY, W.
KING JR, RAY, W.
KINGSBURY JR. WILLIAM, C.
KIRILUK, NICK
KIRKHUFF, DEBORAH, A.
KIRKHUFF III, MILES, L.
KITTLESON, KENNETH, E.
KJOSA, ROYCE, M.
KLEIN, EVERETT, H.
KLEPPS, GERHARDT
KLEVENS, SIDNEY, S.
KLOTZ, FRANK, G.
KNAPP, JAMES, W.
KNAPP, WILLIAM, A.
KNAPP, J. C. D.
KNOLL, RICKELL, D.
KNOX, ALLAN, C.
KOBBEMAN, JOHN, F.
KOLLER, MICHAEL, G.
KOPERSKI, MIKE
KOTTAS, WILLIAM, M.
KOVICH, ANDREW, S.
KOWALSKI, STOSH
KOZA JR. FRANK
KOZAK, DAVID, C.
KOZLOWSKI, PETER, P.
KRAUSE, PAUL, H.
KRAUSE, ARTHUR, F.
KREPP, DENNIS, L.
KROSKEY, JAMES, A.
KUENNING JR. THOMAS, E.
KUNDIS, WALTER
KUTULIS, ROBERT, W.

L

LAFFERTY, JOSEPH, A.
LAGERBERG, TY, M.
LAIRD, ROBIN, R.
LAMB, DOUGLAS, E.
LANDER, JACK, D.
LANDRY PE, PAUL, L.
LANGEY, MICHAEL, B.
LANGSTON, JOSEPH, L.
LARGENT, ROBERT, E.
LASHER, JOHN, L.
LAVIGNE, RONALD, J.
LAWRENCE, DAVID, J.
LAYMAN, PHILLIP, A.
LAZZARO, JOSEPH, A.
LEACH, JACK, A.
LEAMING, STEVEN
LEATHERS, JACKIE, P.
LEAVITT, BARBARA
LEBER, JOHN, C.
LEE, JOHN, A.
LEE, ROBERT, M.
LEE II, CHARLES, J.
LEGAMARO, SALVATORE, J.
LEHMAN, HARRY, A.
LEHNERTZ, MICHAEL, F.
LEIGHNINGER, DAVID, E.
LEONARD, HAROLD, W.
LETTEER JR, CHARLES, E.
LEWANDOWSKI, DANIEL, P.
LEWIS, DAVID, R.
LIBURDI, SAMUEL, A.
LIEBENGUTH, EDWARD, D.
LIEN, DAVID, A.
LINDAHL, CARL OSCAR
LIPKA, PATRICK, T.
LIPSCOMB, CHARLES, L.
LITTLE, MICHAEL, O.
LITZLER, ALBERT, F.
LIVINGSTON, ROBERT, D.
LOGAN, THOMAS, L.
LONDON III, JOHN, R.
LONG, JERROLD, D.
LONG JR. JAMES, E.
LORD, LANCE
LOWE, CLINTON, V.
LOZE, MICHAEL, B.
LUCAS, BEN, A.
LUTMAN, CHARLES, C.
LYNN, DALE, R.
LYON, DENNIS, R.
LYONS, CLARENCE, E.

M

MACCRACKEN, JAMES, C.
MACK JR. PHILIP, G.
MACLAREN, ALLAN, J.
MADDEN, DONALD, E.
MADTES, DONALD, L.
MAGILL, JAMES, B.
MAHADOCON JR, FREDERICK, R.

MAHEUX, DANNY, P.
MALLORY, PAUL, G.
MALONEY, PATRICK, W.
MALONEY, PATRICK, T.
MANSOLILLO II, NICHOLAS, W.
MANTON, EDGAR, J.
MANTOVANI, THOMAS
MARKLINE, CHARLES, K.
MARKS JR, HENRY, P.
MARLER SR, JERRY, L.
MARSH, GARY, E.
MARSHALL, F. JOHN
MARSHALL III, DOUGLAS, M.
MARTIN, FRANK, K.
MARTIN, LARRY, D.
MARTIN, JOSEPH, S.
MARTIN, JEREMY, L.
MARTIN, JOHN, R.
MARTINO JR, FRANK, E.
MATHESON, HUGH, M.
MATHIS, JOSEPH, M.
MATHIS, PAUL, C.
MATTSON, SCOTT, D.
MATTSON, ROBERT, C.
MATTUS, ROBERT, J.
MATZKE, VANCE, G.
MAXON, ALICE
MAYHEW, JASON, F.
MCANSH, BEN, W.
MCCARTHY, CHARLES, J.
MCCARTHY, THOMAS, R.
MCCASKY, ROBERT, R.
MCCAW, TIMOTHY, L.
MCCLELLAND, WOODROW
MCCORMACK, RANDY
MCCORMICK, ROGER, L.
MCCORMICK, THOMAS, H.
MCDERMOTT, FRANCIS, D.
MCDONALD, JOHN, D.
MCDONALD, MICHAEL, C.
MCELHINNEY, RAYMOND, A.
MCFARLANE, FARLANE, J.
MCGEE, MICHAEL, W.
MCGIBENY, ARTHUR, D.
MCGILLVRAY, CHARLES, E.
MCGOWAN, SAMUEL, M.
MCGRATH, THOMAS, M.
MCGUIRE, KATHRYN, M.
MCHUGH, JAMES, P.
MCKEE, WILLIAM, C.
MCKEE, MICHAEL, C.
MCKINNEY, HERBERT, H.
MCKNIGHT JR. WALTER, R.
MCMAHON, TIMOTHY, J.
MCNIEL, WENDY
MCNIEL, SAMUEL
MCPHERSON JR, DANIEL, E.
MCWILLIAMS, KENYON, W.
MEAGHER, JAMES, J.
MEDLEY, MARK, A.
MELBERG, RAYMOND, H.
MELHORN, VAUGHN, J.
MELONE, SABATO, J.
MELTON, MARVIN, R.
MERCER, RONNE, G.
MERCER, JOE, D.
MERCER JR, ROOSEVELT
MERLA, ROBERT, C.
MERRILL, RICHARD, L.
MEYER, MAX, L.
MICKUS, KURT, T.
MIDDLETON, SCOTT
MIEDZIAK, STEVEN, E.
MIKUTIS JR, ALBERT, P.
MILLER, GERALD, E.
MILLER, RONALD, W.
MILLER, VINCENT, B.
MILLER III, CHARLES, G.
MILLOTTE, RALPH, A.
MILLS, RANDALL, M.
MILLS, JOHN, W.
MINARDO, THOMAS, J.
MINER JR, NORMAN, E.
MINO, ANTHONY, P.
MINTON, GEORGE, R.
MITCHELL, RONALD, S.
MITCHELL, ELLSWORTH, L.
MITCHELL, EUGENE, R.
MITCHELL, JOHN, F.
MITCHELL JR, ROBERT, E.
MITNAUL, HENRY
MONTEITH, WAYNE, R.
MOODY, TERRY, W.
MOORE, EVANS, W.
MOORE, ROBERT, F.
MOORE, JOHN, R.
MOORE, RICHARD, I.
MOOREHEAD, GAIL, P.
MORAN JR. EDWARD, F.
MORGAN, JOE, P.
MORGAN, MARK, L.
MORGAN, RICHARD, M.
MORLEY, CHARLES, J.

MORNEAU, MIKE, J.
MORRIS, JAMES, L.
MORRIS, JEFFREY
MORRIS III, JACKSON, A.
MORRISON, KENNETH, R.
MORRISON, RONALD, A.
MORRISSEY, DAVID, M.
MOSER, JOHN, T.
MOULTHROP, ROSCOE, E.
MOWERY, ROBERT, A.
MUELLER, JAMES, H.
MULLADY, JOHN, R.
MULLIN, ROBERT, E.
MULLINS, DON, E.
MULREADY, MICHAEL, J.
MUNDT, BRIAN, D.
MURPHY, PAUL, F.
MUSSELMAN, ARCHIE, L.
MUZZIO, C. PAUL
MYERS, ANDREW, J.
MYRES, BILLIE, L.

N

NADOLSKI, WILLIAM, F.
NAGY, GEORGE, R.
NAILEN, JOHN, T.
NANNING, RON
NAPOLITANO, JOHN, J.
NAVARRO, GUY, A.
NEARY, THOMAS, H.
NEBEN, ROBERT, H.
NEIGHBORS, GARY, L.
NEIGHBORS, WILLIAM, A.
NELSON, MARK, W.
NELSON, WESLEY, F.
NICHOLS, MATTHEW, A.
NICKERSON, CHARLES, L.
NOEL, RICHARD, J.
NORDMANN, BRIAN, D.
NORTHRUP JR, EDGAR, A.
NORTON, MARTIN, C.
NOTMAN, ROBERT, V.
NUESSLE, WALTER
NUNEZ, NELSON

O

O'BANION, CHARLES, C.
O'BRIEN, JOHN, N.
O'CONNOR, ROBERT, K.
O'CONNOR JR, JOHN, E.
O'HERN, FRED, H.
O'SULLIVAN, CHARLES, P.
OAKES, KENTON, F.
OAKMAN, JACK, A.
OELSCHLAGER, ALAN, L.
OGLETREE, GREGORY, W.
OHNEMUS, JOHN, J.
OLIVER, JOHN, R.
OLIVERIO, RONALD, D.
OLMSTED, LEWIS, L.
OLSEN, LAURA, L.
OLSON, TIMOTHY, C.
OLSON, BRUCE, D.
ORESKOVICH, RON
ORNE, WILLLIAM, H.
OSBORNE, EDWARD, W.
OSTRANDER, MICHAEL, M.
OTY, JAMES, L.
OUTLAW, GLEN, R.
OWEN, CLYDE, W.

P

PADDEN, DAVID, T.
PADGETT, KENNNETH, E.
PAEPCKE, JOHN E. C.
PALIN, JUDITH
PALKO, CHARLIE, F.
PALMER, CHESTER, D.
PALUMBO, GERALD, J.
PANGRAC, ELIZABETH, A.
PANSARI, SKELLY
PAQUETTE, CONRAD, A.
PARKER, ROBERT, W.
PARKER, GRAYDON, K.
PARKS, STEVEN, R.
PARSON, THOMAS, J.
PASEKA, RAY
PATTERSON, MICHAEL, J.
PAUSZ, ANTON, L.
PAYNE, EDWARD, D.
PEARSON, EUGENE, LD.
PECK, ORRIN, R.
PECK III, MAXWELL, C.
PEDLIKIN, HOWARD, L.
PEGUESE, H. ANTHONY
PEIRSON, WALTER, V.
PELLETIER, ROBERT C. M.
PELZER, JAMES, L.
PEPE, ROBERT, J.

125

PEPINO, EDWARD, J.
PERKINS, JOE, C.
PERO, JAMES, W.
PERONE JR, CARMEN, F.
PERRYMAN, GERALD, F.
PESEK, KENNETH, M.
PETERS, NICHOLAS, J.
PETERSON, ELMER, L.
PETERSON, ROBERT, G.
PETITO, JOHN, P.
PETRILLO, DAVID
PETTIT, DONALD, P.
PFEIFFER, THOMAS, J.
PFLEPSEN, GARY, N.
PHELAN, STUART, D.
PHELPS, THOMAS, L.
PHILLIPS, JOHN, E.
PHILLIPS, RODGER, S.
PHILLIPS, EARL
PHILLIPS, CHARLES, E.
PICKETT, JOHN, M.
PICKOFF, JULIUS
PILONETTI, DENNIS, M.
PLANTE, RONALD
PLETCHER, BRIAN, S.
POIRIER, MICHAEL, J.
POLITI, JOHN, J.
POND, GARY, W.
PORTER, PHILLIP, R.
POSEY, JAMES, N.
POWERS, JAMES, J.
PREBECK, STEVEN, R.
PRICE, JACK, C.
PRINSTER, RICHARD, R.
PROUDFOOT, JERRY, L.
PUGSLEY III, CHARLES, S.

Q

QUARLES, ROSS
QUINTON, BILLY, E.
QUIST, PETER, P.

R

RACHEL, ALLEN, K.
RADAZ, F. C.
RADEMACHER, WILBERT, W.
RAISON, WILLIAM, T.
RALEIGH, BRUCE, W.
RAMIREZ-TORRES, LUIS, R.
RAMSAY, STEVEN, T.
RANDLEMAN, MIKEL, D.
RANDOLPH, JOHNSON, V.
RAUSCH, EDWARD, W.
RAWLINGS, DONALD, G.
RAWSON, DAVID, R.
RAY, DOUGLASS, D.
RAYA, JOSEPH, F.
RAYHO, SCOTT, A.
RECKER JR ROBERT, I.
REE, HARRY, A.
REED, JOHN, T.
REED, DONALD, C.
REED, ROSS, D.
REESE, HERMAN, L.
REFFNER, JACK, F.
REINHARD, EDWARD, O.
REITER, RICHARD, B.
RENDON, RENE, G.
RENEHAN, JEFFREY, N.
RENNINGER, HAROLD, J.
RENTERIA, RICHARD, R.
RESTEL JR, JAMES, D.
REYES, VICTOR, J.
REYNOLDS, CHARLES, L.
REYNOLDS, JAMES, L.
RHOADES, KEVIN, M.
RHOADS, BILL
RIBBACH, JOHN, A.
RICCI, JOHN
RICE, HOWARD
RICE, RICHARD, A.
RICE, RICHARD, N.
RICE, HOWARD, L.
RICE, CLARENCE, R.
RICHARDSON, FLOYD, D.
RICHTER, EDGAR, C.
RIDENOUR, DAVID, E.
RIDER, THOMAS, E.
RIEMER, MICHAEL, R.
RINER, WILEY, A.
RIZZO, PHILIP, A.

ROADARMEL, STANLEY, B.
ROBB, HARRY, W.
ROBBINS, JACK, B.
ROBERSON, W. F.
ROBERT, EDMUND, L.
ROBERTS, TIMOTHY, A.
ROBERTS, DOUGLAS, J.
ROBERTS, MILDRED, L.
ROBERTS, GORDON, S.
ROBERTS, CARL, H.
ROBERTS PHD, JACK
ROBINSON, GEORGE, W.
ROBINSON, JAMES, L.
ROBINSON, MICHAEL, C.
ROBINSON JR ALROY, I.
RODEFFER, FRANK, E.
RODIE, BRUCE
ROEGGE, CHARLES, B.
ROETCISOENDER, ROBERT, J.
ROGERS, CHARLES, E.
ROGERS, JAMES, R.
ROLLINS, JAMES, G.
ROLSTON, GORDON, C.
RONEY, KEN
ROSS, DONALD, A.
ROSSKOPF, JAMES, D.
ROUSH JR, MILFORD, T.
ROWLAND, LARRY, A.
ROYSTER JR JACK, A.
RUDY, JAMES, D.
RULE, RONALD, L.
RUMPLE, JACK
RUVOLO, SAM, J.
RYAN, JAMES, H.
RYAN, TIM
RYAN, TIMOTHY, R.
RYAN JR, EDWARD, N.

S

SACHSE, BILLY, E.
SAFRAN, DAVID, J.
SAFRENO, ROBERT, R.
SAGE, ARLYN, H.
SALAS, ROBERT, L.
SALVADOR, BEN
SANCHEZ JR, PABLO
SAND, GARY, H.
SANDERS, NUEL, E.
SANDERS, NEIL, D.
SANDS, WILLIAM, H.
SANKS, JULIUS, F.
SAWYER, VIRGINIA, M.
SCANNELLI, VINCENT, J.
SCHACHT, OTTO, B.
SCHAFFERS, JOSEPH, J.
SCHANKEL, RICK
SCHEMENAUER, ROBERT, G.
SCHLASNER, STEVEN, M.
SCHOOLFIELD, CLIFTON, C.
SCHOONMAKER, RICHARD, W.
SCHOONMAKER, RICHARD, L.
SCHRAMM, K. RANDALL
SCHREPEL, MILTON, F.
SCHUUR, DAVID, H.
SCHWENK, LARRY, D.
SCOTT, RALPH, D.
SCOTT, ROBERT, R.
SCREDON, S. RICHARD
SCRUITSKY, WILLIAM, R.
SCUDDER, THOMAS, W.
SEARES, DAVID, F.
SECRIST, GRANT, E.
SEETOO, WILFRED, H.
SEHON, RAYMOND, N.
SEIBER, DALE
SELF, KENNETH, E.
SELMAN, H. PAUL
SENIAWSKI, CHARLES, D.
SERCER, PETER, E.
SESTAK, JOSEPH, G.
SEVERSON, WINTON, E.
SHADE, BARRY, L.
SHAMER, GEORGE, P.
SHAPIRO, MICHAEL, N.
SHARP, DAVID, C.
SHARP, RICHARD, E.
SHARPE, R. C.
SHAW, CHARLES, W.
SHEERIN, JAMES, O.
SHERMAN, DONALD, H.
SHIPLEY, BRUCE, C.

SHOLL, GEORGE, W.
SHORTER, JOHN, W.
SHROADS, JAMES, L.
SHULTS, JOHN, N.
SIEBERT, LEO, N.
SILCOX, BRADLEY, N.
SIMON, WILLIAM, G.
SIMONSON, PAUL STEVE
SIMPSON, CHARLES, G.
SIMS, WILLIAM, F.
SINK, LELAND, R.
SKRAINY, WALTER
SLAGLE, JAMES, H.
SLEGEL, EUGENE, W.
SLOTNESS, GLENN, N.
SMITH, ROBERT, B.
SMITH, PIERCE, L.
SMITH, ROGER, C.
SMITH, H. E.
SMITH, JAMES, H.
SMITH, WALLACE, D.
SMITH, LOREN, R.
SMITH, WALTER, P.
SMITH, JOHN ELTON
SMITH, CHARLES, M.
SMITH, BRIAN, G.
SMITH, CARL, D.
SMITH, MARK, T.
SMITH, RICHARD, V.
SMITH JR LARKIN, B.
SMITH O.D. DOUGLAS, G.
SMITH JR KEITH, R.
SMYTH, DAVID, R.
SNEDDON, JAMES, K.
SNIDE, THOMAS, W.
SNIDER, RICHARD, L.
SNIDER, DAVID, K.
SNODGRASS, MICHAEL, L.
SNYDER, NANCY, D.
SOILEAU, NUJIE, L.
SOMERSET, RICHARD, E.
SONNEFELD, FREDRICK, L.
SOVA, STEVEN, M.
SPAIN, DAVID
SPARKS, JAMES, W.
SPEIDEL, KEN
SPEIGHT, JAMES, D.
SPENCER, RICHARD, B,
SPENCER, ROGER, L.
SPEZIA, JIMMIE, V.
SPIZZIRI, FRANK
SPRABERRY PHD, HUBERT, O.
SPRAKER, RALPH, E.
SPRINGER, LAUREEN, M.
SPRINGER, ALBERT, J.
SPRINGER, WILLIAM, K.
SPURLIN, OGDEN
STACHOWIAK, DANIEL, S.
STAIANO, ANDREW
STANBERY, CHARLES, E.
STANCATI, BERNARD
STANDISH, PETER, M.
STANLEY II, ROBERT, W.
STARRETT, DENNIS, G.
STEARNS, SHERRY, L.
STEFFEN, MICHAEL, W.
STEINKE, GREGORY, V.
STERNS, CHARLES, R.
STEWARD, RICHARD, L.
STIDSEN, CARLTON, A.
STIEVE, TIMOTHY, S.
STIVISON, ROGER, L.
STONE, JOSEPH, L.
STONE, CARL, E.
STOSS III, FRED, B.
STOUGHTON, PETER, W.
STOWE, GERRY, F.
STRACK, HAROLD, A.
STREHLE, STEPHEN, B.
STROHECKER, HARRY, F.
STRONG, JERRY, R.
STUART, HERB
STUBBS JR, EDGAR, D.
STUMP, EUCALYPTUS, T.
STUMPF, DAVID, K.
SUCHECKI JR, THEODORE, J.
SUDDERTH, STEVEN, K.
SUMMERS, ROBERT, P.
SUTTER, JOSEPH, E.
SWANSON JR, THOMAS, R.
SWETT, EDWARD, B.
SWISHER, WILLIAM, S.
SZAFRANSKI, WAYNE, H.

T

TAFFET, HARVEY
TAFT, GERALD, R.
TALCOTT, THOMAS, G.
TANNER, ARTHUR, H.
TANNER, BILL, O.
TARTER, JAMES, W.
TASHNER, RICHARD, E.
TAVENNER, RICHARD, D.
TAYLOR, CLIFFORD, V.
TEAL JR, JESSE, R.
TEALE, LLOYD, V.
TEIGEN, DAVID, H.
TEMPLETON, WILLIAM, J.
TEPFER, DANIEL
TERRILL, RICHARD, J.
THOMAE, STEVEN, A.
THOMAS, GRADY, L.
THOMAS, CLIFFORD, O.
THOMAS, JAMES, P.
THOMAS, MICHAEL, L.
THOMASON, CHARLIE, R.
THOMPSON, RECTOR, A.
THOMPSON, ROBERT, W.
THOMPSON, RICHARD, B.
THOMPSON, KIRBY, G.
THOMPSON, ROBERT, S.
THRESHER JR, RUSSELL, W.
TILL, WILLIAM, A.
TIMBERLAKE, GEORGE, I.
TIMM, CHRIS, V.
TIRCUIT, ELWOOD, C.
TOBIN, WARREN
TOBIN, TERRY, L.
TOLLERUD, ROGER
TOME, ARTHUR, W.
TOPOLOSKI, JOHN
TORRES JR, GEORGE
TOWNSLEY, BRUCE, M.
TOYE, RICHARD, G.
TRAINA, JOSEPH, V.
TRAINOR, PAUL, V.
TRAVIS, RICHARD, W.
TRENT JR, PHILIP, W.
TRIPP, CHARLES, S.
TRULL, FRANK
TUCKER, WILLIAM, T.
TUCKER, JON, L.
TUCKERMAN JR, THEODORE, L.
TULLY, JAMES, E.
TURNER, NELSON, C.
TURNER, HIRAM, G.
TURNER, MARK, C.
TURNER, DENNY, E.
TUTHILL, WILLIAM, S.
TUTTOBENE, ROBERT, T.
TYLER, TIMOTHY, D.
TYMOFICHUK, RANDY, B.

U

UHLAND III, CHARLES, B.
ULVOG, JAMES
URBANSKI, RAYMOND, M.
USUI, FREDERICK, M.

V

VAN HORN, PHILLIP, F.
VAN HOUTTE, MAURICE, J.
VAN JURA, JOHN, G.
VANCE, W. RILEY
VARNADO, LATHAN, A.
VARVI, CHARLES, J.
VAUGHN, MICHAEL, G.
VAUGHN, DARRELL, E.
VENTURA, MYRON, A.
VITITO, THOMAS, E.
VOGEL, MICHAEL, C.
VOGNILD, BRIAN, L.
VON INS, PAUL, R.
VRIEZELAAR, DONALD, W.

W

WADDELL JR, JOSEPH, A.
WADE, CARL, H.
WADE JR, WILLIAM, E.
WAGENER, WAYNE, F.
WALKER, ROBERT, E.
WALKER, DUNCAN, E.
WALKER, JAMES, C.
WALLACE, STANLEY, H.

WALSTON, WILBUR, J.
WALTER, RODNEY, W.
WALTON, TRACY
WAPLE JR, CHARLES, G.
WARD, CLARK, W.
WARD, WILLIAM, L.
WARD, KENNETH, W.
WARD, KELLY, M.
WARD JR, WILLIAM, A.
WARKASKE, STEVEN, A.
WARNER, JON, S.
WASHINGTON, GLENN, E.
WASSMANN, EDWARD, E.
WATERFIELD, JACK, D.
WATSON, JEFFREY, L.
WATTS, MONTE, S.
WEBB, KENNETH, D.
WEBBER, JAMES, E.
WEBBER, RICHARD, E.
WEBER, WALDON, R.
WEBSTER, ALAN, D.
WEDIN, GUSTAVE, A.
WEIDEMAN, ROBERT, F.
WEIHS, GORDON, J.
WEINSTEIN, JACK
WELCH, RANDY
WELCH, GUY, F.
WELLS, DANIEL, G.
WELSH, FRANK, J.
WELSH, KEVIN, D.
WEST, WILLIAM, H.
WESTFALL, C. R.
WESTWOOD, EDWARD, P.
WHETSTONE, MARK, J.
WHIPPLE, ALAN, P.
WHITAKER, MICHAEL
WHITE, JAMES, R.
WHITE, GAYLE, C.
WHITE, WILLIAM, G.
WIATROWSKI, RONALD, F.
WIBLE, DONALD, W.
WIDLAR, JAMES, A.
WIKSTROM, FLOYD, E.
WILCOX, D. H. "SKIP"
WILCOX, JENNIFER, J.
WILKERSON, KEVIN, R.
WILKERSON PHD, R. CHRIS
WILLEY, LES
WILLIAMS, RONALD, C.
WILLIAMS, EUGENE, C.
WILLIAMS, TERRY, L.
WILLIAMSON, DAVID, M.
WILLOUGHBY, DAVID, J.
WILLOUGHBY, STEVEN, C.
WILSON, HUGH, F.
WILSON, JAMES, W.
WILSON, WALTER, P.
WILSON, P. J.
WINGATE, JAMES, M.
WINSLOW JR, EDWARD, F.
WITT, DAVID, A.
WITTKOFF, JULIE
WOLFE, KENNETH, W.
WOLFE JR, DONALD, P.
WOMACK, JOHN, H.
WOODLAND, TEDDY
WOODS, DAVID, M.
WOOSLEY, HIRAM, E.
WORDEN JR, RAYMOND, A.
WRIGHT, ALAN, S.
WRIGHT, LANCELOT
WYANT, FRED, O.
WYCKOFF, ROBERT, A.

Y

YAROSH, DENIS
YASENOVSKY JR, ANDREW
YATES, MILES, A.
YOUNG, KENNETH, W.
YOUNG, FREDERICK, D.
YOUNG, JAMES, B.
YOUNG, LONNIE
YOW, DAVID, E.
YUHAS, JOHN, J.

Z

ZACHARY, FRANK, E.
ZACHMANN, ROBERT, F.
ZARZOUR JR, FRANCIS, D.
ZEAK, PAUL, W.
ZENK, ARVID, C.

Former Member Roster

A

ADAMS JR., CHARLES. A.
ANDERSON, CARSON, E.
ANDERSON, JAMES, S.
ANDREWS, JOHN, F.
ANDREWS, ARDEN, P.
ARRINGTON, DAVID, A.
ASHBERRY, EDGAR, A.
ASSID, MICHAEL, A.
AYARS II, SAMUEL, A.
AYER, KENNETH, R.

B

BAILEY, CHARLES, S.
BALDWIN, GREGORY, T.
BANNISTER, LAWTON, K.
BARNBY, LEE, M.
BARNES, WILLIAM, E.
BARNETT, HOBERT, A.
BATCHELDOR, ROBERT, E.
BATTERTON, LARRY, K.
BEACH, PAUL, R.
BELL, LESLIE, A.
BELL, JAMES
BENEDICT, MICHAEL, F.
BENNETT IV., CHARLES, M.
BERGER, HARRIS, A.
BLACK, SHELTON, L.
BLAKE JR., EARL, G.
BOLLER, RICHARD, D.
BOND. DAVID, A.
BOOTH. RULON
BOUQUET JR, VICTOR, H.
BOWEN, CRAIG, H.
BOWLES, ALBERT, G.
BOYD JR., BILLIE, B.
BRAMLETT, JAMES, D.
BRANZELL, RUSSELL, P.
BRINDLEY, GORDON, C.
BROOKS, PHILLIP, F.
BROWN, HOWARD, O.
BROWN, KEN
BROWNE, LEWIS, C.L.
BRYANT, CHARLES, E.
BURDIS, GEORGE
BUTCHER, STANLEY, R.
BUZAN, ROBERT, E.

C

CADENA, RICHARD, R.
CANESTRARI, BRIAN, L.
CARRAGHAN, KEITH
CASTLE, GARY, R.
CHACE., HENRY, V.
CHAMBERS, ROSALIE, F.
CHRISTIANS, WILLAM, A.
CLOUSE, LELAND, F.
CODY, GEORGE, E.
CONGDON, NORMAN, B.
CONLEY JR, DONALD, M.
CONNOR, JOHN, H.
COUNSMAN, RANDALL, F.
CRENSHAW, OTTIS.
CROWLEY, RICHARD, J.
CUMMINGS, ROBERT, L.
CUMMINS JR., NOAH, R.

D

D'ALBERTIS, MICHAEL. A.
D'ANTONIO, JACK
DAVIS, STEPHEN, L.
DAY, WILLIAM, R.
DEGASIS, THEODORE, J.
DELAURA, ERNEST
DEMATTEIS, JOHN, A.
DENNARD, SUEANNE. E
DERSHEIMER, GEORGE
DICK JR, ORMOMD, H.
DICKSON, HARRY, H.
DRABNIS, ALFRED, J.
DROZD, ROBERT, S.
DURHAM, WILLIAM. G.

E

EASON, PRENTISS, E.
EGGLESTON, KEITH, C.
EILERTS, RICK
ENDICOTT, RICHARD. E.

F

FARMER, WALTER, L.
FARRELL III, CLAUDE, H.
FEIN, ALBERT, L.
FINNER, CLYDE, D.
FITCHETT, ROBERT, F.
FLOWERS, LEWIS, W.
FORTENBURY, JOHN, S.
FRANCZAK, JAMES, J.
FRAZIER, MARCUS, T.
FRESHWATER., MARJORIE, L.
FRIEDLANDER, TOM, A.
FRIEDMANN, ROBERT
FRITZ, ROBERT

G

GARRISON, CHARLES, N.
GENNARO, EDWARD, J.
GERRINGER, L T.
GILL, JAMES
GORDON, LAWRENCE, T.
GOSS, DAVID, A.
GRAHAM, KEITH, A.
GRAHAM JR., WILLIAM, M.
GRAY, CARY
GROHN, DAN, A.
GROSVOLD, HARRY

H

HAILE, ALLEN, C.
HALL, JAMES, H.
HALLORAN III, JAMES, V.
HALM, FRANK, N.
HAMILTON, SHERRY, L.
HARDESTY, EDWIN, E.
HATLESTAD, JAMES, S.
HAUN, ROBERT, L.
HENNING, CHARLES, T.
HERRING, THOMAS
HINSON, TRAVIS, L.
HIROHATA, DEREK, K.
HODGES JR., WILLIAM, R.
HOLDEN, KENNETH, L.
HOLLAND, DALE, A.
HOUSTON, JIMMY, R.
HUGE, HARRY, D.
HUGHES, ROBERT, L.
HUNTER JR., HERBERT
HUSS, WILLIAM, G.

I

ISAACSON, RONALD, A.
IVEY JR., JAMES, C.

J

JAMES JR., CLIFTON, C.
JOHNSON, LAWRENCE, E.
JOHNSTEN, RICHARD, A.
JOHNSTONE, CHARLES, E.
JONES, MICHAEL, H.
JONES II, TONISH, E.

K

KAIN, REECE
KENNEDY, STEVEN, L.
KENT SR, JOHN, H.
KILGORE, CHARLES, B,
KILLIAN, JACOB, J.
KLECKNER, RICHARD, D.
KLONOSKI, FRANK, J.
KNEECE, JOHN, E.
KOZIMOR, JOHN, P.

L

LACKEY, PAUL, K.
LANGER, MATTHEW, P.
LARKINS III, JOHN, B.
LATTEN II, AEDEE
LAVEN, TRACY
LEE, ROBERT, M.
LEEMING, EDWARD, R.
LOOSE, THOMAS, A.
LOVELAND, GLENN
LUCKENBACH, CARL, A.
LYLE, LEWIS, E.
LYON, ARMSTRONG

M

MAGNUSON, LEO, W.
MAHER, FRANCIS
MAILLET, STANLEY, M.
MARTIN, DOUG
MARTIN, FRANCIS, J.
MARTINO, JOSEPH, P.
MAXIM, STANLEY
MAYEUX II, GILBER, E.
MCDANIEL, CHARLES, B.
MCLAUGHLIN, JOHN, J.
MEHARG, WILLIAM, B.
MIHALIK, RICHARD, P.
MIKITA JR., JOSEPH, V.
MILLER, RICHARD, L.
MILLER JR., HARRY, C.
MILLS, BRUCE, D.
MOFFITT, MELVIN, H.
MOORE, JAMES, W.
MORROW, ROBERT, T.
MORTON, MELINDA, S.
MURRAY, ROBERT, P.
MURRAY, RAYMOND, J.
MUSTAINE, JAMES, J.

N

NAPIER, CHARLES, E.
NELSON, DAVID
NEWSOME, JERRY, L.
NOONE, RICHARD, F.

O

OLSEN, JAMES, P.

P

PARKHURST, WAYNE, L.
PEACE, JAMES, D.
PFEIFFER, JOHN, L.
PIEPENBRINK, JOHN, O.
PIKE, WILLIAM, F.
PINKHAM, ROBERT, W.
PISCHKE, NORMAN, A.
POPE, CHARLES, F.
PULS, LAWRENCE

Q

QUINN, STEPHEN, H.

R

RAMAGOS, DAVID, G.
RAYNER, BENJAMIN, A.
REED, WILLIAM, R.
REGAN, THOMAS, J.
REYNOLDS, WILLIAM, R.
RIDDLE, DONALD, K.
ROBERTS, WILLIAM
ROBERTS, TIMOTHY, K.
ROBINSON, HUNTER, R.
ROGGERO, MICHAEL, J.
ROSSI, JAMES, M.
ROTHWELL, LYMAN, W.
RUDZIANSKI, JOHN
RUNT, DAVID, J.
RUSSELL, JAMES, F.
RYAN, JOHN, A.

S

SACKMAN, DENNIS, R.
SAHLFELD, JOSEPH, A.
SANDERS, CHRISTOPHER, C.
SANTUS, BOB
SARGENT, LOUIS, P.
SARTE, VICTOR, J.
SCHOENBERG, IRVING, B.
SCHROEDER, LA VERNE, L.
SCOTT, ART
SCOTT, WILLIAM, K.
SEARS, JOE
SENTER, THOMAS, E.
SHAFER, JOHN, A.
SHARPE, CHRISTOPHER, C.
SHAVERS, MICHAEL, E.
SHERIDAN, JOHN, W.
SIDERS, JERRY, D.
SKIRKO, ANTHONY
SKRUBER, RICHARD
SKUNDA, DALE GEORGE.
SLOAT, THOMAS, L.
SMITH, LAWRENCE, E.
SMITH, B. DEAN
SMITH, SCOTT, W.
SMITH, GENE, L.
STAMM, THOMAS, D.
STEVENS, ROBERT, E.
STEWART, WILLIAM, A.
STROMWALL, CHARLES, L.
STURGEON, LYLE, L.
SYLVIA, WILLIAM, A.

T

TABOR. LARRY, S.
TALLEY JR, NOLAN, A.
TAYLOR, THOMAS, L.
THEADO, THOMAS.
THORNE, DONALD, M.
TILLEY, LLOYD, L.
TIMPE, JONATHAN, E.
TITUS, DONALD, A.
TOLLERUD, GORDON.
TONGUE, WILLIAM, L.
TOURINO, RALPH, G.
TUBBS, JACK, W.
TURNER, LELAND, E.

U

UHRHAMMER, DALE, E.

W

WALKER, LARRY, R.
WALKER, LEONARD, R.
WALKER, JAMES, C.
WALLIN, EVERETT, D.
WALSH., MARTIN, R.
WATKINS, WARREN, S.
WEGNER, PHILIP, A.
WEIL, GEORGE, E.
WERNETTE, CLARENCE, M.
WHITEHURST, LARRY, S.
WIEDERRECHT, TOM.
WILLIAMS, STEVEN, D.
WILLIAMS JR, THOMAS, A.
WILSON, SETH, J.
WILSON, JOSEPH, D.
WILSON JR., WILLIE, J.
WOODWARD, WILLIAM, B.
WUEST, WILLIAM, A.

Z

ZEIGLER, TIMOTHY. R.
ZIEGLER, GEORGE, F.
ZOLCZYNSKI, STAN, P.

Other Former Members

BAILEY, EDWARD, D.
CHRISTICH, BRUCE, K.
COX, SHAWN, ,
PARSLEY, JOSEPH, M.
RAY, DANNY, L.
SCHLACHTER, ALBERT, J.
TACKETT, ARDY, L.
WALDMANN-BOHN, JOHN, E.
MCCOMBS, FRANCIS, J.
COFFIN, JERRY, R.
STAYTON, WILLIAM, H.
WESTHAUSEN, GARY, H.

Other Missileers Roster

A

ABATANGELO, NICK
ABBOTT, DARWIN
ACEVES, BILL
ACKERMAN, GARY
ACQUISTO, RONALD
ADAM, MICHAEL
ADAMS, JERRY
ADAMS, GAIL
ADAMS, RODNEY
ADAMS, KEN
ADAMS, MARK
ADAMS, RICH
ADAMS, JERRY
ADAMS, BOB
ADCOCK, WALTER
ADDISON, ROBERT
ADKINS, JAMES
AFFLERBAUGH, DON
AGAR, RALPH
AHERN III, JAMES, D.
AHONEN, DAVID, A.
AHRENS, EUGENE
AIKEN, JAMES
AKERS, DAN
ALBERT, WILLIE
ALCORN, STEVE
ALDERMAN, ORBA
ALEXANDER, LARRY
ALLEE JR, BERT
ALLEN, ROLAND
ALLEN, ROBERT
ALLEN, JAMES
ALLEN, GEORGE
ALLEN, JOHN
ALLISON, BOB
ALLWERDT, JIM
ALMASY, RONALD
ALTENES, DAN
ALTIERE, LAUREN
ALTUS, JAMES
AMOS, LARRY
AMOS, CHARLES
AMOS, JIM
ANDERL, ROBERT
ANDERS, PERCY
ANDERSON, KENNETH
ANDERSON, VINCENT
ANDERSON, ROBERT
ANDREASEN, EARL
ANDRUS, TOM
ANGRY, BOBBY
ANTERSON, BILL
ANTHONY, ROSEMARY
ANTHONY, KENNETH
ANTHONY, RON
ANTHONY, TONY
APACHE, DON
APPLEGATE, RICHARD
ARAGON, ANTHONY, J.
ARANCE, ROBERT
ARCHER, JOHN
ARCHIBALD, CHARLES
ARCHULETA, THOMAS
ARCIERO, ANTHONY
ARMSTRONG, BILL
ARNOLD, WALTER
ARNOLD, JONATHAN
ARNST, WAYNE
ARRELL, VIRGIL
ARRINGTON, DAVE
ARROYO, HARIS
ARRUDA, ROBERT
ARTHUR, JIM
ARTMAN, WILLIAM
ASHBROOK, DONALD
ASHBY, LOU
ASHMAN, STEPHEN
ASHTON, RICHARD
ASLAENDER, ADAM
ATCHLEY, GORDON
AUFDERHEIDE, HOWARD
AUSPELMYER, RON
AXTEL, STEPHEN

B

BABCOCK, BOB
BACKES, KEN
BACON, JOE
BACS, JOHN
BADEN, JAMES
BAER, ARTHUR
BAGGENSTOSS, JOHN
BAGLEY, EDWIN
BAIER, KEN
BAILEY, PENNY
BAILEY, JOE
BAILEY, CHARLES
BAIN, BUD
BAINBRIDGE, DONALD
BAIRD, FRED
BAKER, ROB
BAKER, STANLEY
BAKER, DONALD
BAKER, BRYCE
BAKER, HOLLIS
BAKER, HAROLD, L.
BAKKE, HARVEY, J.
BALDASSANO, ROBERT
BALDWIN, EDWARD
BALDWIN, BOB
BALDWIN, JERRY
BALDWIN, IRA
BALIN, LARRY
BALL, THOMAS
BANCALE, MIKE
BANES, BOB
BANGHART, JOHN
BANKHEAD, STEVEN
BANKS, ERNEST, S.
BANKS, WILLIE
BARANOWSKI, BRONISLAW
BARBER, ROBERT
BARBOZA, ALPHONSE
BARFIELD, DOUGLAS
BARLOW, CHARLES
BARLOW, TIM
BARNES, JOHN
BARRETT, EWING
BARRETT-SMITH, R
BARRY, WILLIAM
BARRY, ANDREW
BARTLOW, BOB
BARTMAN, GILBERT
BARTU, DON
BARWICK, RICHARD
BASS, JACKIE
BATEMAN, DENNIS
BATES, RONALD
BATES, ROBERT, D.
BAUMAN, TED
BAVUSO, ANTONIO
BAXTER, ERNEST
BEACH, ROBERT
BEACHEN, EUGENE
BEADLING, CHARLES
BEAUREGARD, NORMAN
BEAUVAIS, FRANK
BECK, RICHARD, W.
BECK, STUART
BECKER, FLOYD
BECKER, RICHARD
BECKETT, DARWIN
BEE, JAMES
BEERS, SUZANNE
BEERY, HAROLD
BEESON, VERLE
BELLING, LEW
BENDER, BRUCE
BENDER, LARRY
BENNER, LEW
BENNETT, RICHARD, M,
BENNETT, RUSSELL
BENNETT, ROBERT, E.
BERBERICH, BILL
BERGANDI, LOU
BERGELIN, WILLIAM
BERGESON, MICHAEL
BERMAN, HERBERT
BERNTH, TERRY
BERRETH, ELAINE
BERRY, ARNIE
BERRY, MARTY
BERTAM, VAN
BERTHELSEN, LARRY
BERTHOLF, DAVE
BERTRAM, KENNETH
BESBRIS, DAVID, C.
BIALAS, STAN
BIALAS, GEOF
BICKERSTAFF, RODGERS
BIDLACK, HAL
BIEHL, DONALD
BIGELOW, RICHARD
BIGELOW, MAX PAUL
BILSBARROW, DON
BINFORD, ARRON
BIRCHARD, DAVID
BISEL, ROBERT
BITHELL, WAYNE
BITNER, JOHN
BITTLE, JOHNNY
BIXBY, TOM
BLACK, KEN
BLACK, CLAIR
BLACK, RICHARD
BLACK, TERENCE
BLACKBURN, ALAN
BLACKMER, DOUGLAS
BLACKWELDER, HOWARD
BLAKELY, NORMAN
BLAKENEY, WILLIAM
BLALOCK, LAMBERTH
BLASCOVICH, LEONARD
BLASI, FRED
BLAYLOCK, CHARLES, R.
BLEVINS, WILLIAM
BLISS, JODIE
BLISS, FRED
BLONDIN, JIM
BLOSS, LARRY
BLUBAUGH, WILLIAM
BODOM, EARL, T.
BOHANNON, CHARLES
BOHN, HAROLD
BOHNER, HARRY, R.
BOHNET, ROLAND
BOLAND, TOM
BOLING, JAMES
BOLLINGER, PETE
BOMBA, NELSON, L.
BOMBECK, WALTER
BON TEMPO, JOHN, C.
BOND, RODNEY
BOSEMAN, JERRY
BOSSHART, ROBERT
BOSTON, WILLIAM
BOSWELL, GORDON, F.
BOTHELHO, RAY
BOTHWRIGHT, HAROLD
BOURRETT, GARY
BOWE, DONOVAN, K.
BOWEN, ROY
BOWEN, THOMAS
BOWER, GLEN
BOWERSOCK, DAVID
BOWLES, DAVID
BOWMAN, PETER
BOWMAN, RANDY
BOZEMAN, PHIL
BRABEC, JAMES
BRACE, DONALD
BRADFORD, MICHAEL
BRADISH, ART
BRADLEY, JEFF
BRADY, MARK
BRAGG, CLOYCE
BRAHAN, MIKE
BRAMLEY, PETE
BRANCH, ROBERT
BRAND, ERNEST
BRAND, JEFF
BRANSON, GRANT
BRANTLEY, MIKE
BRANTLEY, SHADY
BRAOZIER, J. C.
BRASHEAR, MIKE
BRAXTON, M. E.
BRAZEALE, ED
BREAUD JR, ALFRED
BRECK, ROBERT
BREER, LAWRENCE
BREMNER, RICHARD
BRENNAN, WILLIAM
BRENNAN, JEMIMA
BREWEN, CHENEY
BREWER, JAMES, L.
BREWER, ALLEN
BRICKELS, MICHAEL
BRIDGES, JAMES
BRIGGS, LEON
BRIGHT, RAY
BRINKLEY, DONALD, E.
BRISENO, DOMINGO
BRITT, BRIAN
BROADBENT, ART
BRODERICK, GEORGE
BROLLIER, GRANT
BROOKS, SONNY
BROOKS, DONI
BROOKS, DENNIS
BROTHERTON, DAWN
BROWER, ED
BROWER, WALTER, L.
BROWN, ALAN
BROWN, ROY
BROWN, GALE
BROWN, JERRY
BROWN, JOSE
BROWN, DONALD
BROWN, DANIEL
BROWN, MARK
BROWN, KEITH
BROWN, CLYDE
BROWN, CHARLES
BROWN JR, JAMES
BROWNFIELD, ROYCE
BROWNING, BILL
BROWNLEE, JIM
BRUBAKER, HOWARD
BRUEREN, QUENTIN
BRUM, TERRY
BRUMMER, DAVID
BRUMMETT, KEVIN
BRUNER, ROBERT
BRUS, TERRY
BRYANT, CHAUNCEY
BRYANT, LEO
BRYANT, JACK
BRYDEN, JAMES
BUCHANAN, FRANKIE
BUEKER, JIM
BUGG, RAOUL
BULEY, PETER
BUMGARNER, BOBBY
BUNDY, E. L.
BUNKER, KENNETH
BUNKER, KENNETH
BURBRIDGE, DOUGLAS
BURCH, TOM
BURDICK, BARRY
BURGESS, ROBERT
BURINSKAS, WALTER, G.
BURKE, ROBERT
BURKE, WILLIAM, J.
BURKEY, ROY
BURKHARDT, JOHN
BURLING, JOHN
BURLINGHAM, JAMES
BURNS, JACK
BURNS, EDWARD
BURNSTAD, BASIL
BURTON, BRADLEY
BURTON, ROBERT
BURZENSKI, TED
BUSH, DAVE
BUSHEE, KENITA
BUSHNELL, GARY
BUTLER, WILLIAM
BUYS, EARL
BUYTAS, STEPHEN, J.
BYERTS JR, WILLIAM
BYNUM, JOHN
BYRN, JAMES
BYWATER, M. A.

C

CABRERA, DENNIS
CADE, RONALD, L.
CAFFREY, EDWARD
CAHELO JR, GEORGE
CALDWELL, JOHN
CALLIGAN, PAT
CALLIGAN, PATRICK
CAMPBELL, THOMAS
CAMPBELL, TERRY
CAMPOS, GERALD
CAMUNEZ, JOSEPH
CANIPE, RICKY
CANNION, MIKE
CANNON, WALTER, G.
CANNON JR, JOHN, H.
CANTRELL, JIMMY
CAPLES, BUDDY
CAPOFERRI, LOUIS
CAPRA, TONY
CARAWAY, CHARLES
CARAZO, MANUEL
CARDEN, JAMES
CARDOSI, JOHN
CARDOZA, THOMAS
CAREY, MARK
CARLOW JR, CARL
CARLSON, MILO, L.
CARLSON, ERIC, D.
CARLYLE, SAMUEL
CARNES, SANFORD
CARNITO, LAWRENCE
CARPENTER, RUSSELL
CARR, ROBERT
CARRIGAN III, ROBERT
CARSON, RICHARD
CARTER, FREDERIC
CARTER, DONALD
CARTWRIGHT, JAMES
CARTWRIGHT, CHARLES
CARTY, A
CARVER, DEWEY
CASBEER, ART
CASE, TERRY
CASEY, MAUREEN
CASSIDY, PETE
CATON, JEFFREY
CAUFFIEL, ROBERT
CAUSBY, JIMMY
CAUTHEN, JOHN
CAVANAUGH, JAMES
CERAOLA, ANTHONY
CHADDOCK, DONALD
CHADWELL, MARGOT
CHAFFEE, TIMOTHY
CHAFFEE, RANDY
CHALFONT, ALAN
CHAMBERS, STEVEN
CHANEY, NEWTON
CHANG, LEWIS
CHAPMAN, DAVID
CHAPMAN, ROBERT
CHAPPLE, WILLIAM
CHAPUT, GARY
CHARCZAK, GLENN
CHASE, HENRY
CHEATHAM, RICHARD
CHEEK, JIM
CHERRY, CARLOS
CHESKO, ROBERT
CHESTER, STEVEN
CHEVESS, DAVE
CHILDERS ,
CHIPMAN, JOHN "CHRIS'
CHITTUM, WILLIAM
CHOATE, TROY
CHRISTIANSEN, JAMES
CHRISTOFF, GEORGE
CHRISTY, MARLENE
CHRISTY JR, HARRISON
CHUN, CALVIN
CIACCI, LARRY
CISNA, CRAIG
CIVICK, RICHARD
CIVIOK, RICHARD
CLARK, HOWARD
CLARK, GARY
CLARK, LARRY
CLARK, ALEX
CLARK, EARL
CLARK, DENNIS
CLARK, CLARENCE
CLAUSEN, ROBERT
CLAY, ROGER
CLEARY, WALT
CLEMENTS, CAL
CLEMENTS, JIM
CLEMONS, GERALD
CLERMONT, GARY
CLIATT, ED
CLINTON, ALLEN
COCHRAN, MICHAEL
COCHRANE, MIKE
COFFEY, GEORGE
COFFIN, RICHARD
COIGNARD, JULES, B.
COIN, MIKE
COLBERT, LARRY
COLBURN, HUGH
COLBY, RHONDA
COLE, DANIEL
COLE, LEWIS
COLE, RONALD
COLEMAN, DONALD
COLEMAN, FORREST
COLEMAN, GEORGE

COLEY, JON
COLLINGS, MICHAEL
COLLINS, LOUIS
COLOSIMO, PHIL
COLSON, WILLIAM, B.
COLSON, JOHN
COMBS, ROBERT
COMERFORD, BILL
CONANT, ROGER
CONE, NORRIS
CONELL, LONNIE, R.
CONGDON, DAVID
CONLEY, SCOTT
CONLEY, SCOTT
CONNELLY, GRANT, J.
CONNER, JOSEPH
CONROY, JAMES
COOK, BOB
COOK, PHIL
COOKE, GEORGE
COOKE, DAMON
COOKSY, DAVID
COOLEY, DAVID
COONEY, LARRY
COOPER, BILL
COPELAND, CHARLIE
COPPINGER, STEVE
CORK, LARRY
CORL, DEWEY
CORSO, ANDREW
COSNER, DENNIS
COTTERMAN, RONALD
COUCH, DAREN
COUNTRYMAN, PERRY
COURTNEY, RAYMOND
COURVILLE, NOLTON
COUVILLION, JERRY
COUZINS, RICHARD
COWAM, RAYMOND
COX, BILL
COX, ROBERT, E.
COX, BILLY
COX, AL
COX, STU
COY, ED
CRAIG, DONALD
CRAMER, SUE
CRANK, FRANK
CRANMER, BRUCE
CRAVEIRO, RICHARD
CRAW, KENNETH
CRAWFORD, ROBERT
CRAWLEY, GEORGE
CREECH, ROBERT
CREEL, JOEL
CRITTENDON, C. T.
CROMER, MARK
CROSCEY, TERRENCE
CROSIER, DEL
CROW, JAMES
CROWDER, BENJAMIN
CRUMP, HERSHEL
CRUSSELL, ANDREW, D.
CRUTHFIELD, MARION
CUADROTORRADO, JUAN
CUELLAR, ANDRES, N.
CULLEN, TOM
CULVER, RONALD
CUMMINGS JR, REX
CUNNING, ROGER
CUNNINGHAM, PAUL
CUNNINGHAM, TOM
CUNNINGHAM, JAMES
CUNNINGHAM, LARRY
CURRAN, EUGENE, M.
CURREY, MICHAEL
CURRIN, THOMAS
CURRY, TONY
CURTIS, RON
CURTIS, JACK
CUSENBARY, CHARLES
CUSTER, ED
CYOSKI, DAVID
CZAJKA, RAYMOND

D

DAACKE, ROBERT
DAHLE, SIMEND
DAHLGREEN, THOMAS, C.
DAHMEN, BOB
DALE, THOMAS
DALRYMPLE, GARY
DALTON, RANDALL
DALY, JOSEPH
DALY, WILLIAM
DALZELL, LEONARD
DAMEWOOD, CHRISTINE
DAMON, KARL
DANFORD, MAX
DANIEL, EDWARD
DARR, STEVE

DAVIDSON, CULLEN
DAVIDSON, JAMES
DAVIDSON, GARRY
DAVIDSON, WALTER
DAY, BENJAMIN
DECLOEDT, DEREK
DEES, BUDDY
DEGENERES, FREDERICK
DEICHERT, LEROY
DEINKEN, JOHN
DEITRICK, ROBERT, L.
DEKAW, PETE
DEKOWSKI, RONALD
DELEON JR, EDWIN
DELISA, ROBERT
DELORENZO, FELIX
DEMOSS, MICHAEL
DEMOULLY, MIKE
DEMPSEY, JOSEPH
DENBOW, JAMES
DENCHY, THOMAS
DENEKA, RICHARD
DENHAM, TRACY
DENMAN, EDWIN
DENNY, WAYNE
DERUYTER, MARK
DESIGIO, JAMES
DEVOSS, PHIL
DEWEY, WILLIAM
DEWITT, LEIGHROY
DEXTER, R
DEZARRAGA, MIGUEL
DIAMOND, LEO
DIAMOND, ERNEST
DIAS, MELVIN
DICKENS, JACKIE
DICKERSON, ORVILLE, E.
DICKINSON, STEVE/ LINDA
DICUS, CRAIG
DIEKMAN, JOLENE
DIETZ, PAUL
DILELLO, TONY
DILLMORE, JAMES
DIXON, JAY
DOCKERY, NOEL "DOC'
DODGE JR, JOHN
DOERFERT, GARY
DOHERTY, JOSEPH
DOLLAR, KEN
DOLLARD JR, JOHN
DOME, MARTY
DOMINGUEZ, MARK
DOMINGUEZ, MANUEL
DOMINUEZ, CONRAD
DONAHUE, PAT
DONES, ERNEST
DONEZ, SALVADOR
DONLEY, DEREK
DONNELLY, JOHN
DONOHOE, RICHARD
DONOVAN, JOHN
DORMAN, RICHARD
DORR, ROBERT, F.
DORR, RAYMOND
DOUBERLY, JOHN
DOUGHERTY, BRIAN
DOUGHTY JR, BENJAMIN, T.
DOW, FRANK
DOWELL, BUCKY
DOWNEY, RICHARD, F.
DOWNING, KEN
DRAKE, CARL
DRAKE, CLARENCE
DRAMER, MICHAEL
DREESEN, VICTOR
DRISCOLL, KEVIN
DRISKELL, LOREN
DRISKILL, MELVIN
DUBOIS, TOM
DUCK, TRACY
DUDLEY, RAY
DUGO, TODD
DUHON, ALAN
DUNKER, GENE
DUNN, LUCKY
DUNN, JOSE
DUNN, K
DUNNING, BRAD
DUNSHEE, RICHARD
DURHAM III, WILLIAM, S.
DUSEK, RICHARD
DUYCK, ROBERT
DWYER JR, MICHAEL
DYE, DON
DYER, JON
DYER, DAVID

E

EARNSHAW, MIKE
EARY, LOY
EASTEP, GARY

EATON, JAMES
EATON, JAMES
EBERHARDT, RICHARD
EBERLING, GEORGE
ECK, RAYMOND
ECKERT, BILL
ECKERT, LEO
ECKERT, TOM
ECOPPI, J. B.
EDDINGTON, ROBERT
EDDY, WILLIAM
EDGERTON, RICHARD, W.
EDGEWORTH III, THOMAS
EDGIN, CHUCK
EDMOND, JOHNNIE
EDSON, JAMES, W.
EDWARDS, LARRY
EDWARDS, LYNN
EDWARDS, DONOVAN
EDWARDS, SCOTT
EGGLESTON, TIMOTHY
EGLER, JOHN
EHMER, JACK
EIGENMANN, JOHN
EISCHEID, JAMES
EISENHAUER, MICHAEL
EISNER JR, JIM
EKLUND, LESTER
ELBRACHT, WILLIAM
ELDER, JOSEPH
ELDER, SAM
ELDRIDGE, GEORGE
ELLERBECK, DENNIS
ELLINGTON, DUKE
ELLIOTT, DAVID
ELLIS, GEORGE
ELLIS II, JOHN
ELORANTO, JEFFREY
EMOND, LOUIS
EMRICK, TIM
ENGELKEN, MARTIN
ENGER, ROLF
ERWIN, CARL
ESPARRA JR, PORFIRIO
ESTELL, TIM
ESTES, RONNIE
ESTRELLA, FERNANDO
ETTE, STEPHEN
EVANOFF, JON
EVERETT, HAROLD
EVERETT, WILLIAM
EVERETT, CLINT

F

FAAS, DAVID
FAAS, DAVID
FABBRI, MICHAEL
FAGNER, J. LOGAN
FAISCA JR, JESUS
FALL, JOHN
FALLON, JIMMY
FAMILITON, ROBERT
FARLEY, JAMES
FARMER, BERYLE
FARNELL, LARRY
FARNELL JR, LELAND, B.
FASANO, DOMINIC
FAUST, WILLIAM
FEENSTRA, GERTRUDE
FEIGHAN, MICHAEL
FEIST, SCOTT
FELDMAN, LAURA
FELDSTEIN, KEN
FENNELL, ROB
FERAN, PETER
FERBER, MARIANN
FEREBEE, MIKE
FERENZ, JOHN
FERGUSON, JAMES
FEWER, DANIEL
FIELD, JEFF
FILLER, JIM
FILLETTE, JOHN
FINK, BARRY
FINNAMORE, RONALD, P.
FIRTH, FRANK, H.
FISCHER, BRUCE
FISHER, JOHN
FISK, DAVID
FITE, C.W.
FITZPATRICK, RONALD
FLETCHER, JACK
FLINN, ROBERT
FLOERCHINGER, KEVIN
FLOOD, STU
FLOTTEN, EDWARD
FLOWERS, JAMES, A.
FLYNN, ROYCE
FLYNN, LARRY
FOLEY, JAMES
FOLTS, EDWARD

FONKEN, AL
FORRISTALL, RICHARD
FORSEMAN, DENNIS
FOSTER, THOMAS
FOSTER, EARL, E.
FOULGER, KEITH, H.
FOUST, GARY
FOWLER, LEONARD
FOWLER, STEVEN
FOX, BEN
FOX, ALAN
FOX, ROBERT
FRANCIS, ROBERT
FRANKLIN, JAMES
FRANZ, GORDON, G.
FRAZIER, CARL
FRAZIER, CHARLES
FRECHE, EDWIN
FREELAND, GRANT
FREEMAN, OTIS
FREEMAN, KARMA
FREEMAN, WILLIAM
FREEMAN, LORAINE, W,
FREIER, DONALD
FRENCH, CHRISTOPHER
FRIEHAUF, FRITZ
FRIESEN, NYLE
FRUEHAUF, DUANE
FRY, ROBERT, H,
FULDERSON, RICHARD
FULDS, REAFORD
FULKERSON, PAUL
FULKS, JOHN
FULLER, ED
FULLWOOD, BILL
FULTZ, J. D.
FUNDIS, TOM
FUNKHOUSER, GARY

G

GABERT, PAUL
GAFFNEY, JOSEPH
GAINES, GREG
GALATI, VINCE
GALDEANO, JESSIE
GALE, MARK
GALLARDO, C. L.
GALLETTA, PAT
GALLIGAN, CHRIS
GAMBLE, GAYLE
GANGL, GENE
GANT, JACK
GARBER, GARES
GARBER, MEYER
GARCIA, FREDDY
GARDINER, CAROLYN
GARDNER, SAMUEL
GARLAND, EUGENE
GARNER, MARVIN
GARREN, WALTER
GARRIGUES III, CASPAR
GARVER, HOWARD
GASAWAY, WILLIAM, F.
GASIOR, KEN
GATES, ROBERT
GATLIN, DONALD, L.
GAUTIERI, JOHN
GAVICH, ROBERT
GAY, TERRY
GAYLOR, DON
GAYLORD, DONALD
GAYZUR, HANS
GEBHARDT JR, FREDERICK
GEHRIS, GEORGE
GEIGLE, STEVEN
GENTRY, CALVIN
GEORGE, DIANE
GETZ, RANDALL
GIEGLER, SAMUEL
GILBERT, JOHN
GILCHRIST, MIKE
GILKESON, TOM
GILLIKIN, STEVEN
GILLIS, JACK
GIMMI, RICHARD
GIRARD, WILLIAM
GIST, JAMES
GIUFFRE, CHARLES
GLADSTONE, THOMAS
GLANTZ, DONALD, L.
GLAUSH, GEORGE
GLENN, ROY
GLOSHEN, JAMES
GLOVER, KAZUMI
GNIADEK, EDWARD
GOEN, ROBERT
GOERTZ, JEFF
GOETZINGER, THOMAS
GOFF, GERALD
GOLD, BARON
GOLDBERG, SAUL

GOLDBLATT, JOSEPH
GOLDMAN, HAROLD
GONZALEZ, CRUZ
GOOD, CHARLES
GOOD, THOMAS
GOODRIDGE, THOMAS
GOODSON, EDWARD
GORCZOK, PETER
GORDON, DAVID
GORE, STEVEN
GORNELL, DANIEL
GORT, JACK
GORYCHKA, TED
GOSNEY, CHARLES
GOSSETT, WILLIAM
GOTNER, NORBERT
GOTT, GEORGE
GOULD, TONY
GRABLE, DONNA
GRAHAM, ROBERT
GRAMMER, CHARLOTTE
GRANT, HARRY, F.
GRANT, ROBERT
GRANT, JACK
GRAVES, RUSS
GRAVITT, VICTORIA
GRAY, TOM
GRAY, JIM
GRAY, RUSS
GREAVER, AARON, R.
GREEN, DAN
GREEN, DON
GREEN, MURICE
GREEN, DANNY
GREEN, WARREN
GREENE, SKIP
GREENE JR, ALBERT
GREENWOOD, CRAIG
GREER, MARK
GREY, GORDON
GRIFFEN, JOHN
GRIFFIN, RALPH, O.
GRIFFIN, WAYNE
GRIMSLEY, KEN
GRIPP, BERNIE
GROSS, TONY
GROTE, HERB
GROVE, JACK
GRUBB, ROBERT
GRUWELL, WALLACE
GRYCZKO, STANLEY
GRZEBINIAK, VAL
GUEDNOW, WILLIAM
GUERTLER, CARLTON
GUIDRY, MURREL
GUILFORD, CLIFFORD
GULA, MIKE
GULL, RALPH
GUNDERSON, ROBERT
GUNDY, ISAIAH
GUPTIL, LEROY
GURTIS, RICHARD
GUSTAFSON, LOWELL
GUSTAFSON, B. G.
GUTIERREZ, JOE
GUTIERREZ, OSCAR
GUYER, KURT
GWINN, GEORGE

H

HAAN II, ALBERT
HACKETT, HELEN
HAGAN, MARC
HAGEN, WILLIAM, H.
HAGERNES, ROBERT
HAIGHT, LARRY
HAIGHT, WALTER
HAILEY, JAMES
HAINES, GARY
HAIRSTON, MICHAEL, A.
HAITHCOAT, K. E.
HALE, FRANK
HALL, WILLIAM
HALL, STANLEY
HALL, RALPH
HALT, CHARLES
HALTER, JERRY
HAMBELTON, WILLIAM, A.
HAMBLETON, GENE
HAMILTON, GEORGE
HAMMACK, GLENN
HAMMOND, CIINTON
HAMPTON, JOHN
HAMPTON JR, LUTHER
HANCOCK, TOM
HAND, THOMAS
HANDY, ART
HANDY JR, WALTER, R.
HANKS, CHARLES
HANKS, MICHAEL
HANN, ED

HANNA, ROY
HANNON, JOSEPH, E.
HANSEN, CRAIG
HANSEN, BARRY
HANSON, GAIL
HANSON, BILL
HANUS, RUSSELL
HANUSHEK, KENNETH
HARBOUR, KENNETH
HARDENBROOK, MILO
HARDING, HOWARD
HARFORD, MICHAEL
HARGROVE, NORVILLE
HARGROVE, DONALD
HARKINS, WARREN
HARNEY, JOHN
HARPER, LAWRENCE
HARPOLE, G
HARRELL, WILEY
HARRINGTON, MAHLEN
HARRINGTON, MARK
HARRIS, WILLIAM
HARRIS, ERNIE
HARRIS, PAUL
HARRIS, CHARLES
HARRIS, CHARLES
HARRIS, JIM
HARRIS, ALEXANDER
HARRISON, WILLIAM
HART, WILLIAM
HART, LESTER
HARTILL, SCOTT
HARTING, WILLIAM
HARTJE, ANGIE
HARTLY, HOLMES
HARTMAN, HAROLD
HARTMAN, DONALD
HARTNEY, JOHN
HARUTUNIAN, DAVID
HARVAN, FRANK
HASKELL, FREDRICK
HASSELL, ED
HATHAWAY, MILTON
HAWK, TROY
HAWKINS, ALAN
HAWKINS, JARED
HAWORTH, RICHARD
HAWORTH, WENDELL
HAYNES, JAMES, W.
HAYNES, CHARLES
HAZEL, ROBERT, A.
HEADLEY, GORDON
HEATH, CHESTER
HEDAHL, DUANE
HEGNER, WILLIAM
HEINSCH, PAUL
HEISER, SHERWOOD, W.
HELGESON, LARRY
HELLYAR, DENNIS
HEMMERT, RALPH
HEMPHILL, ROBERT
HENDERSON, JON
HENDERSON, ROGER
HENDERSON, JAMES
HENDRIX, GLEN
HENNEY, DONALD
HENNING, JERRY
HENNING, DAVID
HENNINGS, MAXWALTON
HENNY, DONALD
HERMANN, OSCAR
HERRIN, LEXIE
HERTEL, EDWARD
HESS, BARRY
HESTER, CLIFF
HETRICK, TIMOTHY
HEWITT, DONALD
HICKLIN, KENT
HICKMAN, VERL
HICKMAN, TERRY
HICKS, CECIL
HICKS JR, LONNIE
HIGGASON, BOB
HIGGINS, MICHAEL
HIGGINS, JACK
HIGHT, FRED
HILDERBRAND, DEL
HILL, DONALD
HILL, MIKE
HILLIARD JR, JESSE, E.
HILLS, JACK, E.
HILLYER, WILLIAM
HILT, DANIEL
HIMMER, JOHN
HINCK, WALTER
HINES, WILLIAM
HINKLE SR, GEORGE, G.
HINTON, MARK
HINTON, JAY
HIXON, J. VERN
HOAG, DAVID
HOARD, DUANE

HOBBS, WILLIAM
HOBENSACK, CORWIN
HODGERSON, KEVIN
HOEPPNER, THOMAS
HOFFMAN, HAROLD
HOGER, JAMES
HOLDEN, RANDY
HOLLAND, MAURICE
HOLLAND, GEORGE
HOLLENSTAIN, ANDREW
HOLLERICH, KENNETH
HOLLINGA, PAUL
HOLLMAN, JOHN
HOLMAN, WILLIAM, S.
HOLSKE, CLIFFORD
HOLSTROM, ERIC
HOLT, RICHARD
HOLT, CHARLES
HOLZMAN, WAYNE
HONAKER, HARRY
HOOD, CHARLIE
HOOD, ROBERT
HOOD, BOB
HOOD, DAWSEY
HOOVER, MARVIN
HOOVER, JACK
HOPKINS, ROBERT
HOPPER, TERRENCE
HOPUN, JAVIER
HORNE JR, AZIE
HORNER, FOREST
HORNEY, JIM
HORSCH, JOHN, T.
HORTON, CLARENCE, F.
HORTON, DONALD
HORTON, JIM
HOSKINS, THOMAS
HOSTERMAN, JACK
HOSTETTER, CECIL
HOSTETTER, DONALD
HOUGHTON, HOMER
HOWE, JACK
HOWELL, DAVID
HOWELL, ROBERT
HOWELL, DAVID
HUBBARD, WARREN
HUBBARD, RONALD
HUBBARD, RONALD
HUBER, SCOTT
HUDDLESTON, JAMES
HUDDLESTON, DALLIS
HUDSON, BRUCE
HUDSON, RICK
HUEY, JAMES
HUFF, MITCHELL
HUFFMAN, BILL
HUGHES, SHARON
HULT, LEROY
HUMBERT, STEVE
HUMMER, ROBERT
HUMPHRIES, BUFORD
HUNTER, JAMES, D.
HUNTER, EDWARD
HUNTER, BRUCE
HUNTER, CEDRIC
HUTCHINS, BETTYE
HUTCHISON, HOWARD
HUTTO, KENNETH
HUTTON, HARRY

I

INGHAM JR, JOSEPH
IRELAND, OSCAR, R.
ISAACS, CLAUDE
ISEBRAND, ROBERT
IVES, DONALD
IVY, JAMES

J

JACKSON, KENNETH
JACKSON, DEAN
JACOB, CAROLYN
JACOBS, WILLIAM
JACOBSEN, KEVIN
JACOBSON, JAMES
JAKUBOWSKI, JAMES
JAMES, RONALD
JAMES, CHET
JANICKE, JERRY
JANKE, MICHAEL
JANSEN, JERRY
JARRELL, VERNON, H.
JAY, DEAN
JEFFRIES, LAVAY
JEFFRIES JR, HASKELL
JENKINS, DAVID
JENKINS, KIRK
JENKINS JR, ROBERT
JENNINGS, PAUL
JENNINGS, JAMES

JENNINGS, JIM
JENNINGS, MIKE
JENSEN, MARTY
JENSEN, RONALD
JESSUP, DAVID
JINE, GENE
JOHANNES, JOHN
JOHNS, MERVYN
JOHNSON, GROVE, C.
JOHNSON, VAUGHN
JOHNSON, HARVEY
JOHNSON, JOHNIE
JOHNSON, MERLE
JOHNSON, LARRY
JOHNSON, LAWRENCE, H.
JOHNSON, JAY
JOHNSON, KEVIN
JOHNSON, CLYDE
JOHNSON, CHAN
JOHNSON, LEO
JOHNSON, RICHARD
JOHNSON, RUSSELL
JOHNSON, RONALD
JOHNSON, ELWOOD
JOHNSON, ROBERT
JOHNSON, ED
JOHNSON, GENE
JOHNSON, JIM
JOHNSON, WILLIAM
JOHNSTON, CARL
JONES, JOSEPH, W.
JONES, HARTZELL
JONES, GARY
JONES, LEONARD
JONES, RALPH
JONES, DALE
JONES III, GRINNELL
JORDAN, RICHARD
JORDAN, CHARLES
JOSH, WALTER
Joyner, Buck
JUDA, FRANCIS
JUDD, KENNETH, D,
JUDY, JACK
JUNEAU, MURLEY
JUSTIS, PERRY
JUVETTE, KENNETH

K

KADERKA, RICHARD
KAMPA, RICHARD
KANE, EDMUND
KANEKO, BOB
KAPLAN, LEON
KAPPELER, JACK
KARCEWSKI, DONALD
KAYLOR, DON
KAYS, THE
KAYSON, MATTHEW
KAZMIERCZAK, RONALD
KECK, NICK
KEE, TERRY
KEE, TODD
KEENAN, TOM
KEENER, JIM
KEENEY, ROGER
KEITH, DAVID
KELLER, K
KELLER, DAVE
KELLEY, MICHAEL, O.
KELLEY, REGINALD
KELSEY, LETA RAE
KELSO, RICHARD
KENNEDY, MIKE
KENNEDY, RUSS
KENNEDY, JEFF
KENNON, JOHN
KENRICK, JAMES
KERMICK, DOUG
KES, PETER
KETCHUM, REX
KIBLER, RONALD
KIECHLIN, KEVIN, P.
KIEKLAK, RON
KILPATRICK, WILLIAM, J.
KIMBELL, JAMES
KIMMEL, PHILIP
KIND, BOB
KING, WILLARD
KING, B. A.
KING, CHRIS
KING, TERRY
KING, JAMES
KING JR, RAYMOND, J.
KINNEY, JAMES
KINSER, JOHN
KINSTEDT, RICHARD
KIRBY, GEORGE
KIRCHHOEFER, ARTHUR
KIRCHNER, NATE
KIRCHOFF, WILLIAM

KIRKBRIDE, JIM
KIRSCHNER, RIKKI
KIRTLAND, MICHAEL, A.
KITCHENS, DANIEL
KLATT, JOHN
KLEIN, JOSEPH
KLINE, DONALD, P.
KLINE, RICHARD
KLOES, KARL
KNAPP, DONALD
KNIGHT, BEN
KNIGHT, MICHAEL
KNORRE, MICHAEL
KNUDSON, ROGER
KOBLE, DALTON
KOCH, CHARLES
KODALEN, FLOYD
KOELBEL, LEONARD
KOELLNER, ROBERT, E.
KONDZIOLKA, EDWARD
KOONTZ, DAVE
KOONTZ, DAVE
KORGER, HARRY, F.
KORNREICH, JAMES
KOTERAS, FRANK
KOTWICKI, GEORGE
KOWALCHUK, CHUCK
KOZIEL, DENNIS
KRACKE, GENE
KRAFT, BRYAN
KRAHENBUHL, DAVID
KRESGE, CLETE
KRIEGER, TOM
KRINER, DENNIS
KRUEGER, ROBERT
KUFTA, NICK
KUJAWA, WALTER
KUNKEL, JAMES
KUNKLE, ROBERT
KURDY, EDWARD
KYLE, JAMES
KYRIOPOULOS, FRANK
KYSER, LINVEL

L

LA PORTE, CHRIS
LABAR, GLENN, E.
LABARRE, PAUL
LACY, WALT
LADD, ROY
LAFFERTY, ROBERT
LAING, HERBERT
LAKE, DAN
LAKE, JOHNNIE
LAKEY, RICHARD
LALUMONDIER, JIM
LAMB, FELIX
LAMB, ALLEN
LAMBERT, STEWART
LAMBERT, DWIGHT
LAMPE, GERRY
LAMPSON, TOM
LANCASTER, WILLIE
LANCE, BILL
LANE, STEVE
LANEY, ALVIN
LANGFORD, RON
LANGLEY, LEROY
LANTZ, ZANE
LAPORTE, RICHARD, J.
LARMER JR, JOHN
LARSON, BRENT
LASHBROOK, LARRY
LASKEY, RICHARD
LATSHAW, ROBERT, J.
LAVENDER, STEVEN
LAVIGNE, RICHARD
LAWRENCE, DANIEL
LAWSON, ERIC
LAYFIELD, JEFFERY
LAYTON, GREGORY
LAYTON, STEVE
LEAHY, CORNELIUS, M.
LEAR, JOE, B,
LEARN, HAYWARD
LEBAK, MIKE
LEBLANC, WILBUT
LEE, GEORGE
LEE, MELVIN
LEEPER, JOSEPH
LEFFELMAN, FRANCIS
LEFTEROFF, CHRISTOPHER
LEGER, LEOMARD
LEGGETT, DAVID, C.
LEHR, RICHARD
LEHRMAN, MARK
LEMIIEUX, WILLIAM, L.
LEONARD, CHRIS
LEONHARDT, JACK
LESSARD III, ADELARD
LESTER JR, DAVID

LEVENTEN, GEORGE
LEVESQUE, LEO
LEVIN, GERALD
LEVINE, HOWARD
LEWIS, JESSE
LEWIS, WILLIAM
LEWIS, BOBBY
LEWIS, J, D,
LEWIS, JAMES
LEYNES, DAVID
LIDARD, ARTHUR
LIDDY, STEVEN
LIENEMANN, JAMES
LIF, HAROLD
LIGHT, JOHN
LILEVJEN, MARK
LILLEY, EARL
LILLO, ERNEST
LIMOGE, ALBERT, H.
LIND, STEVEN
LINDEMUTH, DAVE
LINDSAY, JON
LINE, DESMOND
LINER, WILL
LINX, JACK
LIPINSKI, JAMES
LIPSEY, CHARLES
LIPTON, ROBERT
LITSINGER, DAVID
LITTLE, ERNEST
LITTLEPAGE, OTIS
LITTLEPAGE III, FRED
LITZSINGER, O. J.
LIVINGSTON, JAMES, W.
LIVINGSTON, ROBERT
LIVINGSTON, NEAL
LIVINGSTONE, DOUGLAS
LOCKHART, FRED
LOFTUS, JERRY
LOLEAS, MONICA
LOMAYESVA, GEORGE
LONG, VINCE
LORENZ, EDWARD
LOTZ, CY
LOUGHRAN, HAROLD, R.
LOURIE, JOE
LOVERN, JOHN
LUDLUM, BILL
LUDWIG, CHARLES
LUDWIG, SANDRA
LUFFMAN, CHARLES
LUND, RAY
LUNDY, JOHN
LUNSFORD, STEPHEN
LUNSFORD JR, GEORGE
LUPOLT, RANDY
LUTERMAN, ANDY
LUTES, PAUL M V B
LYKINS, JAMES
LYNCH, JIM
LYNCH, RICHARD
LYNCH, GERALD
LYNN, DONALD
LYNN, RICHARD
LYONS, JOHN, P,

M

MACARNEY, MICHAEL, J.
MACDONALD, JAMES
MACFARLANE, FARLANE
MACINTYRE, SCOTTY
MACKOWSKI, JOHN
MACMILLIAN, ROBERT
MACOMBER, TOM
MADDEAUX, TERRANCE
MADDOCK, M, S,
MADDOX JR, VERNICE
MAFFRY, RICHARD
MAGNESS, WILLIAM
MAGUIRE, WILLIAM
MAHAN, ALLEN
MAKINSON, JIM
MALONE, ED
MALONE, ELBRIDGE
MALONE, JAY
MANN, RAYMOND, L.
MANNING, KENNETH
MANNING, MICHAEL
MANOR, JERRY
MANS, WALTER
MANSON, DENNIS
MANSON, TOM
MANTOOTH, BILL
MARKS, CONRAD
MARSH, ROBERT
MARSH, KENNETH
MARSHALL, DARYLL
MARTIN, RICHARD
MARTIN, ROB
MARTINEZ, JOEL
MARTINEZ JR, HENRY

MASICA, STEVE
MASON, CECIL
MASON, ROBERT
MASSEY, VERN
MASTAL, JEROME
MATEJKA, LYNN
MATHEWS, BILL
MATHIS, JAMES
MATTERN, JAN
MATTERN, GLEN
MATTESON, ALAN
MATTHEWS, MIKEL
MATTIOLI, JOHN
MATTOX, CHARLES, A.
MATZAT, ALBERT
MAXWELL, ULMA
MAYARD, JEROME
MAYER, VAL
MAYS, MICHAEL
MAZUR, MIKE
MC GUIRK, GEORGE, W.
MCALISTER, DAN
MCANASPIE, DUANE
MCANNALLY, ROB
MCCABE, WILLIAM
MCCALLUM, CHESTER
MCCARTHY, JAY
MCCARTHY, JOSEPH
MCCARTNEY, ROBERT
MCCAY, THOMAS
MCCLENAHAN, MALCOM
MCCLURE, DAVID
MCCOLL, HUGH
MCCOLL, BILL
MCCORD, ROBERT
MCCORD, FRED
MCCORMICK, MELVIN
MCCRACKEN, JAMES, D.
MCDANIEL, WALTER, R,
MCDERMOTT, DON
MCDONALD, MIKE
MCDONALD, JOHN
MCDOWELL, TIM
MCELFRESH, JIM
MCFARLAND, ROBERT, L,
MCGAUHEY, J. F.
MCGEE, RICHARD
MCGEE, CHARLES
MCGOLDRICK, CLARENCE
MCGOWAN JR, ROY
MCGUIRE, JAMES
MCHARVEY, H. F.
MCINNIS, FRANCIS
MCINTYRE, ROBERT
MCKEAN, JOEL
MCKENNA, RALPH
MCKENZIE, HOWARD
MCKINNEY, TED
MCKNIGHT, THEODORE
MCLAUGHLIN, JAMES
MCLAURIN, ALLEN
MCLEAN, THOMAS
MCLEOD, FRED
MCMAHAN, JERRY
MCMAHON, RICHARD
MCMILLAN, LEROY
MCMILLAN, VERNON
MCMILLEN, PHILIP
MCNEEL, KENNETH
MCNORTON, MARCELLAS
MCPHERSON, JOHN
MCPHIERSON, MELANIE
MCWATTERS, ROBERT
MEAD, STEVEN
MEADE, LAWRENCE
MEAUX, STEVEN
MECCA, FRANZ BERT
MEDDRESS, JIM
MEEK, STEVE
MEIER, GARY
MEJIA, RICHARD
MELENOVICH, TERRY
MELGHEM, DOUGLAS
MELKO, EDWARD
MELLAND, NORMAN
MELVIN JR, ROBERT, T.
MENCH, CARL
MENZEL, R
MERCER, BOB
MERCIER, ROBERT, G.
MERKLE, ELLEN
MERTEN, DONALD
MERZ, RONALD
MESSER, ROBERT
METCALF, GARY, R.
METHENEY, FRANK, W.
METZGER, DAVE
MEUTH, HUGO, C.
MEYER, MELVIN
MEYER, ROGER
MEYER, JERRY
MEYERS, ELWYN

MICHAUD, THOMAS
MIDDLETON, PHILLIP
MIGNEAULT, ELEANOR
MILANESE, JOHN
MILLER, DAN
MILLER, DONALD
MILLER, WYATT
MILLER, ROBERT
MILLER, RAY, E.
MILLER, LUCIENNE
MILLER, JOHN
MILLER, RICHARD
MILLER, JERRY
MILLER, JOHN
MILLER, MURN
MILLER, FRED
MILLER, KENT
MILLER, NANCY
MILLER, LEONARD
MILLER, GEORGE
MILLER, G. WAYNE
MILLER, BILL
MILLER, BOB
MILLER, MICHAEL
MILLER, JAMES
MILLER, SIDNEY
MILLER, WILLIAM CARL
MILLER JR, TYRUS, F.
MILLHOLLEN, STEVEN
MILLIN, JACK
MILLS, EDWARD
MILLS, MELVIN
MILLS, ARCHIE
MINDERMAN, ERVIN
MINDLING, GEORGE
MION, LOUIS
MITCHELL, DAVID
MITCHELL, J. C.
MITCHELL, ROBERT
MIZE, JOHN
MOCHEL, MICHAEL
MOCKLER, HUBERT, B.
MODE, GORDAN
MOHON, BOB
MOHR, DEWAYNE
MONACO, ANTHONY
MONEYMAKER, JAMES
MONNETTO, JIM
MONTGOMERY, THOMAS
MONTGOMERY, ROY
MOON, ERNST
MOORE, BILL
MOORE, CARROL
MOORE, RAYMOND
MORE, DON
MORGAN, ED
MORGAN, DANIEL
MORGAN, LEROY
MORGAN, DENNIS
MORIN, JANET
MORRIS, TOM
MORRIS, PHILIP
MORRIS, GARY, S.
MORRIS, BRETT
MOSER, GREG
MOSS, RICHARD
MOSS, EDWARD, P.
MOYERS, JAMES
MOYES, WARREN
MUCIA, DAVE
MUCKEY, JOHNNY
MUELLER, KURT
MUNIZZA, DOM
MURPHY, JAMES
MURPHY, DEWEY, V.
MURPHY, RICHARD
MURRAY, JOHN
MURRY, MICHAEL
MUSHKIN, NATHAN
MYERS, DON
MYERS, CAROL
MYERS, STEVE
MYERS, CARL
MYRVOLD, RICHARD, K.

N

NACEY, ALICE
NAST, NANCY
NEAL, GORDON
NEEDLEMAN, LEONARD
NEFF, JOSEPH
NEIGER, NORM
NEIL, CHARLES
NELLER, HAROLD, E.
NELSON, GERRY
NELSON, NILS
NELSON, LYNN
NELSON, PHYLLIS
NELSON, JEFFREY
NELSON, BOB
NELSON, BILLY

NELSON JR, THOMAS
NETZ, ROY
NEUENDORF, CHARLES, A.
NEUMANN, RONALD
NEW, JOHN
NEWBOLD, RONALD
NEWELL, THOMAS
NEWELL, BILL
NEWKIRK, LOREN
NEWSOM, JOSEPH
NEWTON III, RICHARD
NIBLETT, CHARLES
NICHOLS, WILLIAM
NICHOLS, DANNY
NICOLINI, JOSEPH
NIEMEYER, ROBERT
NILSSON, CHERYL
NINNEMANN, TOM
NISHIDA, DARREL
NOBLE, RAYMOND
NOBLE, TED
NOBLES, JIM
NOGUS, JOHN
NOHMER, FREDERICK
NOLAN, HOWARD
NOLAN, PATRICK
NOLAN, PAT
NORCROSS, MIKE
NORDSTORM, DAVE
NORMAN, MURRAY
NORMAN, FRENCHIE
NORMANDO, JOE
NORRIS, LEONARD
NORRIS, STEVE
NORROD, RICHARD
NORTHEY, CLARENCE
NORTON, JAMES
NORVELL, JOHN
NUNN, ROBERT
NYBERG, DAN
NYHUS, DALE

O

O'BRIEN, RAY
O'CONNELL, CRAIG
O'CONNER, JOHN
O'DELL, KENNETH
O'DONNEL, JACK
O'DONNELL, JOSEPH, D.
O'HEARN, THOMAS
O'NEILL, STEVE
O'TOOLE, RANDY
OGDEN, ZIBA
OGLE, FRED
OKLAND, RONALD
OLIVA, BRIAN
OLIVER, SAMUEL
OLIVER, RICHARD
OLSEN, RICK
OLSON, NORM
OLSON, GLENN
OLSON, BRENDA
ONDERS, CARRUS
OPDAHL, OWEN
OPERHALL, RICHARD
OSBUN, GREGORY
OSSWALT, RICHARD
OSTERBERG, CHRISTOPHER
OSTRUM, THOMAS
OTT, PAUL
OTT JR, CHARLES
OVERALL, WAYNE
OVERLING, WILLIAM
OVERSTREET, JOHN, H,
OWENS, FRED, J.
OWNBEY, HENRY

P

PAASCH, JOHN
PACE, RUSSELL
PADBURY, NEAL
PADGETT, WINSTON
PADILLA, MANUEL
PAGE, DAVID
PAGE, PHILLIP
PAGE, WILLIAM
PAGE, JACK
PAGE, JOE
PALANIUK, OREST
PALMER, BILL
PALMER, DAVE
PALMER, ROBERT
PALMIER, PAT
PANNAGE JR, CHARLES
PAPENFUSS, LOREN
PARDECK, RON
PARDILLA, JERRADO
PARENT, CHARLES
PARENTE, HENRY
PARK, JIM

PARKER, FRANK
PARR, WILLIAM
PARR, GERALD, E.
PARRAMORE, WOODY
PARSONS, LARRY
PARSONS, MICHAEL
PARTIN, ALAN
PARTRIDGE, STEVE
PASANEN, RON
PASSAMANI, ANDREW
PATINKA, ROBERT
PATTERSON, JOHN
PATTERSON, GEORGE
PATTERSON, JOHHNY
PATTON, JERRY
PAUL, RICHARD
PAULSON, JAMES
PAVLOVICH, JOHN
PAWLOW, TOM
PAYNE, BILL
PAYNE, GLENN
PAYNE, CALVIN
PAYNE, WARREN
PAYTON, LOWELL
PEARSON, ED
PEARSON, DAVID
PECK, BILL
PECK, GEORGE, E.
PEDERSEN, LESLIE
PEDLEY, JAMES
PEEK, KENNETH
PEHAN, TERRY
PENMAN, DAVID
PERALTA, LOUIS
PERENOVICH, RODNEY
PERETTI, HOWARD
PEREZ, JOHN
PEREZ, HARRY
PEREZ, P
PERKIN, PAUL
PERKINS, JOE, C.
PERKINS, RALPH
PERRY, GEHARDT
PERRY, CHARLES
PERRY, GREG
PERRYMAN, HARRY
PERSONS, DOUG
PETERS, JOSEPH
PETERSON, DENNIS, E.
PETERSON, SCOTT
PETERSON, ALGER
PETERSON, MICHAEL ERIC
PETRIE, WILLIAM
PETTINGILL, MIKE
PETTY, RICHARD
PETZOID, JOHN
PFEUFFER, GEORGE
PHEE, MICHAEL
PHILBRICK, GENE
PHILLIPS, JAMES
PHILLIPS, BILL
PHIPPS, GLENN
PICCONATTO, AL
PICKENS, MAURY
PICKETT, TOM
PIERCE, DOUGLAS
PIERCE, MICHAEL
PIETILA, JOHN
PIGEON, SCOTT
PIKE, LAURA
PINDER, JAMES
PINNER, CLAY
PISARCHICK, BERNIE
PISHNEY, JOHN
PISLE, STANDFORD, C.
PLAKORUS, GEORGE
PLETCHER, BRIAN
PLOSZAY, STEVE
PODAGRASI, ERNEST
POLITTE, V
POOLE, LYNN
POOLE, RICHARD
POOS, HANK
PORTER, MICHAEL
PORTER, PHILLIP
POSTELNEK, BILL
POSTEMA, MARK
POTTER, ROBERT
POWELL, LEE
POWERS, JOHN
POYNTER, HARRY
PRATT, ALBERT
PRENTICE, JIM
PRESLEY, DELLY
PRICE, HERBERT
PRIEST, DICK
PRIMME, D
PRING, JAMES
PRINGLE, ELZA
PRITCHARD, THOMAS, F.
PROCTOR, STEPHEN
PROKOP, JEROME

PRUITT, WILL
PRZYBLOWSKI, RONALD
PUGH, IKE
PUGH, MILTON
PUGH, FLOYD
PUGNET, RONALD
PUND, KEVIN
PUTNAM, CHUCK

Q

QUAY, ROBERT
QUEDNOW, WILLIAM
QUINN, KENNY
QUINN, JIM
QUINONES, DANIEL

R

RAABE, FRITZ
RAFFERTY, PETE
RAMSAY, NEDOM
RAMSEY, CRAIG
RAMSTINE, KURT
RANDERSON, DAVID
RANDLES, FRED, C.
RANDOLPH, ROSS
RANKIN, TOM
RASMUSSEN, RICHARD
RAUPP, TIM
RAUSCH, EDWARD
RAYMOND, FRANK
REED, BRUCE
REED, DON
REED, ROBERT
REESE, JOHN
REESE, H. L.
REILLY, RON
REINKOESTER, EDWARD
RENFRO, MAYO
RENNE, JERRY
RESER, JAMES
RESKE, FRED
REST, DAVID
REUSS, THOMAS
REYHER, FRANKLIN
REYNIOLDS, JIM
REYNOLDS, JASON
REYNOLDS, WILLIAM, E.
RHOADS, JOHN
RICE, DENNIS
RICE, GENO
RICHARD, GEORGE
RICHARD, SCOTT
RICHARDS, STEPHEN
RICHARDSON, LYNN
RICHARDSON, BOB
RICHIE, ROGER
RICKETTS, KEITH
RIEKER, HERMAN
RIGGS, FRANK
RILEY, HARRY
RIMKUS, HENRY
RIORDAN, JERRY
RIPP, MASON
RISING II, EDDIE, J.
RIVERA, RANDY
ROACH, RONALD
ROARK, JOHN
ROBB, FRANCIS, J.
ROBERSON, CHARLES
ROBERT, DAVID
ROBERTS, KENNETH
ROBERTS, DON
ROBERTS, KENNETH
ROBERTSON, DAVE
ROBERTSON, PAUL
ROBILLARD, WILLIAM
ROBINSON, STEPHEN
ROBINSON, CLARENCE
ROBINSON, JACK
ROCHELAU, CHRIS
RODD JR, WILLIAM
RODEN, MICHAEL
ROEGGE, ROBERT
ROGERS, MARK
ROLAND, JAMES
ROMANOWISKI, BRIAN
ROMERO, TOMAS
ROMERO, RYAN
RONCHETTO, JOSEPH
ROOS, GEORGE
ROPER, ROBERT
ROSENFELDER, PETER
ROSS, LARRY
ROSSA, ROBERT
ROSSEL, EUGENE
ROSSMEISSL, TOM
ROTENBERGER, RICHARD IVER
ROTZKO, R
ROULSTON-WARREN, NANCY
ROWE, LLOYD

ROWLAND, M. O.
ROWLEY, HOWARD
ROY JR, FRANCIS
ROZNER, MORRIS
RUANA, RAYMOND
RUDIG, JIM
RUIZ, ELIAS
RUMLER, WILLIAM
RUNKLE, JAMES, G.
RUNNER, HERBERT
RUOTOLO, MICHAEL
RUPP, JOHN
RUPP, KENNETH, L.
RUSH, LARRY
RUSNAK, BARB
RUSS, HENRY
RUSSELL, HORACE
RUSSELL, EDGAR
RUTH, JIM
RUTHERFORD, DON
RUTLEDGE, DAN
RYGG, LOWELL
RZEWUSKI, JOSEPH

S

SACHES, RICHARD
SACONE, FRANCIS
SADLER, RICHARD
SAGER, WILL
SAILAR, BUD
SAKAKIDA, DAVID
SALATA, WILLIAM
SALAVA, GARY
SALISKI, THOMAS
SALLY, CLARENCE
SALYERS, JAMES
SALZMAN, JOHN
SAMPLES, LOUIS, L.
SAMUELS, JON
SANBORN, DAVE
SANDERS, SIDNEY
SANDERS, FORREST
SANDERS, FLOYD
SANDOVAL, JERRY
SANGER, ROBERT
SANSONE, DAVID
SANTOS, DEJESUS
SAPP, GLENN
SAUNDERS, PAUL
SAURER, RUDOLPH
SAVAGE, DOC
SAVAGE, MACK
SAWAYA, EDWARD
SAWYER, CHARLES
SCALLORN, BEN
SCANLON, WILLIAM
SCARCELLA, JIM
SCHADE, DAVE
SCHAEFER, WARREN
SCHAEFFER, HERRB
SCHAFF, GARY
SCHARF, RICHARD
SCHEFFEL, REYNOLD
SCHERBINSKI, STEVE
SCHEUERMAN, R. LAMAR
SCHEUFLER, STEPHEN
SCHIEBER, RUSSELL
SCHIRMER, GEORGE
SCHLESAK, DON
SCHLUSSEL, NEIL
SCHMID, C. A.
SCHMIDT, STEVE
SCHMIEG, CHARLES
SCHNELL, TOM
SCHOOLER, WILLIAM
SCHOONOVER, PETE
SCHORK, ROBERT
SCHROEDER, ROBERT
SCHROEDER JR, HENRY, A.
SCHULTZ, FRED
SCHWABE, CRAIG
SCHWANKE, OTTO
SCHWARZENBACH, JACK
SCIVALLY, FRANK
SCOTT, BYRON
SCOTT, ROGER
SCOTT, WILLIAM
SCOULAR, EUGENE
SCUDDER, ROBERT
SCURZI, JOSEPH, R.
SEALING, LESTER
SEALS, BILLY
SEARS, JOHN
SEASHORE, PAUL
SEEBY, BILL
SEGALINI, JIM
SELF, GARY
SELMAN, PHILIP
SELPH, FRANCIS, M.
SEMMLER, BILL
SESSOMS-PENNY, SANDRA

SETHER, GILBERT
SEVERO JR, ORLANDO
SEVERUD, TIMM
SHACKLETTE, BILL
SHAFER, JOHN
SHAFF, WILLIAM
SHARPE, GLENDON, P.
SHAW, JOHN
SHELDON, LANCE
SHELLEY, DELBERT
SHELTON, JACK
SHELTON, RICHARD
SHEPPARD, CHARLES
SHERROW, LEONARD
SHIAU, JEN
SHORETTE, CHARLES
SHORT, WILLIAM
SHORTRIDGE, WAYNE
SHULTZ, GEORGE
SHUTE, RICK
SHUTT, JOHN
SHY, GEORGE
SIAU, BOB
SIBCY, ALFRED
SIGNORINO, CHARLES
SIKKINK, GARY
SILK, TODD
SILVA, ANGEL
SILVER, SAMUEL
SIMBECK, JOHN
SIMMONS, WALT
SIMMONS JR, FINNIS
SIMMS, JOHN
SIMPSON, GUS
SIMS, JOHN
SIMS, JOHN
SIMS, JOHNNY
SIMUNAC, JOSEPH
SINCLAIR, JERRY
SISLER, JACK
SITZMAN, YVONNE
SKALICKY, JAMES
SKINNER, TODD
SKINNER JR, THOMAS
SLAGLE, HAROLD
SLAUGHTER, CHARLES, W.
SLAYDEN, VAN, H.
SLIFKA, MIKE
SLOAN, BRETT
SLOAN, SHARON
SLOVIK, JULIAN
SMITH, BILL
SMITH, BILL
SMITH, ROBERT
SMITH, STEVE
SMITH, FRAN
SMITH, PAUL, K.
SMITH, ERNEST, L.
SMITH, DONALD, L.
SMITH, CLYDE
SMITH, WILLIAM
SMITH, IVAN
SMITH, LEO
SMITH, JOHN
SMITH, NORMAN
SMITH, FRANK
SMITH, SCOTT
SMITH, PETER, S.
SMITH, MIKE
SMITH, GLENN
SMITH, GARY
SMITH, STEVE
SMITH, WILLIAM, R.
SMITH III, ELDON, R.
SMYERS, EDWARD
SNEDKER, HAROLD
SNELL, ROY
SNIDER, REX
SNODGRASS, CLYDE
SNODGRASS JR, WILLIAM
SNYDER, JOANNE
SNYDER, RAY
SNYDER, CHARLES, A.
SOBIESZCZYK, NEAL
SOBIK, PAUL
SOLBERG, DARRELL
SOLBERG, GARY
SOLLITT, RONALD
SORENSON, DONALD
SORENSON, JAMES
SOTRINES, FRANK
SOUCY, ROBERT
SOUTHARD, ROBERT, J.
SOVICH, FRANK
SOVINE, CHARLES
SPANN III, WILLIAM, C.
SPENCER, PHILIP
SPENCER, JOHN
SPENGLER, WALTER, F.
SPINNEY, VAN
SPITZ, RAY
SPIVA, JAMES

SPLAWN, JERRY
SPRINGER, CHARLES
SPRIZZI, FRANK
SPROUL, PARKER, D,
SPROUSE, DONNA
ST CLAIR, JEFFERY
ST MYER, JAMES
ST. GEORGE, LEONARD, L.
STACHURSKI, DICK
STANKEWICH JR, ANDREW
STANLEY, D. J.
STANLEY, DARRELL, L.
STANLEY, GREG
STANLEY, ROB
STAYTON, JACK
STEADMAN, SCOTT
STEM, WALTER
STEPHENS, TIM
STEPHENSON, BOB
STEVENS, KENNETH
STEWART, SHIRLEY
STEWART, RICHARD
STEWART, TOM
STEWART, JOHN
STINSON, JERRY
STINSON, WILLIAM
STOCKER, WILLIAM
STODDARD, DAROLD
STOLARSKI, DONALD
STONE, BILLY
STOPPKKOTTE, WARREN
STORM, GARY
STOTLER, BRIAN
STOVAL, ROY
STRAUME, MARK
STROME, DICK
STROUD, ROBERT
STROUD, CONLEY, B.
STROUD, RALEIGH
STROUT, DON
STRUCKHOFF, ANDREW
STRUTHERS, DAVID
STRUTZ, JAMES
STUBBLEFIELD, RICHARD
STUBBLEFIELD, JAMES
STUBBS, GREGORY
STUEBE, THOMAS
STULLENBERGER, EARL
STUMBAUGH, LOUIS
STUPSKI, ED
STURDEVANT, STEVEN
STUTTE, RICHARD
SUBLETTE, KENNETH
SUGARMAN, RICHARD
SULLENBERGER, DONALD
SULLIVAN, JOHN
SULLIVAN, REUBEN
SULLIVAN, RICHARD
SUNDERLAND, HARRY
SUPER, RONALD
SUSMAN, JOEL
SUTHERLAND, RICHARD
SWANSON, HAROLD
SWAP, ARTHUR
SWARTZ, JIM
SWEENEY, JAMES
SWEENEY, RAYMOND

T

TALBOT, LOUIS
TALBOTT, NEIL
TALLEY, KENNETH, R.
TARESH, DARRYL
TAUTJES, RAY
TAYLOR, JIMMY
TAYLOR, RICK
TAYLOR, JAY
TAYLOR, CORT
TAYLOR JR, JAMES
TEAGUE, BENNIE
TEDRICK, JOHN
TELLES, RUDY
TEMPLE, RICHARD
TEMPLE, JAMES
TERHARK, & STEVEN
TERRELL, JAMES, O.
TERRELL, TEDDIE, D,
TESTA, WILLIAM
TEW, LARRY
THAYER, MELVIN
THEODOSS, GREG
THEOHARIS, GEORGE
THODEN, RICHARD
THOMAS, JOEL
THOMAS, ROY
THOMAS, ROBERT
THOMAS, PHILLIP
THOMAS JR, HECK
THOMASSON, DAVID
THOMPSON, ROBERT

THOMPSON, GEORGE
THOMPSON, CHARLES
THOMPSON, RONALD
THOMPSON, KENNETH
THOMPSON, J
THOMPSON, KERMIT
THORELL, ERIC
THORNBURG, WILLIAM
THORNBURGH, ERIC
THORPE, WILLIAM
THRUSH, ALLEN
THURMAN, MARTIN
THURMAN, MARTY/GAIL
TIBBETTS, VACNE
TICE, LARRY
TIDWELL, JAMES, A.
TIETJEN, ROBERT
TILL, WILLIAM
TILMANS, H. M.
TIMS, GREG
TINKLER, RICHARD
TOBIA JR, JOHN
TOM, JERRY
TOMS, JAMES, R.
TONER, SAMUEL
TOPPER, STEVE
TORKELSON, DAVID
TORRINGTON, AL
TOTTEN, EVAN
TRAMM, PETER
TRAMP, ARTHUR
TRAVERS, ALFRED, P.
TREADWELL, CRAIG, R.
TREGRE, CHARLES
TRIMARCHI, VICTORIO, G.
TROGDON, BOB
TROMBLEY, DONALD
TROMBLEY, TOM
TROUT, MICHAEL
TRUHLAR, DOUG
TRUITT, JERRY
TRUMPFHELLER, GREG
TUBB, SAM
TUCHOLSKI, HENRY
TUCKER, TOM
TUCKER, GARY, B.
TUCKER, WILLIAM
TURNER, DION
TURNIPSEED, JON
TUROSKI, RICHARD
TUTKA, JOHN
TWIGG, JACK
TYLER JR, MACEO

U

ULRICH, WALLY
UNDERHILL, STEPHEN
UNDERWOOD, R. E.
UNSELL, RAYMOND
URBANEK, FRANK
URICH, RIELLY, E.
USUI, ED
UTSEY, VINCENT

V

VALENTINE, ROGER
VALERIO, BENJAMIN
VALLE, AUGUSTINE
VAN DORPE, DOUGLAS
VAN HOUTEN, DIETHER, H.
VAN METER, KENT
VAN TRUMP, HERB
VANCE, PAUL
VANCHIERI, VINCENT, S.
VANDAGRIFF, TREADWELL
VANDELL, VERNON
VANDEN BOSCH, PETE
VANKEUREN, JERRY
VANLEAR, JERRY
VANSANDT, JACK
VARGO, RONALD
VARNER, FRANCIS
VASQUEZ, OMAR
VASSAR, ALEX
VAUGHAN, BEN
VAUGHAN, JAMES, E.
VEALE, ROBERT, L,
VEITH, BILL
VERTENTEN, JAMES
VIOLETTE, MICHAEL
VICKERY, CHUCK
VILLANUEVA, ANTHONY
VINING, THEODORE
VIPOND, DANIEL
VIS, BOB
VOLKMER, JOHN
VOORHEES, GARRETT
VOYLES, VALLEY

W

WAAGE, TAYLOR
WADE, TOMMY
WAGNER, WILLIAM
WAGNER, STEVEN
WAGNER, ALLEN
WAGNER, TOM
WAITER, LEN
WAKEFIELD, GERALD
WALKER, EDDIE
WALKER, JAMES
WALKER, BENJAMIN
WALKER, LESTER
WALKER, DAN
WALKER, GREG
WALLER, HOWARD
WALLER, GLEN
WALSH, DENNIS
WALSH, STEVE
WALTERICK, KEN
WALTON, JACK
WALTON, MICHAEL
WALTON, DAVID, L.
WAMSLEY, CHARLES
WAMSLEY, HAROLD
WARD, LEW
WARD, MICHAEL
WARD, JOHN
WARD JR, WILLIAM
WARE, GEORGE
WARE, WILLIAM
WARFIELD, JAMES
WATKINS, LARRY
WATKINS, CHARLES
WATKINS, ANDREW
WATSON, GEORGE
WATSON, JAMES
WATT, JAMES
WATT, GLENN
WATTS, ROBERT
WATTS, BOBBIE
WATTS, ROBERT, E.
WAY, ROBERT, E.
WEADON, RICHARD
WEAVER, JAMES
WEBB, RICHARD
WEBB, JAMES
WEBSTER, JOHN, A.
WEDIG, RAY
WEEKS, BENJAMIN
WEGMULLER, WILLIAM
WEICHEL, STEVEN
WEISS, FRANK
WEISS JR, WALTER, A.
WELCH, ROBERT
WELCH, THOMAS
WELKER, TED
WELLMAN, MICHAEL
WELLS, JACK
WELLY, MARVIN
WENDL, JOE
WERNER, RICHARD
WERTZ, DON
WERTZ, LAWRENCE
WESBROOK, RICHARD
WESSELS, DONALD
WEST, ROBERT, L.
WEST, WILLIAM
WEST, RUFUS
WESTBERG, DEL
WESTCOTT, LAWRENCE
WESTROM, JOHN
WHALEY, JOE
WHEATCRAFT, L. D.
WHEELER, GREGG
WHITE, PENDLETON
WHITE, JOSEPH
WHITE, JOHN
WHITE, JOE
WHITE, CARL
WHITE, WARREN
WHITED, JOE
WHITNEY, JAMES
WHITSON, FRED
WHITT, WESLEY
WHITTENBERG, JAMES
WIEDE, JIM
WIEDER, SIDNEY
WIESNOSKI, JOSEPH
WILBERT, JAN
WILCOXON, JAMES, F.
WILDE, DAVID
WILDER, DANIEL
WILKE, DENNIS
WILKIE, BOBBY, D.
WILKINS, JAMES
WILKINSON, WILLIAM
WILKINSON, LEO
WILKS, JERRY
WILLARD, PAUL
WILLIAMS, ROBERT

WILLIAMS, JOYCE
WILLIAMS, JOHN
WILLIAMS, RONALD
WILLIAMS, CHUCK
WILLIAMS, MARCUS
WILLIAMS, RONALD, H.
WILLIAMS, CHARLES, E.
WILLIAMS, TOM
WILLIAMS, TRENT
WILLIAMS, KENYON
WILLIAMS, EARL
WILLIAMS JR, EVERETT
WILLIAMSON, STANLEY
WILLINGHAM, WILLIAM
WILLIS, ISAAC
WILLYARD, ROBERT
WILSON, HARRY
WILSON, RICHARD, S.
WILSON, DOUG

WIMBERLEY, GARY
WINDHAM, WILLIAM
WINKEL, BILL
WINN, RICK
WISEMAN, TERRY
WITHERITE, SAMUEL
WITZENBURGER, STEPHEN
WOLD, HAL
WOLF, DONALD
WOLFE, MARK
WOLFF, ROBERT
WOLUSKY, TONY
WONER, J. MURRIN
WONG, MICHAEL
WOOD, M. D.
WOOD, PATTI
WOODCOCK, HERBERT
WOODS, MICHAEL
WOODS, RICHARD

WOODS JR, CHARLES
WOODSON, ROBERT
WOOLF, BILL
WORKMAN, JIMMY
WORTH, GRANT
WORTHINGTON, ROY
WOUTERS, PETER
WRIGHT, RON
WRIGHT, AL
WROBLEWSKI, JOSEPH
WRONKOSKI, EDWARD
WUERL, MIKE
WYLAM JR, AUSTIN, J.
WYNN, CHUCK

Y

YAGUCHI, MIKE
YATES, HOMER

YOCUM, JOHN
YOHN, JACK
YOUMANS, KENNY
YOUNG, RICH
YOUNG, FRANK
YOUNG, ROB
YOUNG, JAMES, B.
YOUNG, MICHAEL
YOUNG, BRUCE
YOUNG, KENNY
YOUNG, KENNETH
YOUNGBLOOD, FRANCIS
YUST, GARY

Z

ZALENSKI, ROBERT
ZALESKI, JOSEPH
ZARAMBO, STANLEY

ZEBRON, SAMUEL
ZEBROWSKI, THOMAS
ZIGLI, RONALD
ZIMMER, RICHARD
ZIMMER JR, CHARLES
ZIMMERMAN, ROBERT
ZIMMERMAN, DAVID
ZIMMERMAN, FRED
ZIOBER, ALVIN
ZITZKA, WAYNE
ZOLLINGER, JOE
ZOOK, CARROLL
ZORICH, SAM

New AAFM Members Since First Publication

A

AHRENDT, WILLIAM M.
AHRENSFIELD, CLAYTON E.
ALEKSON, ROBERT F. SR.
ALLMAN, WILLIAM P.
AMBROSE, PETER
ANDERSON, WALTER E. JR.
ASHMORE, CHARLES D.
ASIK, RAYMOND J.

B

BABCOCK, ROBERT M.
BALDNER, ROGER E.
BARDEN, PAUL A.
BASS, AARON C.
BATH, DAVID W.
BAZZEL, BOBBY G.
BEASLEY, CATHY
BEECHLER, CARLTON R.
BERKSHIRE, JOHN H.
BETTIS, ALVIN R.
BLANKS, RANDOLPH M.
BLUME, GEORGE D.
BOLAND, THOMAS R.
BOMBERG, CRAIG L.
BOSCH, HARRY C. JR.
BOSTON, FELIX J.
BOYD, STEVEN H.
BOYER, TIMOTHY J.
BRADFORD, BEN
BRAND, DAVID F.
BRANDE, WENDELL S.
BRANTLEY, SHADY L.
BRIDGES, JAMES R.
BRIDGES, ROBERT
BROOKS, DONALD
BUCKLES, CURTIS L.
BUNCH, DEBORAH L.
BURNETT, DANNY A.
BURNS, JACK E.

C

CAIN, ANDREW S.
CAMPBELL, HARVEY H.
CAMPBELL, THOMAS
CARSON, RAYMOND L.
CARTER, DAVID M.
CARTER, HAROLD B.
CARTWRIGHT, HENRY J.
CASLETON, HAROLD D.
CHALFONT, ALAN C.
CLAUS, JONATHAN C.
CLUBB, FREDERICK R.
COCHRANE, MICHAEL J.
COLEMAN, GEORGE P.
COLON, WILLIAM JR.
COUGHLIN, RAYMOND J.
CRUDUP, HAYWOOD L.
CULLEN, STEPHEN G.
CUNNINGHAM, JOHN S.
CURRIE, GEORGE F. JR.

D

DABROWSKI, SUSAN E.
DALRYMPLE, ROBERT S.
DAVIS, RICHARD W.
DAYHOFF, GEORGE W.

DICK, STANLEY A.
DORSEY, BRUCE E.
DOTY, WILLIAM K. JR.
DUGGAN, CARL A.
DUNCAN, C. MIGUEL
DUNCAN, CLYDE W.
DUNKIN, SHERRILL L.

E

EDDLEMON, JOHN A.
EDWARDS, SCOTT D.
ELA, BERNARD L.
ELICH, ANTHONY M. III
ELLIS, ROLAND D.

F

FABBRI, MICHAEL
FALL, GERALD G. JR.
FICK, MATTHEW P.
FISHER, KENNETH M.
FLYNN, GERALD B.
FORESMAN, DENNIS
FOULGER, KEITH H.
FRANKHOUSER, JEFFREY E.
FREMSTAD, JON R.
FUNKHOUSER, GARY D.

G

GASPAR, CARL L.
GEMMILL, ROBERT L.
GIBEAU, JOHN P.
GOLMITZ, JOHN J.
GOODALL, RICHARD K.
GOODCHILD, GEOFF
GOTTS, RICHARD D.
GRAGG, TEDDY E.
GREENHILL, KARL J.
GRIFFITH, JAMES S.
GRIM, GARRETT D.
GROMAN, ROBERT W.
GROMETER, BRENT A.
GROSS, RICHARD E.

H

HACKER, RANDY W.
HALE, CHRIS
HALE, FRANK
HALE, LEWIS R.
HANSEN, CRAIG A.
HARDEN, THOMAS P.
HARGROVE, DON
HARRIS, MICHAEL B.
HARRISON, GEORGE
HAWTHORNE, GARY F.
HENDRICKSON, DEAN P.
HERNANDEZ, ANDREW A.
HINKEN, DAVID F.
HODGE, CLIFFORD A.
HOKETT, JEFFREY A.
HOOD, GREGORY W.
HOSTETLER, DAVID K.
HUGHES, RICHARD J.
HUNLEY, JOHN D.

I

ISAACSON, STEPHEN A.

J

JACOBSEN, KEVIN J.
JACOBSON, EVANS C.
JEAS, WILLIAM C.
JENSEN, JOHN M.
JONES, THOMAS F.
JONES, WILLIE JR.

K

KELLY, MARTIN C.
KELSO, THOMAS S.
KENWORTHY, WILLIAM G.
KETCHUM, ROBERT A.
KILE, ERIC D.
KLINGE, KRISTIAN C.
KOGGE, STEVEN R.
KOVACH, GEORGE F.
KRYWANY, JOSEPH M.
KUEHL, DANIEL T.
KUFTA, NICK C.
KUGLER, LAWRENCE

L

LA PORTE, CHRIS P.
LAMB, RICHARD B.
LAMPE, TIMOTHY J.
LAYFIELD, J. DAVID
LAYFIELD, JEFFREY A.
LAZIER, KENNETH
LEACH, PHILLIP A.
LEDDY, CARL J.
LEGGETT, DAVID C.
LIEN, RICHARD A.
LINDEAU, MICHAEL
LORENZ, RICHARD F.
LUBBE, GARY L.

M

MARSCHKE, JAY
MARTIN, CHRIS
MASONE, JAMES
MCALPIN, NORMAN B.
MCKEAN, JOEL M.
MELLOR, RICHARD C.
MERZLAK, PAUL M.
MESSMER, DONALD E. JR.
MEURY, KAREN J.
MEZZATASTA, JOSEPH M.
MINDLING, GEORGE G.
MITCHELL, WILLIAM EARL JR.
MOLLISE, RODNEY F.
MONROE, JOHN R.
MOORE, JAMES F.
MOORE, V. DOUGLAS
MORIN, RONALD E.
MORTENSEN, KEVIN S.
MOYNIHAN, TIMOTHY J.
MUESSING, SCOTT K.
MURRAY, WILLIAM H.
MYERS, RICHARD B.

N

NATIONAL AEROSPACE TRUST
NEDIMYER, STEPHEN E.
NELSON, STEVE
NOBLE, TED M.

NOLDER, KEN
NORDHAUS, PAUL M.
NORWOOD, KAI LEE
NYBERG, FREDERICK D.

O

ODEN, LUCIAN B.
OKULICZ, CHARLES J.
OLEYAR, WILLIAM S.
ORGERON, JAY E.

P

PARSONS, MICHAEL B.
PAVLOVICH, J. GREGORY
PERDUE, WILLIAM T.
PERENOVICH, RODNEY
PERKINS, RALPH Q.
PERRY, ALLEN H.
PHILLIPS, KEITH L.
PIPER, DENNIS M.
PISLE, STANFORD C.
POTTER, DOUG
PREBLE, THOMAS A.
PREIDIS, DOUGLAS A.
PRESLEY, FERREL L.
PUFFER, RAYMOND L.

Q

QUAY, ROBERT

R

RAFFIELD, WILLIAM D.
RAINBOLT, HAROLD E.
RAPONE, GENO J.
RAY, JEFFREY S.
REEDY, JOHN S.
REID, JAMES H.
REIMER, PAUL E.
REST, DAVID W.
RICE, KENNETH M.
RIORDAN, JOHN S.
ROBERTS, ROY C.
ROBINSON, BOBBY
ROBINSON, GEORGE W.
ROBISON, RODNEY K.
ROCK, RICHARD W.
RODRIGUEZ, RICHARD JR.
RODRIQUEZ, GERARDO U.
ROEWER, MARTIN W.
ROGERS, CHARLES L.
RUGGIERO, FRANCIS X.
RYAN, GEORGE W. JR.

S

SANDERS, JOHN R. JR.
SCHIFF, JEFFREY M.
SCHMIDT, EDWARD B.
SEARS, ROBERT L. II
SEE, RICHARD M.
SERVANT, ROBERT E.
SHAW, RONALD R.
SHELDON, JOHN H.
SHILLER, DAVID K.
SHINEMAN, KIRBY A.
SHUEMAKER, ROBERT A.
SHUPING, ROBERT F. JR.

SILLIMAN, MARK E.
SMITH, JEFFREY J.
SMITH, LESLIE T.
SMITH, VAL
SMYSER, CARL J.
SNEDKER, HAROLD C.
SOCHA, STANLEY W.
SPIEGELMAN, MICHAEL H.
ST. JOHN, KENNETH
STICKELS, ANGELA
STOCKER, JOHN E. III
STONE, CHARLES JR.
STONE, GUY D.
SULLIVAN, ROBERT L.
SUNDERLAND, HARRY R.
SUSMAN, JOEL D.
SWEDA, GARY H.

T

TALLMAN, GRANT R.
TAYLOR, GREGORY S.
TAYLOR, MALCOLM L.
TENNANT, JOHN E.
THOMAS, EVERETT H.
TIRABASSI, PETER E.
TURNER, THOMAS P.

V

VAN DILLEN, KENNETH B.
VAN SANFORD, JEFFREY
VENARCHICK, KEVIN M.

W

WALKER, CLAYTON H.
WALLACE, BILLIE R.
WALLACE, ERIC O.
WALTERICK, KEN
WALTMAN, GLENN C.
WARNER, JAMES F.
WEIGERT, RICHARD M.
WEIRICK, ANDREW
WELLER, GRANT T.
WELLS, JAMES M.
WELLS, STEVEN C.
WESTPHAL, WILLIAM R. JR.
WETMORE, DANIEL R.
WHITBY, THOMAS I.
WHITE, WARREN E.
WHITMORE, PAUL H.
WILLEY, VINCENT R.
WILLIAMS, DALTON III
WILLIAMSON, STANLEY J.
WILLIS, BRIAN E.
WILSON, ROSS J.
WOEPPEL, DAVID W.
WOODS, HAROLD L.
WOODS, MICHAEL W.
WUEST, ERIC C.

Z

ZACCHERO, CHUCK

Index

Symbols

ALCM 15, 16, 17, 21
Atlas 7, 19, 22, 23, 24, 27, 28, 30, 33, 35, 53, 55
Atlas E 24, 38, 46
Atlas F 11, 25, 43, 67, 68, 71
Bell Rascal I 13
Bell Tarzon 13
Bird Dog 13
BOMARC 3, 17, 20, 21, 36, 42
Bullpup 16
Dragonfly 13
GLCM 11, 21
Delta Dagger 13
Falcon 14
Firebird 13
GAM-77 35
Genie 13, 14
Gryphon 17, 21
HARM 16
Harpoon 16
High Card 13
Hound Dog 13, 15, 16, 19
IRBM 19, 22
Jupiter 22, 23, 35
Mace 17
Mastiff 13
Matador 11, 17, 18, 19, 52, 58
Maverick 16
Minuteman 7, 11, 22, 28, 29, 30, 31, 34, 35, 42, 51, 60, 71
Minuteman I 28, 30
Minuteman IA 28, 30
Minuteman IB 28, 29, 30, 31
Minuteman II 28, 29, 30, 31, 33, 65
Minuteman III 28, 29, 30, 31, 35, 61
Missile-X 31
MX-777 13
MX-778 13
MX-779 13
MX-799 13
Navaho 15, 17, 19, 54
Peacekeeper 31, 36, 49, 65
Polaris 31
Poseidon 31
Quail 14, 15, 16
Rascal 13, 54
Redstone 19
Regulus 44
Regulus II 21, 44
Roc 13
SCAD 16, 17, 21

Shrike 14
SLCM 17, 21
Sidewinder 13, 14
Skybolt 15, 16, 36
Snark 8, 17, 20, 38, 53, 54
Sparrow 13, 14
SRAM 15, 16, 17, 21
T-MX 22
Tarzon 13
Thor 8, 19, 22, 23, 35, 38, 39, 45, 46, 47, 48, 54, 55
Tiamat 13
Titan 7, 22, 28, 31, 35, 53, 55
Titan I 2, 11, 22, 25, 26, 27, 30, 33, 37, 38, 40
Titan II 3, 11, 27, 28, 31, 41, 42, 43, 63, 65, 68, 69, 70, 72
Tomahawk 17, 21
Trident 31

A

Adams, Chic 66
Adams, Dick 3, 42
Allen, Gen. 54
Andrew, Joe 11
Appel, Jack 36
Arnold 53
Aueri, Michael J. 39

B

Babbidge, Mike 42
Bacik, Joseph 69
Bacs, John 33
Bailey, F. Lee 72
Banning III, Thomas M. 36
Bates, Katherine Lee 59
Bieleski, Stan 38
Biggs, Richard J. 71, 72
Bishop Jr., Ronald J. 11, 72
Blair, Edison T. 52
Blanchard, William H. 31
Blaydes 37, 50
Boverie, Richard 6, 35, 36
Bricker, Creight 36
Brixey, Robert D. 69
Brooks, Elmer T. 43, 71
Brooks, Leo C. 71
Bunker, Kenny 67
Burba, Jim 11, 33
Burns, Ken 36
Bush, William 66

C

Campbell, John 39
Carpenter, Bob 38
Carswell Jr., Horace S. 70
Carter, Jimmy 17
Casey, Aloysius G. 35, 36
Castillow, Bennie 39
Castro, Dayna 11
Catton, Jack J. 36
Chamberlen, Jerry 71
Chappell, Brian "Chappie" 58
Cherry, Carlos 46
Clark, Fred 36
Clugston, Col 46
Cofield, William B. 71
Coglitore, Seb 7, 11
Cook, Bruce 46
Counts, Dave 69
Crossley, Bobby L. 69
Crouch, Jim 11

D

Daniel, Lynn B. 69
Davidson, B.H. 36
Davis, Cassel 69
de Zarraga, Miguel 68
Dee, Harry J. 69
Defede, Ed 36
Dempsey, Jack 59
Donnes, John 60
Dorsett, Harry L. 69
Downing 56
Downinl, Darrell 29

E

Emmerling, Bob 52
Erlenbush, William C. 6
Evans, James D. 69

F

Farley, Jack 39
Farmer, Beryle 46
Flores, Paul M. 69
Fogelman, Ronald 8
Fohr, Rodney 39
Foster, Jim 36
Foster, John P. 71

Fowler, John 38
Fox, Kenneth 71
Fox, William J. 69
Freer, Al 36
Fuchs, Klaus 62

G

Gibbons, John 59
Gifford, Charles E. 39
Gill, Stephen C.M. 69
Gloston, Jerry J. 69
Gramley, Dale 36
Gray, Ron 7, 11
Guyer, Jeffrey D. 69

H

Haas, Victor J. 35
Haley, John W. 47, 68
Hall III, Alonzo 65
Hancock, Delbert L. 69
Hannon, John O. 69
Hansbrouck, Larry 42
Hansen, Eileen 65
Harrel, Walter L. 69
Harshbargar, James W. 69
Hartzell, Robert 56
Hasbrouck, Larry 39, 71
Henry, Pat 7, 11
Hepworth, Jan L. 69
Herres, Bob 38
Heth, Darcie 58
Hilden, Jack G. 65, 66
Hodges, William L. 39
Holloway, Bruce 33, 71
Horton, Barry 7, 11
Hughes Falcon 13
Hunnm, Van 39
Hunt, Wells E. 67, 68
Hutchinson, Philip T. 69
Hutstetler, Michael 38

I

Ingram, Dick D. 36

J

Jagger, Donald E. 69
Jameson, Dirk 7, 11
Jankowski, Michael T. 69
Jewell, Larry E. 69

K

Keen, Dick 11
Kelchner, Bob 7, 11
Kelley, Jay 7, 11
Kelly, Mike 46

Kindsfather, Jake 6
Kingsley, Jack 56
Kissel, Lt. 37, 50
Knapp, Jim 11
Knapp, William A. 44
Kopac, Nick 52
Kundis, Walter 33, 37, 50, 69, 70

L

Lander, Jack 7, 11
Langhoff, Myron M. 69
Leber Jr., John C. 68
Lehnertz, Mike 11
Leuty Jr., Elbert J. 36
Liebenguth, Duffy 46
Lightsey, Tann 36
Livingston, R. Douglas 33
Lloyd, James E. 69
Lord, Lance 7, 11

M

Marquis, Jim 47
Marsh, Gary E. 54, 56
Marsh, Robert E. 38
Martin, Glenn L. 25
Marzyn, Richard P. 69
Mayhew, Jason F. 39
McNair, Michael 65
McPeak, Tony 8
Meddress, James W. 69
Meyer, Max L. 39, 46
Miller, Bill 36
Miller, John 36
Mitchell 53
Morris, Tom 64
Mulcany, Joseph 56
Murphy, Paul F. 48

N

Nagy, George R. 48
Napolitano, John J. (Pat) 39
Nassr, Michael A. 35
Neary, Tom 11
Northrup, Ed 11
Northrup, Jack 20

O

Ogletree, Greg 6, 33, 48, 49
O'Hern Jr., Fred H. 51
Osborne, Ed 11
Ott, James A. 69
Ott, Richard L. 69

P

Palmer, William M. 69
Paquette, Conrad 7, 11

Parker, Bob 11
Parker, Hamp 69
Partner, Dan 51
Pate III, Joseph A. 36
Payne, Edward 8
Pepe, Bob 11
Pfeiffer, Rhuena B. 69
Phillips, Samuel 8
Pillman 37, 50
Power 45
Prahalis, Stan 36
Proffitt, Dick 36
Prohaska, Chuck 36
Pyne, Gerald F. 69

Q

Quarles, Donald A. 54

R

Rennigner, Harold 39
Rhiney, R.T. 68
Rice, Richard A. 42, 43
Richie, Roger L. 69
Roberts, Jack 46
Roosevelt, Teddy 38
Roper, Herbert G. 69
Rosenberg, Ethel 62
Rosenberg, Julius 62

S

Sampson, Ry D. 47
Santistevan, Michael 68
Sassone, Dick 36
Schoonmaker, Dick 7, 11
Schriever, Bernard A. 22, 53
Scott, Ralph 66, 68
Shaw, Charles W. 56
Shirley, Lance 39
Simpson 56
Simpson, Charles G. 7, 11, 29, 31, 42
Slegel, Gene 58
Smith, Bruce D. 51
Smith, Doug 43
Soreco, Carl 36
Spaatz 53
Spooner, Jerry 38
Spraker, Ralph 7, 11
Stachowiak, Daniel 3
Stricklin, Loyal E. 69
Strong, Jerry 43
Stults, Richard L. 69
Sutton, Larry 38

T

Taffet, Harvey 38
Taylor, Marvin D. 69

V

Veres, Mike L. 59, 60, 62
Vivian, Edward A. 71
von Braun, Werner 22

W

Wade, David 38
Walker, Samuel 69
Warren, Francis E. 68, 69
White, Jerry W. 69
White, Thomas 6
Widlar, Jim 11
Wilt, Salem 52
Wisniewski, Ron 38
Wittkoff, Julie 11
Womack Jr., John H. 69
Wood, Harold 39
Worthy, Cliff 38
Wright, Gene 46
Wyckoff, Robert A. 63, 64
Wynne, Charles 65

Element of 11th pilotless bomber squadron (Matador)Orlando AFB, Florida. NAFM member Richard Boverie front row, kneeling left.

Titan I on trailers ready for transport to Norton AFB, CA. Access Road to 568 sms, "C" complex.